ORIGINS

THE SCIENTIFIC STORY OF CREATION

JIM BAGGOTT

OXFORD
UNIVERSITY PRESS

OXFORD
UNIVERSITY PRESS

Oxford University Press is a department of the University of Oxford. It furthers the University's objective of excellence in research, scholarship, and education by publishing worldwide.

Published in the United States of America by
Oxford University Press
198 Madison Avenue, New York, NY 10016, United States of America

Library of Congress Control Number: 2015936023
ISBN 978–0–19–870764–6 (hbk.)

1 3 5 7 9 8 6 4 2
Printed by Sheridan Books, Inc., Ann Arbor, MI

For Grace (aged 7),
who wants to be a planetary geologist

CONTENTS

ABOUT THE AUTHOR

Jim Baggott is an award-winning science writer. A former academic scientist, he now works as an independent business consultant but maintains a broad interest in science, philosophy, and history and continues to write on these subjects in his spare time. His previous books have been widely acclaimed and include:

Farewell to Reality: How Fairy-tale Physics Betrays the Search for Scientific Truth (Constable, 2013);

Higgs: The Invention and Discovery of the 'God Particle' (Oxford University Press, 2012);

The Quantum Story: A History in 40 Moments (Oxford University Press, 2011);

Atomic: The First War of Physics and the Secret History of the Atom Bomb 1939–49 (Icon Books, 2009), short-listed for the Duke of Westminster Medal for Military Literature, 2010;

A Beginner's Guide to Reality (Penguin, 2005);

Beyond Measure: Modern Physics, Philosophy, and the Meaning of Quantum Theory (Oxford University Press, 2004);

Perfect Symmetry: The Accidental Discovery of Buckminsterfullerene (Oxford University Press, 1994);

The Meaning of Quantum Theory: A Guide for Students of Chemistry and Physics (Oxford University Press, 1992).

ACKNOWLEDGEMENTS

This has been an extraordinarily ambitious project, and I could not have contemplated it without the support of Latha Menon, my editor at Oxford University Press. I remain forever in her debt. Thanks must also go to Emma Ma, Jenny Nugee, and Kate Gilks at OUP, and Frances Topp and Paul Beversey, who have been instrumental in helping to turn my words and pictures into the book you now hold in your hands. I doubt that the book would be what it is were it not for John Gribbin, whose book *Genesis*, published in 1981, was an early source of inspiration (well, I remember cursing when I read it, because I realized that this was a book I would very much have liked to have written).

I'm never happier than when I'm learning something new. Writing *Origins* has taken me into some familiar territory, about which I've written before, but its wide-ranging scope has also drawn me into some less familiar spaces. I will therefore be eternally grateful to all the academic experts who have given of their valuable time to read through sections of the manuscript and offer helpful advice and criticism. These include Jennifer Clack at Cambridge University; George Ellis at the University of Cape Town; Andrew Knoll at Harvard University; Jeremiah Ostriker at Princeton University; Michael Russell at the Jet Propulsion Laboratory in Pasadena, California; Chris Stringer at London's Natural History Museum; Ian Tattersall at the American Museum of Natural History in New York City; Steven Weinberg at the University of Texas at Austin; and Simon White at the Max Planck Institute for Astrophysics in Garching, Germany. Any errors or misconceptions that remain are, of course, down to me.

Now, you know well enough that happiness doesn't come only from learning things and writing about them. It also derives from a great sense of place and the company of good people. Thanks then to Madeleine and Jur and Guido and Mikael, good people who have created two such fabulous places where some of these words were written, and to my wife Jini for just about everything else.

Jim Baggott

June 2015

LIST OF FIGURES

PREFACE

What is the nature of the material world? How does it work? What is the universe and how was it formed? What is life? Where do we come from and how did we evolve? How and why do we think? What does it mean to be human? How do we know?

These questions, many others just like them, and many subsidiary questions that logically follow are, collectively, the 'big questions' of human existence. They are questions that we have been asking ourselves for as long as we have been capable of rational thought.

We weave what answers we can discover or contrive into a creation story, and such stories have formed an essential foundation for every human culture throughout history. We have a seemingly innate and rather insatiable desire to want to comprehend our own place in the universe, to understand how we and everything around us came to be. This is a desire driven in part by simple curiosity. But I suspect it is also driven by a deeper emotional need to connect ourselves meaningfully with the world which we call home.

There are many different versions of our creation story. This book tells the version according to modern science. It is the result of an enterprise which has involved (and continues to involve) thousands of scientists struggling to piece together parts of the puzzle, constantly speculating, hypothesizing, testing, debating, and revising. These scientists strive to ensure that the puzzle pieces are individually coherent and consistent. But if the story is to be unified and comprehensible, the pieces themselves must also fit together, from the large-scale grandeur of the universe to *Homo sapiens* to the smallest microorganism to the elementary particles from which all material substance is composed. This is a powerful constraint.

That said, there is no such thing as an 'authorized' or 'official' version of the scientific story of creation. But I'm modestly hopeful that if there were, then it *might* look something like the book you now hold in your hands. Without wishing to appear overly melodramatic, I believe all my efforts as a popular science author over more than twenty years have been building up to this. Deep down, this is a book I have always wanted to write.

Now, I'm sure that's all very well, but is this a book you will want to read? This is obviously something only you can judge, but here are a few things you might want to think about.

First, I think the time is right for a book like *Origins*. In the past few years there has been a glut of popular physics titles telling us about new theories of everything or arguing that we live in a universe which is but one of a multiplicity of universes. In truth, although this stuff is advertised as science, none of it is accepted outside of a relatively small community of theorists, and it actually explains nothing relevant to our story. I have written elsewhere about such unsubstantiated 'fairy-tale' physics and I believe (or, at least, hope) that readers are growing increasingly wary of it.*

And what do we do when we are assaulted by news headlines screaming of another dramatic breakthrough in our understanding of our origins, only for the conclusions to be retracted months later when it emerges that the analysis was faulty and the announcement premature? In these circumstances, it's all too easy to lose sight of what's regarded as accepted scientific fact. And then what do we do when scientists publish books arguing in favour of their pet theories—theories that perhaps very few of their colleagues in the scientific community buy into? It's difficult to know what to make of these. Should we believe them?

In *Origins* I've tried to distinguish unquestioned fact from majority explanation from debatable interpretation from pure speculation. This is a book for readers who want a reasonably clear, balanced, and (hopefully) unbiased perspective on what we think we know and can explain. Yes, there are gaps and

---

* See my book *Farewell to Reality*, published in 2013.

things we don't understand clearly, and there are places in the story where scientists have had no choice but to indulge their inner metaphysician. I've been sure to point these out.

Second, anyone interested in tracing the scientific account of creation from the big bang to the transition to behavioural modernity in humans will need to range over many different scientific frontiers. These include (deep breath) aspects of modern cosmology and particle physics; the primordial synthesis of hydrogen and helium; the formation of stars and galaxies; stellar evolution and nucleosynthesis; planetary formation and differentiation; the chemistry of life; the evolution of the genetic code and simple, single-celled organisms, complex cells, and multicellular organisms; the sequence of mass extinctions and radiations that profoundly shaped the evolution of marine and land animals; the rise (and fall) of the dinosaurs; the emergence of mammals, primates, early hominins, the genus *Homo*, the species *Homo sapiens*, and the evolution of human consciousness.

Now that's asking a lot, and there's only so much time in a day. There are many popular, accessible, and highly readable books on all these subjects but *Origins* is, I think, unique in attempting to pull the contemporary scientific story together in a single volume.

Telling this story will take us on a journey from the very 'beginning' of the universe to the origin and evolution of human consciousness, 13,820 million years later. Whatever your reason for joining me, you should be aware of the path we will take. I'm going to try to explain what makes the scientific story of creation so different, and why I believe it is more reliable and compelling and ultimately more satisfying than other versions you might have heard of.

I want to reassure you that this is a satisfaction born not of smug certainty, from some sense of scientific triumphalism or authoritarianism. Far from it. I recognize that, while answers are very interesting, it is often the unanswered *questions* that fascinate. The satisfaction comes from acknowledging that although we know (or think we know) much, we don't yet know everything. And the answers we do have are likely to be modified or replaced entirely as we discover more things about the world and about ourselves and our place in it.

Some may view this as a weakness, but over a lifetime I've learned to be very wary of people who lay claim to certain knowledge. Perhaps we can be reasonably certain of one thing. Just ten years from now the story will be different, in some subtle and some not-so-subtle ways. And I guarantee you that the new version will be *better* than the old.

THIS THING CALLED SCIENCE

The physical chemist and novelist C. P. Snow famously wrote about the 'two cultures'—essentially science and the humanities—in an article published in the *New Statesman* magazine in October 1956, just a few months before I was born. Alas, this schism is in some ways wider now than ever before. There remains a persistent perception that science is not about 'us', that it somehow lacks human empathy. Science, so the argument goes, involves the application of a methodology derived from a rather cold, inhuman logic. It is obsessed with materialist mechanisms which demean our human spirituality. Science, the argument continues, can tell us some things about the physical, chemical, and biological mechanics of 'how?' but it can't address our deeper, equally compelling questions concerning 'why?'

I personally believe that *Origins* is very firmly about 'us'. This is *our* story. It is about how the world on which we live came to be, how life began and evolved to produce us: conscious, intelligent beings capable of a scientific investigation of their beginnings.

Science works because when we apply it we impose on ourselves a fairly rigorous discipline. We demand things from ourselves that are often counter-intuitive and counter-cultural and which require considerable intellectual effort. Obviously, we impose exacting standards in the elucidation and reporting of scientific facts. We demand scientific theories that broadly fit these facts, which hopefully make predictions that are potentially accessible to observation and experiment and which provide genuine insight and understanding (although there are lots of grey areas).

But science also demands that we adopt an attitude or perspective which is generally referred to as the 'Copernican Principle'. Although scientists are fundamentally interested in 'us', they presume that we are not particularly *special*. When we step outside the sphere of our distinctly human preferences and prejudices, as *Origins* will recount, we do indeed discover that we are not uniquely privileged observers of the universe we inhabit. To all appearances, the universe is not designed with us in mind. As we will see, science tells us that we are very much a naturally evolved *part* of this reality, but we are not the *reason* for it.

In other words, the universe has no *purpose*, at least if we seek to interpret purpose in a specifically human context. If teleology concerns the search for purpose or evidence of design in nature, then by definition science is firmly anti-teleology. This, in my view, doesn't make science inhuman. What it means is that it is not possible for humans to apply the scientific method without first adopting more than a little *humility*.

And herein lies the rub. Many of our 'why?' questions are typically driven by our singularly human search for purpose and meaning in life. It should therefore come as no surprise to find that science will struggle to answer them if, indeed, they have answers at all. Science is simply not set up this way and, if needed, we must reach for other belief systems to provide such answers as we can find.

Does this make science any less valid? No, it does not. Even though it cannot address many of our 'why?' questions, the simple truth is that we do not know its limits. One of the most remarkable aspects of modern science is its relatively new-found capacity to provide answers to questions that not so long ago might have been considered the preserve of high priests, priests who historically have had a habit of conflating 'why?' with 'how?'.

There is much we can still learn.

THE STORY IN OUTLINE (SPOILER ALERT!)

Origins tells the scientific story of creation in a chronological sequence, beginning with the origin of the universe—of space, time, and energy—in a 'big bang'. The first three chapters trace the origin and expansion of the universe through to something called the moment of recombination, about 380 000 years later, which releases the flood of hot electromagnetic radiation we now identify with the cold cosmic background. By this stage in the history of the universe, the basic building blocks are all available—space, time, energy, matter (dark matter and hydrogen and helium atoms), and light.

Much of our understanding of this early stage in the history of creation is derived from the fusion of so-called inflationary cosmological models and something called the standard model of particle physics. For the earliest stages in this history, we are obliged to reach for educated guesses and extrapolate theories to energy regimes well beyond their domain of applicability or validity. There's a lot we really don't understand about the very earliest stages in the evolution of the universe.

Nevertheless, the discovery of something that looks very much like the standard model's Higgs boson, at CERN in July 2012, suggests that our ignorance of the real state of affairs in the history of the universe might be limited to the first trillionth of a second of its existence. Now, I would respectfully suggest, that's not so bad.

In choosing to tell the story chronologically, I've had to accept some challenges. It means that the opening chapters involve us in some fairly heavy and demanding physics and, since much of the observational evidence that shapes our understanding of the early universe is derived from events that occur later in its history, there are places in the story where I have to ask you temporarily to suspend your demand for the evidence.

Chapter 4 continues the story with the formation of the first stars and galaxies, some time between 300 and 550 million years after the big bang. It describes the fundamentally important roles played by tiny inhomogeneities in the distribution of matter, thought to be imprinted on the large-scale structure of the universe by inflation, and the mysterious dark matter believed to account for almost 27% of the total mass-energy of the universe today. Chapters 4 and 5 detail what we currently understand about the evolution of stars and of stellar nucleosynthesis, which produces a range of chemical elements from hydrogen to iron. Cataclysmic supernova explosions are required to explain the existence of elements heavier than iron.

The sprinkling of a broad variety of chemical elements in the dust and vapour ejected from exploding stars leads to the production of interstellar molecules, many of which are now recognized to be important in the chemistry of life. These molecules seed the interstellar clouds which slowly gather together and eventually collapse to form new third-generation stars with associated planetary systems. Chapter 6 describes the formation of the solar system about 4.6 billion years ago, from a spinning giant molecular cloud which contracts and condenses. Dust in the outer parts of the cloud condenses to form first rock and metals. These accrete to form planetesimals, which combine eventually to form the inner planets.

Our attention switches in Chapter 7 to the early history of the Earth, its differentiation into core, mantle, crust, ocean, and atmosphere and the factors influencing surface composition and temperature. The Earth acquires a Moon in a collision with another planet-size body, which we call Theia. Earth's fluid mantle convects, and the continents start to move. Realignment of the outer planets precipitates the Late Heavy Bombardment, as billions of billions of tonnes of rock and ice crash to the planet's surface. We trace the subsequent evolution of the Earth to the point where conditions are right for the emergence of life.

Chapter 8 opens with a relatively stable warm wet Earth sprinkled with a variety of organic chemicals known to be essential to life and with natural deep-ocean geological systems that may act as factories for the conversion of simple inorganic chemicals into complex biochemical systems. The fossil record shows that primitive, single-celled organisms existed on Earth about 3.5 billion years ago, just a billion years after the Earth is first formed. How abiogenesis—the spontaneous generation of living from non-living matter—happens remains essentially mysterious. We have some compelling theories but there is as yet no 'standard model' for the origin of life.

Irrespective of how it comes about, we do know that the basic biological structures that emerge about 3.5 billion years ago establish a template that will be replicated in all subsequent life forms. Chapter 9 describes how early single-celled organisms use photosynthesis to terraform the planet by pumping their waste oxygen into the atmosphere. Oxygen opens up new opportunities for evolution to experiment with. Some single-celled organisms now merge with each other to form complex cells, and go on to form larger and larger multicellular creatures. After 2.8 billion years of evolution, the first primitive animals appear.

The story of the last 540 million years or so of Earth's history is a song of ice and fire. As Chapter 10 recounts, the planet swings violently between inhospitable ice ages and periods of deadly volcanic eruptions and at least one asteroid impact. Life clings on, both in the oceans and now on the land. Evolution drives frantic periods of diversification of different animal species, only for these to be cruelly pegged back in a series of planet-wide mass extinctions. In the mass extinction event 252 million years ago, called the 'Great Dying', almost 95% of all marine animals and a substantial proportion of all land animals are wiped out. But each extinction is followed by an evolutionary 'radiation' in the near-empty ecology that results. From the ashes of the Great Dying comes the age of the dinosaurs, until they too are destroyed by an asteroid, 66 million years ago.

Among the creatures that survive this last calamity are small mammals. Chapter 11 describes the evolutionary radiation that now occurs, paying particular attention to the primates. As primate species diversify, they follow different evolutionary lines, much like the branches of a tree. The first to branch away are tarsier monkeys, followed by New World monkeys, Old World monkeys, gibbons, orangutans, and gorillas. Finally, about 5–7 million years ago, chimpanzees part evolutionary company with the line of hominins.

Tracing the lineage of modern humans from this point is fraught with difficulty and much is contested. But the fossil evidence supports a line that brings us through early hominins dated 4–7 million years old, to species of the genus *Australopithecus*, 2–4 million years old, to early *Homo* species 2 million years ago. Of these, *Homo sapiens* appears in Africa just 200 000 years ago.

There can be little doubt that human consciousness is what sets humans apart. The ability to have abstract thoughts, imagine concepts, and develop intelligence has allowed humans to break free from the prison of the present, and is the basis for the proliferation of a humanity of 7.2 billion across the Earth. Chapter 12 attempts to describe the origin and evolution of consciousness, its foundation in genetics, and its close fundamental ties to the development of language and society.

I've mapped out a 'timeline of creation' in Table 1. This lists the origins of its various singular components, from space, time, and energy all the way to human consciousness, measured in time from the present (2015) and from the big bang as 'time zero'. To put this into some kind of perspective that we mere mortals might grasp, I've also mapped the time measured from the big bang onto a single hypothetical 24-hour 'day of creation'.

On this reckoning, the universe 'begins' at midnight. Particles with mass appear the merest whisper of a fraction of a second afterwards, and the universe is bathed in light at the moment of recombination two seconds later, as primordial electrons latch themselves to primordial hydrogen and helium nuclei. Stars and galaxies first appear between 12:30 and 1:00 a.m., with complex molecules starting to make their appearance sometime between 3 a.m. and 6:30 a.m., in time for breakfast.

We're then obliged to sit on our hands for most of the day—9–10 hours—as we wait for the Sun and Earth to appear, at nearly 4 p.m. At some time during this wait, the expansion of the universe flips. The matter in the universe that has thus far been slowing down the rate of expansion becomes so dilute that 'dark energy'—the energy of 'empty' space—takes over and starts to accelerate the expansion once again.

Life emerges around 6 p.m., and complex cells and multicellular organisms around 8:30 p.m. A few hours later we see the beginnings of the diversification of animal species in the 'Cambrian explosion'. Modern humans make their first appearance at about 1 second before midnight. Human consciousness

TABLE 1: **The Timeline of Creation**

Chapter	The Origin of...	Measured from the Present (2015)*	Measured from the Big Bang	Mapped to a Single 'Day of Creation'
1	Space, Time and Energy	13.8 Ga	'0'	Midnight
2	Mass	13.8 Ga	10^{-12} s	A fraction after midnight
3	Light	13.8 Ga	380 000 Yrs	2 seconds after midnight
4	Stars and Galaxies	13.5-13.3 Ga	300–550 million Yrs	Between 12:30 to 1:00 am
5	Molecules	10-12 Ga	1.8–3.8 billion Yrs	Between 3:00–6:30 a.m.
6	Solar System	4.6 Ga	9.2 billion Yrs	About 4 p.m.
7	Earth			
8	Life	3.5 Ga	10.4 billion Yrs	Almost 6 p.m.
9	Complex Cells and Multicellular Organisms	~2 Ga	11.8 billion Yrs	About 8:30 p.m.
10	Species (Animal Species Diversity)	540 Ma	13.4 billion Yrs	A little after 11:00 p.m.
11	*Homo sapiens*	200 ka	13.8 billion Yrs	About 1 second to midnight
12	Human Consciousness	50 ka	13.8 billion Yrs	About 300 milliseconds to midnight

* Ga = billions of years ago, Ma = millions of years ago, ka = thousands of years ago, Yrs = years, s = seconds. Based on an estimate of the age of the universe of 13.82 billion years established by recent results (21 March 2013) from the European Space Agency's Planck satellite.

develops throughout this second, but has begun to realize its full potential with the transition to behavioural modernity in the 'Great Leap Forward' which happens with just 300 milliseconds (thousandths of a second) left on the clock.

As *Origins* will, I hope, make abundantly clear, these have been a very busy few hundred milliseconds.

1

IN THE 'BEGINNING'

The Origin of Space, Time, and Energy

Don't be fooled. No matter what you might have read in some recent popular science books, magazine articles, or news features, and no matter how convincing this might have seemed at the time, be reassured that *nobody* can tell you how the universe began. Or even if 'began' is a word that's remotely appropriate in this context.

There's a good reason for this. As we will see in what follows, we know the universe is expanding. By extrapolating backwards, we know therefore that there must have been a moment in its history when all the energy in the universe was compacted to an infinitesimally small point, from which it burst forth in what we call the 'big bang'.

How do we know? This chapter will provide some of the answers to this question, and I'll provide the scientific evidence for the big bang and the expanding universe as subsequent chapters relate the story of its evolution through its first 380 000 years. Suffice to say that something like the big bang *must* have happened, and our best estimate is that it happened about 13.8 billion years ago, give or take a few hundred million years.

Describing the very 'beginning' of the universe is problematic because, quite simply, none of our scientific theories are up to the task. We attempt to understand the evolution of space and time and all the mass and energy within it by applying Albert Einstein's general theory of relativity. This theory works extraordinarily well. But when we're dealing with objects that start to approach the infinitesimally small, we need to reach for a completely different structure, called quantum theory. Now, the general theory of relativity can't handle some things in ways that quantum theory can, and vice versa. But when we try to put these two venerable theories together to create some kind of unified theory that could do the work of both, we find that they really don't get along, and the whole structure falls apart.

So far, nobody has been able to figure out how to fix this.

And there's another problem. Insofar as our extrapolations from the present day universe can be trusted, they tell us that the energy of a hot big bang must have been much, much greater than any energy we could ever hope to re-create on Earth in a particle collider. So, even if we could one day build a theory that could be applied with some confidence, we will simply never be able to build an apparatus to perform the experiments and make the observations that would be required to test such a theory's predictions.* We have no alternative but to rely on what we can discover from the observable universe, and use our theories to infer what *might* have happened in the distant past.

What this means is that the very beginning of the universe (if this is indeed the right word) is beyond the reach of science for the foreseeable future and, quite possibly, for all time. Of course, this doesn't stop us from speculating, and there are many contemporary theories that provide various origin-of-the-universe stories. In some of these, the universe emerges 'from nothing' in a quantum fluctuation.† Or the universe is simply one of a large number (possibly an infinite number) of expanded bubbles of spacetime in a 'multiverse' of possibilities. Or the big bang results from the collapse of the universe that went before, as the cosmic reset button is pressed once again in a cycle that has lasted for all eternity.

There is no empirical evidence for any of these different ideas. It's perhaps possible that some of these theories can be developed to the point where they can predict subtle physical phenomena that might be detectable in our own universe, using Earth-bound or satellite-borne instruments (although, frankly,

* Which is probably just as well. I'm not entirely sure what would happen if the conditions that prevailed during the big bang were ever to be re-created on Earth.

† 'Nothing' is a philosophically loaded concept, best avoided if you're unprepared to argue semantics.

I'm inclined to think that this possibility is remote). Even then, as I've said, the prediction of phenomena observable today still allows us only to infer what might have happened during or before the big bang. Choosing what to believe about this moment will still require something of an act of faith.

The Austrian philosopher Ludwig Wittgenstein once famously cautioned: 'Whereof one cannot speak, thereof one must be silent'.[1] That's probably good advice, but I've promised you a book about origins, so in this chapter I'm going to try to walk the fine line between accepted science—what we do know and can prove—and speculative theorizing: what we can only make moderately educated guesses at, based on scientific principles that have at least some validity. I'll hang warning signs in the appropriate places, so we don't inadvertently stumble and fall down a metaphysical rabbit hole.

Our creation story begins with the origin of space, time, and energy, and it is here that we meet our first challenge, even before we can properly start to tell the tale. For what are space, time, and energy? How should we conceive of these things?

THE NATURE OF SPACE AND TIME

I'm sitting at the desk in my study, typing these words on a keyboard which is wirelessly connected to a docked laptop, watching my sentences take shape on a large-screen monitor. If I take my eyes away from the monitor and look around me, I see a room with the architecturally favoured number of walls—four. Two of these, to my left and behind me, are decorated with shelves on which sits a modest collection of books. Against another wall to my right I have a sofa-bed which is used occasionally when I have sleep-over guests (and which today, unusually, is not piled with yet more books).

Like you, I have no hesitation in concluding that the things in this room are objects in space.

But what, precisely, *is* space? I can move through it, but I can't see it and I can't touch it. Space is not something that we perceive directly. We perceive objects (such as monitors, books, and sofa-beds) and these objects have certain relations one with another that we call spatial relations: this here on the left, that over there on the right. But space itself does not form part of the content of our direct experience. We interpret the objects as existing in a three-dimensional space as a result of a synthesis of electrical signals in our brains translated into visual perceptions by our minds.

Similarly, I move through time (in one direction, at least) but I can't see it and I can't reach out and touch it. Time is not a tangible object. My sense of time would seem to be derived from my sense of self and the objects around me changing their relative positions (this *was* on the left, *now* it's on the right), or changing their nature, from one type of thing into another.

Does the space in this room exist independently of the objects in it? Does time exist independently of the things that happen here? In other words, are space and time 'absolute' things-in-themselves?

In developing the theory of mechanics that he described in his great work *The Mathematical Principles of Natural Philosophy*, first published in 1687, Isaac Newton was willing to acknowledge the essential relativity of space and time, in what he called our 'vulgar experience'. He was willing to accept that objects move towards or away from each other, changing their relative positions in space and in time. This is relative motion, which can be defined simply in terms of the relationships between the objects.

But Newton's theory required an absolute motion which, he argued, must imply an absolute space and time that forms a kind of container within which objects exist and things happen. Take all the objects out of the universe and, Newton's theory demanded, the empty container would remain: there would still be 'something'.

Einstein begged to differ. He pondered this question while working as a 'technical expert, third class' at the Swiss Patent Office in Bern more than two hundred years later, in 1905. He concluded that absolute space and time cannot exist. This conclusion follows from Einstein's special theory of relativity.

This theory is based on two fundamental principles. The first, which became known as the *principle of relativity*, states that observers who find themselves in states of relative motion at different (but constant) speeds *must* observe precisely the same fundamental laws of physics.

This seems perfectly reasonable. Suppose I make a set of physical measurements here on Earth which allows me to deduce some underlying physical law. You make the same measurements on board a distant spaceship moving away from the Earth at high speed. The conclusions we draw from both sets of measurements must surely be the same. There can't be one set of physical laws for me and another set for space travellers. Otherwise they wouldn't be laws.

We can turn this on its head. If the laws of physics are the same for all observers, then there is no measurement we can make which will tell us which observer is moving relative to the other. To all intents and purposes, you may actually be stationary, and it is me who is moving away at high speed. We cannot tell the difference using physical measurements.

The second of Einstein's principles concerns the speed of light. At the time that he was working on special relativity, physicists had rather reluctantly concluded that the speed of light is constant, completely independent of the speed of the source of the light. If I measure the speed of the light emitted by a flashlight held stationary on Earth and you measure it again using the same flashlight on board a spaceship moving at high speed, we expect to get precisely the same answers.

Instead of trying to figure out *why* the speed of light is independent of the speed of its source, Einstein simply accepted this as an established fact. He assumed the speed of light to be a universal constant and proceeded to work out the consequences.

One immediate consequence is that there can be no such thing as absolute time.

Here's why. Suppose you observe a remarkable occurrence. During a heavy thunderstorm you see two bolts of lightning strike the ground simultaneously, one to your left and one to your right. You're standing perfectly still, so the fact that it takes time for the light from each of these lightning bolts to reach you is of no real consequence. Light travels very fast so, as far as you're concerned, you see both bolts at the instant they strike.

However, I see something rather different. I'm travelling at very high speed—half the speed of light, in fact—from left to right. I pass you just as you're making your observations. Because I'm moving so fast, the time taken for the light from the lightning bolts to reach me now has measureable consequences. By the time the light from the left-hand bolt has caught up with me, I've actually moved quite a bit further to the right, and so the light has further to travel. But the light from the right-hand bolt has less ground to cover because I've moved closer to it. The upshot is that I see the right-hand bolt strike first (Figure 1).

You see the lightning bolts strike simultaneously. I don't. Who is right?

We're both right. The principle of relativity demands that the laws of physics must be the same irrespective of the relative motion of the observer, and we can't use physical measurements to tell whether it is you or me who is in motion.

We have no choice but to conclude that there is no such thing as absolute simultaneity. There is no definitive or privileged frame of reference in which

FIGURE 1 The stationary observer in (a) sees the lightning bolts strike simultaneously, as the light from both travels so fast as to appear instantaneous. But the observer in (b), who is moving at a considerable fraction of the speed of light, sees something different. He's moving at half the speed of light from left to right, so by the time the light from the left-hand bolt catches up with him, he's moved a bit further to the right. The light from the right-hand bolt now has less far to travel. Consequently the observer in (b) sees the right-hand bolt strike first.

we can declare that these things happened at precisely the same time. They may happen simultaneously in this frame or they may happen at different times in a different frame, and all frames are equally valid. Consequently, there can be no 'real' or absolute time. Something's got to give. We perceive events differently because time is relative.

Einstein developed a similar set of arguments to show that space is relative, too. The bizarre consequences of special relativity are reasonably well known. Demanding that the laws of physics appear the same for all observers in a universe in which the speed of light is fixed means that time intervals (durations) can dilate and spatial intervals (distances) can contract. This means that durations and distances will be measurably different for different observers travelling at different speeds.

But all is not lost. Time dilation and distance contraction are like two sides of the same coin. They're linked by the speed of the observer making the measurements relative to the speed of light. If we now combine space and time together in a four-dimensional *spacetime*, then intervals measured in this spacetime are unaffected by relativity. In spacetime intervals, time dilations are compensated for by distance contractions, and vice versa.

Does this mean that, although space and time are relative, spacetime is absolute? Some contemporary physicists think so. Others disagree. What's important for us to realize is that we must abandon our simplistic, common-sense notions of an independent space and time and accept that in our universe these are inextricably connected.

MASS AND ENERGY

Einstein's 1905 research paper on special relativity was breathtaking in its simplicity yet profound in its implications. But he wasn't quite finished. He continued to think about the consequences of the theory and just a few months later he published a short addendum in the same journal.

In this second paper he considered the situation in which a moving object emits two bursts of light in opposite directions. The initial energy of the object is entirely in the form of its energy of motion (which is called kinetic energy). Each burst of light carries the same amount of energy away from the object, $\frac{1}{2}E$. As the total energy must be conserved, the object's kinetic energy must therefore fall by a total amount E. This makes perfect sense. The light carries energy away, and the energy must come from somewhere.

He then imagined what two different observers might measure, with one observer moving along in the 'rest frame' of the object (keeping pace with the object so that it appears as though at rest) and the second observer moving at a constant speed relative to this frame. These observers measure the difference in the energy of the object before and after emission of the light. He found, perhaps not altogether surprisingly, that the different observers get different results.

After a little bit of algebraic manipulation, he arrived at a mathematical expression which allowed him to draw an extraordinary conclusion. The energy of the light bursts comes from the object's kinetic energy of motion. This can be calculated from the mass of the object and its speed.[2] From the difference in the two sets of results, Einstein deduced that the energy carried away is derived not from the object's speed (as might be anticipated), but from its *mass*.

If the total energy carried away by the light is E, Einstein concluded that the mass (m) of the object must diminish by an amount E divided by c^2, where c is the speed of light. It doesn't matter what kind of object we might be referring to: this is a general result, universally applicable. The inertial mass of an object (a measure of its resistance to acceleration) is also a measure of the amount of energy it contains.

Today we would probably rush to rearrange the equation in Einstein's paper to give the iconic formula $E = mc^2$. But Einstein himself didn't do this. Although he was uncertain that this was something that could ever be tested experimentally, he was prepared to speculate that the conversion of mass to energy might one day be observed in radioactive substances, such as radium.

The special theory of relativity blurs our commonsense conceptions of a physical reality of space and time, matter, and energy. Space and time are relative, they are defined by the things they contain and events that happen, though spacetime *might* be absolute. Mass is energy, and from energy can spring mass. These modifications of our common conceptions are important to acknowledge if we are to understand precisely what it was that originated in the big bang.

But there is yet a further modification we need to make. Spacetime and mass-energy do not themselves exist completely independently of one another. They are locked together in an elegant dance described by Einstein's general theory of relativity.

GRAVITY AND GEOMETRY

Gravity is a familiar 'everyday' kind of force. When I drop something, it falls to the ground. It does this because it experiences the force of gravity. We struggle against this force every morning when we get out of bed. We fight its effects every time we lift a heavy weight. When we stumble to the ground and graze a knee, it is gravity that causes the hurt. Such is its familiarity that it's tempting to assume that science must have long ago answered all our questions about it.

And, indeed, we learn in school of Newton's law of universal gravitation. Bodies of material substance are attracted to one another, with the force of attraction increasing with the product of their masses and inversely with the square of the distance between them.[3] But, although Newton's law was a great achievement, there are some real problems with its interpretation, problems that were obvious in Newton's time but for which he could offer no solutions.

In the mechanical universe described by Newton's laws of motion, we interpret force to be something that is exerted or imparted by objects impinging on each other. A stone does not move unless we kick it or throw it, thereby accelerating it to some final speed as it sails through the air. But precisely what is it that grasps the Moon as it swoons in Earth's gravitational embrace? How does the Moon push the afternoon tide up against the shore? When a cocktail glass slips from a guest's fingers, what grabs it and forces it to shatter on the wooden floor just a few feet below?

Newton was at a loss. His force of gravity seems to imply some kind of curious action-at-a-distance. Objects influence each other over great distances through empty space, with nothing obviously transmitted between them. Critics accused him of introducing 'occult elements' in his theory of mechanics.

Part of the solution to this riddle would come to Einstein during an otherwise average day at the Patent Office in November 1907, by which time he had been promoted to 'technical expert, second class'. As he later recalled: 'I was sitting in a chair in my patent office at Bern. Suddenly a thought struck me: If a man falls freely, he would not feel his weight'.[4]

In this stunningly simple observation, Einstein realized that our local experiences of gravity and of acceleration are the same. He called it the equivalence principle. Working out what this meant would take him another eight years and would require another extraordinary connection, between gravity and geometry.

The geometry we learn about in school is Euclidean geometry, named for the ancient Greek mathematician Euclid of Alexandria. In this geometry, parallel lines never cross, the angles of a triangle add up to 180 degrees, and the circumference of a circle is twice its radius multiplied by π. This is a geometry associated with a kind of three-dimensional space that mathematicians call 'flat'. We learn about Euclidean geometry because the spacetime of our universe happens to be a flat spacetime.

In a flat space the shortest distance between two points is obviously the straight line that we can draw between them. But what is the shortest distance between London and Sydney, Australia? We could look up the answer: 10,553 miles. But this distance is not, in fact, a straight line. The surface of the Earth is curved, and the shortest distance between two points on such a surface is actually a curved path called a *geodesic*.

Now comes a rather breathtaking leap of imagination. What if the spacetime near a large object isn't 'flat'? What would happen if it were to be curved? Einstein realized that he could get rid of the action-at-a-distance implied by Newton's gravity by replacing it with curved spacetime. An object with a large mass-energy warps the spacetime around it, and objects straying close to it follow the shortest path determined by this curved spacetime.

American physicist John Wheeler summarized the situation rather succinctly some years later: 'Spacetime tells matter how to move; matter tells spacetime how to curve'.[5] In general relativity, gravity is not a force that matter exerts on matter. It is a force that matter (or, strictly speaking, mass-energy) exerts on spacetime itself. Objects do experience a mutual gravitational attraction, but the attraction is *indirect*, mediated by the curvature of the spacetime between them.

These are subtle, but real, effects. The general theory of relativity correctly accounts for some peculiarities in the orbit of the planet Mercury that originate in the curvature of spacetime near the Sun, something that Newton's theory of gravity fails to predict correctly. And, although light from a distant star that passes close to the Sun on its way to Earth follows a straight-line path, the curvature of spacetime near the Sun makes the path appear to bend. When this phenomenon was first demonstrated during a total eclipse in 1919, Einstein became a household name.

On 24 April 2004, an exquisitely delicate instrument called Gravity Probe B was launched into polar orbit, 642 kilometres above the Earth's surface. The satellite housed four gyroscopes, designed to measure the effects of spacetime curvature around the Earth. To eliminate unwanted torque, the satellite was rotated once every 78 seconds and thrusters were used to keep it pointing towards the star IM Pegasi in the constellation of Pegasus (Figure 2).

Two effects were measured. The curvature of spacetime causes the gyroscopes to precess* in the plane of the satellite's orbit (that is, in a north–south direction), an effect known as *geodetic drift*. The second effect is *frame-dragging*. As the Earth rotates on its axis, it drags spacetime around with it in the plane perpendicular to the plane of the satellite orbit (in a west–east direction). This gives rise to a second precession of the gyroscopes.

The results were announced at a press conference on 4 May 2011. Although an unexpected wobble in the gyroscopes had resulted in some significant uncertainty, the measurements of both geodetic drift and frame-dragging provided a very powerful experimental vindication of general relativity.

* This results in a shift in the axis of rotation of the gyroscopes which, though very small, can be measured with great accuracy. I remember playing with a toy gyroscope when I was young. You would set it spinning with a pull string, sit back and watch (these were simpler times). It taught me about precession, although I didn't know that was what I was seeing at the time.

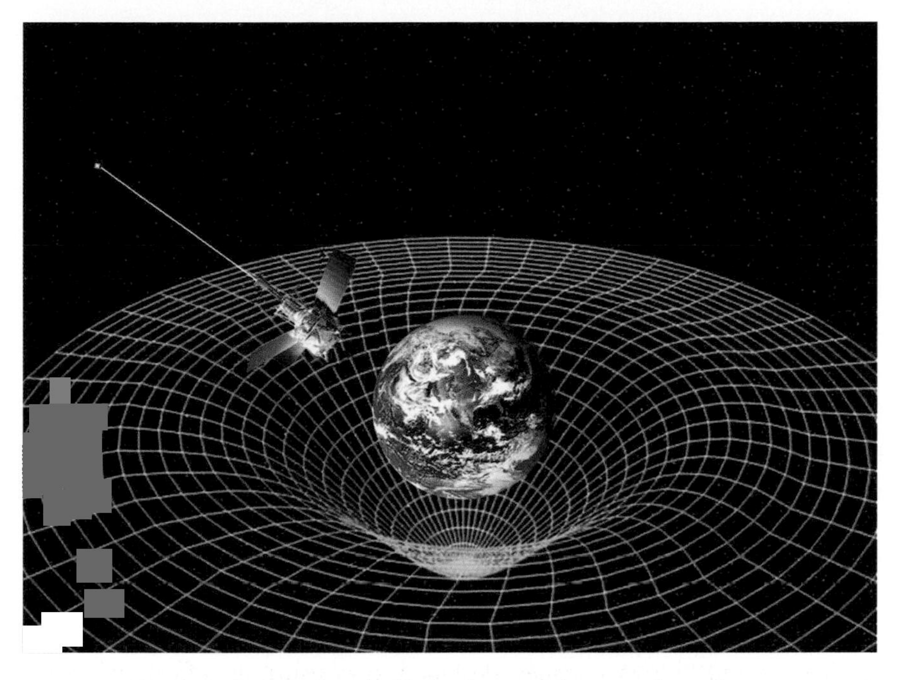

FIGURE 2 Gravity Probe B was launched in April 2004 and measured two phenomena associated with the curvature of spacetime around the Earth. The results were announced in May 2011, and provided a powerful vindication of general relativity. This picture shows the satellite moving in the curved spacetime around the Earth, pointing towards the star IM Pegasi, in the constellation of Pegasus.

THE EXPANDING UNIVERSE

Einstein presented his new general theory of relativity to the Prussian Academy of Sciences in Berlin in 1915. Two years later he applied the theory to the whole universe.

At first glance, this seems impossibly difficult. How can a single set of equations describe the whole universe? The answer is: by making a couple of simplifying assumptions. Einstein had to assume that the universe is uniform in all directions, containing objects that have the same kind of composition. He also had to assume that the universe we observe from our vantage point on Earth is no different from the universe as observed from any and all such vantage points. In other words, observers on Earth occupy no special or privileged position. What we see is a 'fair sample' of the universe as a whole.

What Einstein got was singularly appealing, a universe that is finite but nevertheless 'unbounded', without edges. We know that the ground on which we walk appears to be flat, but if we walk far enough in one direction we also know that we will eventually circumnavigate the Earth, without falling off the edge. So in Einstein's universe, spacetime curves back on itself like the surface of a sphere. At any one point in this universe spacetime looks flat but it is, in fact, gently curved.

But Einstein then quickly ran into a big problem. He had anticipated that the universe that should emerge from his equations would be consistent with prevailing scientific prejudice—a universe that is stable, static, and eternal. What he got instead was a universe that is dynamic—either expanding or contracting depending on the initial assumptions. The field equations suggested that a static universe is impossible.

Gravity is the weakest of nature's forces (we'll meet the others later in this chapter). But it is cumulative and inexorable and acts only in one 'direction'—it serves to pull objects together but it doesn't push them apart. Einstein realized that the mutual gravitational attraction between all the material objects in the universe would cause it to collapse in on itself. This was a troubling result, quite inconsistent not only with prevailing opinion but also arguably with simple observation. After several centuries of astronomy there is no evidence that all the stars in the universe are rushing towards each other in a catastrophic collapse. Quite the opposite, in fact, as we will see.

But there was nothing in his gravitational field equations to stop this from happening. As he later explained: 'We admittedly had to introduce an extension to the field equations that is not justified by our actual knowledge of gravitation'.[6]

The left-hand side of the most general form of this equation describes the curvature of spacetime and hence the strength of the force of gravity that will act on all the mass-energy, which is summarized in the right-hand side of the equation. Einstein chose to modify the equation by subtracting from the left-hand side a term containing a 'cosmological constant', usually given the Greek symbol Λ (lambda).

In essence, this extra term imbues spacetime with a kind of odd, anti-gravitational force, a kind of negative pressure which builds in strength over long distances and counteracts the effect of the curvature caused by all the mass-energy in the universe. By carefully selecting the value of the cosmological constant, Einstein found that he could balance the gravitational attraction that tended to pull everything together, caused by all the mass-energy on the right-hand side, with a spacetime on the left-hand side that has a tendency to push everything apart. The result was perfect balance, a static universe.

It was quite a neat solution. Introducing the cosmological constant didn't alter the way general relativity works over shorter distances, so the successful predictions of the orbit of Mercury and the bending of starlight were preserved. But it was, nevertheless, a rather unsatisfactory 'fudge'. There was no evidence for the cosmological constant, other than the general observation that the universe *seems* to be stable and static.

Freed from prejudices about the kind of universe that *should* result, Einstein's field equations actually yield many different kinds of possible solutions. In 1922, Russian physicist and mathematician Alexander Friedmann offered a number of different solutions of Einstein's original equations. He was not particularly interested in trying to represent our own universe, preferring instead to explore the different possibilities allowed by the mathematics. Consequently, while he retained the cosmological term that Einstein had introduced, he assumed that it could take any value, including zero.

Friedmann discovered a range of different possible model universes, with properties and behaviour that depend on the relationships between the amount of mass and the size of the cosmological constant. He focused his attention on solutions with positive spacetime curvature, showing that they could expand or contract. He was particularly taken with solutions based on an assumed cosmological constant of zero which oscillate back and forth, alternating between expansion and contraction, the period of oscillation depending on the amount of mass.

A universe in which the density of mass-energy is high (lots of objects in a given volume of space) and the rate of expansion is modest is said to be 'closed'. It will expand for a while before slowing, grinding to a halt and then turning in on itself and collapsing. Spacetime in such a universe has a positive curvature. A few years later Friedmann examined universes in which spacetime is negatively curved. Such universes are infinite, they are said to be 'open' and will expand forever.

Tragically, Friedmann died of typhoid fever in 1925. But his expanding-universe solutions were independently rediscovered in 1927 by Belgian theorist (and ordained priest) Abbé Georges Lemaître. Einstein was initially dismissive of the idea of an expanding universe but when, in the early 1930s, the evidence from observational astronomy* suggested rather strongly that the universe is indeed expanding, he accepted that he had been wrong, and expressed regret for fudging his equations.[7]

* We will take a closer look at this evidence in Chapter 4.

The evidence in favour of an expanding universe became overwhelming in 1965, with the discovery of the cosmic background radiation. This is the cold remnant of hot radiation that spilled into the universe soon after its birth. We will encounter it in Chapter 3.

But what *causes* the universe to expand? In a paper published in 1933, Lemaître suggested that expansion is triggered because empty spacetime is not, in fact, empty. Einstein had introduced his cosmological term on the left-hand side of his field equation, as a modification of spacetime itself, designed to offset the effects of the curvature caused by all the mass-energy in the universe. But it takes just a moment and a little knowledge of algebra to move the cosmological term from the left-hand (spacetime) side of the equation to the right-hand (mass-energy) side. Now it represents a *positive* contribution to the total mass-energy of the universe. This is not the familiar mass-energy we associate with stars, planets, and people. Rather, it takes the form of an energy of 'empty' spacetime, sometimes called *vacuum energy*.[8]

It seems that Lemaître's paper had little impact at the time. But it will be useful to remember this connection between the rate of expansion of spacetime and vacuum energy, as we'll need it again quite soon.

TIME, SIZE, AND TEMPERATURE

We're now finally in a position to get this party started. If the universe is expanding, then simple logic suggests that we can 'wind the clock back' and conclude that it must have had an origin at some point in time. For reasons that will become apparent in later chapters, this origin is currently fixed at about 13.8 billion years ago.

Studying the evolution of the universe from this moment forward is much like studying the evolution of a homogeneous ball of hot dense matter that has just exploded into existence. The variables that we need to pay particular attention to are time, size, and temperature. As time elapses, the universe expands, growing in size. As it expands, it cools. And as it cools, the form of the stuff inside it changes.

Figure 3(a) is fairly typical representation of the expansion of the universe from the big bang to the present day. In this diagram, the three-dimensional universe is pictured as a two-dimensional slice, and the expansion is reflected in the area of this slice as time passes. The stuff that exists within it (stars, galaxies, and planets) is then pictured inside the slice as it expands.

(a)

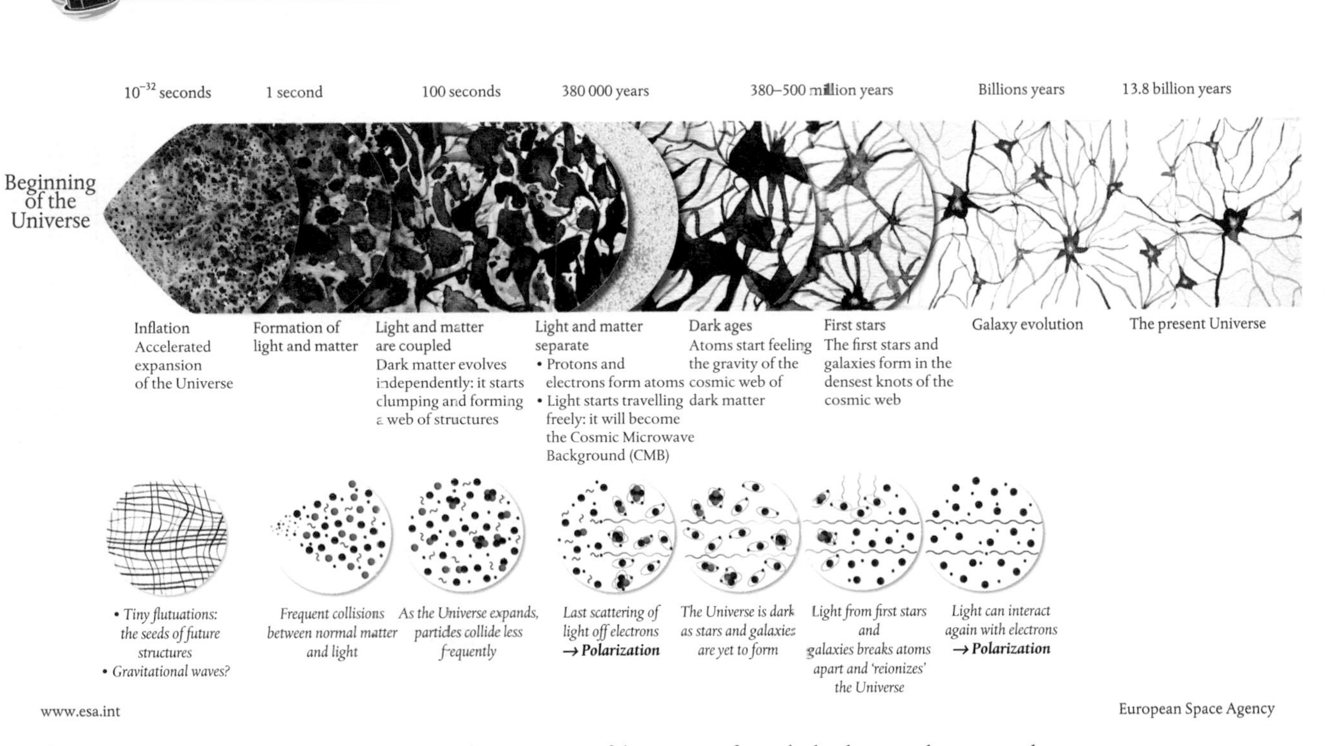

FIGURE 3 (a) A fairly typical way of representing the expansion of the universe from the big bang to the present day.

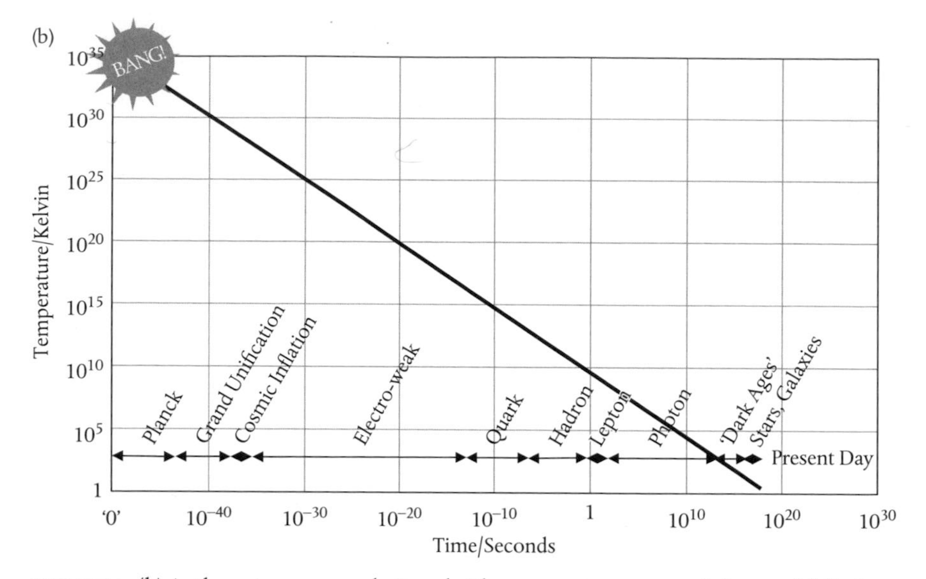

FIGURE 3 (b) As the universe expands, it cools. The average temperature defines and delimits a series of 'epochs' as the nature of forces and particles changes. Note that both time and temperature scales in this figure are logarithmic, increasing in orders of magnitude. This places much greater emphasis on the earlier moments in the evolution of the universe.

Notice that there are some subtleties—this isn't simply a linear expansion from some notional 'time zero'. The universe starts out infinitesimally small, then suddenly undergoes a short burst of exponential expansion as a result of something called cosmic inflation (more on this later). The universe then settles down, the rate of expansion declining as all the mass-energy in the universe applies the brakes. What Figure 3(a) doesn't show is that, about 9 billion years after the big bang, the rate of expansion starts to accelerate again.

As the universe expands, its average temperature falls. Figure 3(b) shows this temperature decline from the point at which it is remotely sensible to talk about 'temperature', shortly after the big bang, to the present day.* This straight-line diagram looks absurdly simple, but note that both time and temperature scales are logarithmic—they increase in powers of ten. Present day is 13.8 billion years after the big bang, or a little over 400 000 trillion (4×10^{17}) seconds.†

* Figure 3(b) gives temperature measured in units of kelvins. The kelvin scale of temperature can be derived from the more familiar celsius (centigrade) scale by adding another 273 degrees. Thus, the freezing point of water is 0°C, or 273 kelvin. The boiling point is 100°C, or 373 kelvin. Zero kelvin is 'absolute zero', the temperature of a (hypothetical) body with absolutely no energy.

† If you're uncertain about the representation of very small or very large numbers as numbers multiplied by a power of ten (as in 4×10^{17}), then please consult the Appendix.

The average temperature of the universe is fundamentally important, as it determines what kinds of things and what kinds of forces can exist within it. So, the average temperature is used to define and delimit a series of 'epochs', from the big bang itself, through the evolution of forces and the evolution of elementary particles, atomic nuclei, atoms, stars, and galaxies.

These epochs are classified according to the dominant themes that shape the physics of the universe. The first few epochs are dominated by physical forces. Subsequent epochs are dominated by material particles, starting with quarks and ending with photons. In these first three chapters, I will attempt to tell the story of this evolution from the big bang to the end of the Photon Epoch.

THE PLANCK EPOCH: 'ZERO' TO 10^{-43} SECONDS

I've already explained why it's not possible for science to provide a definitive description of the big bang in its very earliest moments. If we extrapolate backwards using the general theory of relativity then we will arrive at a *singularity*, a situation characterized by a rather hostile outbreak of infinities in the maths. This tells us that we're pushing the theory too far beyond its domain of applicability. The theory simply can't handle it. We should be able to fix this by mixing in quantum theory but—so far—that hasn't worked.

But what do we *think* might have happened? Remember that the big bang heralds the origin of space, time, and energy, which is everything there is, at least in *our* universe. We don't have a theory that we can apply to the moment of the big bang, but perhaps we can make some educated guesses about what might have happened very shortly afterwards, by extrapolating backwards from the properties of the universe as it appears to us today.

I often buy ice from my local supermarket. It comes in clear plastic bags weighing a couple of kilos. The cubes of ice inside are roughly the same size, but they tend to clump together and can sometimes be difficult to separate without an ice pick. Suppose I empty the contents of such a bag into a large saucepan. The pile of ice is jumbled—it isn't particularly uniform, or symmetrical. It looks different in different directions: there are some smaller lumps over here and bigger lumps over there. This non-uniformity persists at the microscopic level of the water molecules that make up the ice. The molecules form a lattice that looks different in different directions. There's a difference between up, left, and front, and down, right, and back.

What's going to happen if I now put the pan on the hob and apply some heat?

That's easy. The differently shaped cubes and the clumps will start to melt. Very soon I'll have a much more uniform (and much more symmetrical and homogeneous) volume of liquid water in the pan. At the microscopic level the lattice is now replaced with water molecules moving about but linked together in short-range networks. It's now much more difficult to see a difference looking up, left, and front, and down, right, and back.

I now place the lid on the pan and seal it so that nothing can escape. If I continue to apply heat then eventually the liquid will boil and I'll have an even more uniform (and symmetrical) distribution of steam (water vapour) in the pan. At the microscopic level, individual water molecules are now flying past each other and colliding, and the short-range networks are gone. It's now impossible to tell the difference between up, left, and front, and down, right, and back. The vapour looks the same in all directions.

And this is not so very different from how we imagine that lumps of universe would behave if we could heat them to the kinds of temperatures that prevailed shortly after the big bang.

The great variety of particles and forces that shape our universe are expected to disappear as they become more and more homogenized and symmetrical. Today we experience four forces of nature. We've already encountered gravity, which works at the scale of large objects, such as cocktail glasses, books, planets, and stars. The other three forces work at the microscopic scale of molecules, atoms, and subatomic particles.

There are two types of subatomic particle, called quarks and leptons. Quarks have three different kinds of properties: electrical charge, colour and flavour.* The strong force acts on quark colour, and is transmitted from one quark to another by massless particles called gluons. The weak nuclear force acts on quark flavour, and its most common manifestation is in the form of a type of radioactivity called beta radioactivity. The weak force is transmitted by 'heavy photons', called W and Z particles, of which there are three: W^+, W^-, and Z^0. Quarks of different colours and flavours combine to form the more familiar protons and neutrons, which were once thought to be elementary particles.

* Of course, these aren't the 'everyday' properties of colour and flavour that we experience. In the context of quarks, it's best to think of colour and flavour in much the same way we think of electrical charge. There are two kinds of electrical charge—positive and negative. There are three kinds of quark colour—red, green, and blue, and six kinds of flavour—up, down, strange, charm, top, and bottom.

The fourth force, electromagnetism, should be quite familiar—it's the force that passes between electrically charged particles, such as protons and electrons, transmitted by photons, the particles that make up ordinary light. It's the force you feel when you try to push the north poles of two bar magnets together. It's also the force for which you pay every quarter when you receive your electricity bill.

Heat particles and the forces they experience to a high enough temperature and, so some physicists believe, we'll end up with a single 'vanilla' particle and a single primordial force, as the electromagnetic, weak force, then strong force and eventually gravity 'melt' into one another.

If this is correct, then the earliest period in the life of the universe is characterized by the spontaneous emergence of spacetime and energy (although it's not clear if at this stage there's a meaningful difference between these things), governed by a single universal, primordial force. It's tempting to define this as some kind of ultimate time zero, the moment we set the clock ticking. But time itself is being created here. Consequently, 'time zero' is rather poorly defined, and all references to time 'after the big bang' have to be regarded as somewhat ambiguous.

The temperature of the universe in these earliest of moments is incalculable (and, indeed, temperature as a physical quantity has no real meaning at this point). But the universe is expanding and cooling. The universe expands from an infinitesimally small (and so undefinable) point to something we can at least begin to get our heads around.

This is the Planck Epoch. It is characterized by measures of distance and time derived by combining a selection of the fundamental constants of nature, combinations first identified by the German physicist Max Planck.[9] These represent measures beyond which science has nothing really meaningful to say. Theoretically, there can be no distance shorter than the *Planck length*, 1.6×10^{-35} metres. There can be no time shorter than the *Planck time*, about 5×10^{-44} seconds, the time it takes for light to travel the Planck length.

The universe expands and cools to a temperature of about 10^{32} kelvin. To put this in perspective, the temperature at the Sun's core is about 16 million kelvin, so the universe at this early stage in its evolution is about 6 trillion trillion times hotter than the centre of the Sun.

What happens next can be likened to a *phase transition*. If I remove the hot pan of steam from the hob and set it to one side, it will cool down. As it cools, the steam will begin to condense, forming droplets of water which slowly dribble

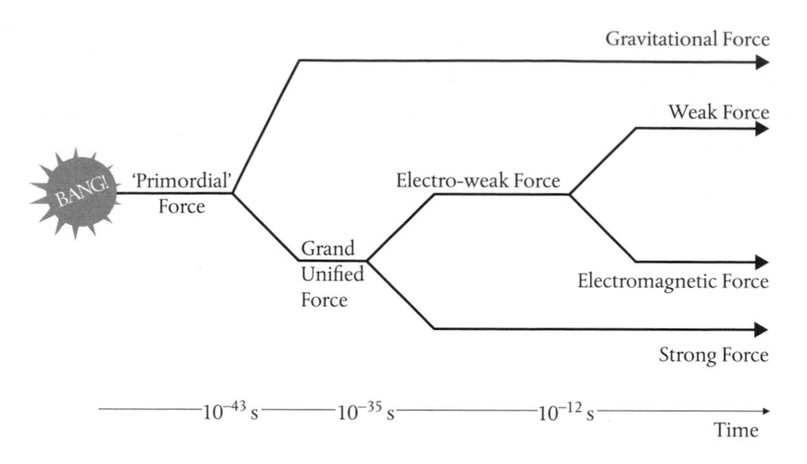

FIGURE 4 At the earliest moments of the big bang, all four forces of nature are thought to be unified in a single, highly symmetric primordial force. After 10^{-43} seconds, the universe undergoes a phase transition, and the force of gravity separates.

down the insides of the pan and gather in the bottom. The phase of the water molecules changes from gas to liquid, and the symmetry is said to be spontaneously broken or reduced.

Likewise, as the early universe approaches the venerable age of 10^{-43} seconds, the primordial force 'condenses'. What we will later recognize as the force of gravity leaches out, alongside a unified 'electro-nuclear' force formed from the strong and weak nuclear forces and electromagnetism (Figure 4).

The universe is well under way.

THE GRAND UNIFICATION EPOCH: 10^{-43} TO 10^{-35} SECONDS

There is as yet no accepted theory that can be used to describe the primordial force. Likewise, there is no accepted theory that can describe the force that would be created through the unification of the strong and weak nuclear forces and electromagnetism. There have been many attempts to derive such theories, which are referred to as 'grand unified theories', or GUTs, but these have been frustrated by a general lack of consistency with experimental observations.

I should quickly point out that these experiments do not re-create the conditions that prevailed during what's known as the Grand Unification Epoch in the early universe, as the energies and temperatures are still way beyond the reach of any terrestrial apparatus we could build. However, GUTs tend to leave their fingerprints on the particles and forces that emerge *after* this period, and which we can observe today at more modest energies and temperatures. To date, no GUT has yielded empirical fingerprints that are consistent with what we observe.

What does the fledgling universe contain? This is really hard to say. Most cosmological models assume that the early universe behaves much like a hot uniform gas, without being specific about the nature of the gas. If there are particles of some kind, then we can safely assume that they are massless and highly energetic, and bear scant resemblance to any particle with which we are familiar today.

The universe continues to expand and cool, to a temperature of about 10^{27} kelvin. About 10^{-35} seconds after the big bang we hit another phase transition, as the strong nuclear force condenses and separates from the residual 'electro-weak' force.

But this time, something is different.

COSMIC INFLATION: 10^{-35} TO 10^{-32} SECONDS

When the ice melts in the pan, the symmetry increases and the water molecules are freed from their prison in the crystal lattice. A considerable amount of energy in the form of heat is required to complete this phase transition. This heat is used to free the water molecules from their restraints, but it does not change the temperature during the transition (which remains fixed at the freezing point of water: 0°C, or 273 kelvin). Such a phase transition is said to be *endothermic*—it requires an input of thermal energy to make it happen.

Conversely, when liquid water freezes to form ice, this energy or 'latent heat' is released and the transition is said to be *exothermic*—there is an output of thermal energy. This heat is released in a short burst as the ice forms. When we make ice in the refrigerator we don't normally notice this, but it can be easily measured using a calorimeter.

In my kitchen experiment, I allow the hot steam to condense to liquid before popping the pan into my freezer, where the liquid will condense further to form ice. This is a conventional freezer, which will cool the liquid in its own good time. But let's now suppose that the water in the pan is ultra-pure, and the inside of the pan is ultra-smooth. This greatly reduces the number of possible sites at which the first ice crystals can nucleate. If I now cool the pan very rapidly, I find that I can reduce the temperature of the liquid well below its freezing point, to almost −50°C (223 kelvin), in fact, without forming any ice. This is *supercooling*.

In supercooled water, the latent heat is not immediately released. It remains latent.

In 1979, young American postdoctoral researcher Alan Guth and his colleague Henry Tye were trying to resolve some of the puzzles of early big bang cosmology. They asked themselves what might happen if, instead of undergoing a smooth phase transition at the end of the Grand Unification Epoch, the universe had instead persisted in its grand unified phase at temperatures several orders of magnitude *below* the transition temperature.

This did indeed resolve their problem. But when Guth went on to explore the wider effects of the onset of supercooling, he found much more than he had initially bargained for. He found that the latent heat of the supercooled universe is stored in the vacuum, in spacetime itself. Remember, vacuum energy does much the same job as Einstein's cosmological term, forcing the rate of expansion of spacetime to increase. Shovelling a very large amount of energy into spacetime results in a short burst of extraordinary exponential expansion, doubling the size of the universe every 10^{-34} of a second. One hundred such doublings would have increased the size of the universe by a factor of 10^{30}.

'I do not remember ever trying to invent a name for this extraordinary phenomenon of exponential expansion,' Guth later wrote, 'but my diary shows that by the end of December [1979] I had begun to call it *inflation*.'[10]

Is this really the way it happened? Theories of cosmic inflation are based on the assumed properties of GUTs and, as there is no generally accepted grand unified theory, then the status of inflation remains rather questionable. But there are some features of the observable universe that are quite handily explained if we assume that something like inflation actually did happen.

Friedmann had shown in 1924 that different assumptions about the density of mass-energy would give rise to different kinds of universe. In a closed universe, spacetime would have a positive curvature, like the surface of a sphere. Parallel lines will cross and the angles of a triangle will add up to more than 180°. In an open universe, spacetime would have a negative curvature, looking something like a saddle. Parallel lines would again cross and the angles of a triangle would add up to less than 180°.

It would seem that we live in a universe in which there is a very fine balance between the density of mass-energy and the rate of expansion, such that spacetime is flat (Figure 5). This is why we learn Euclidean geometry in school.

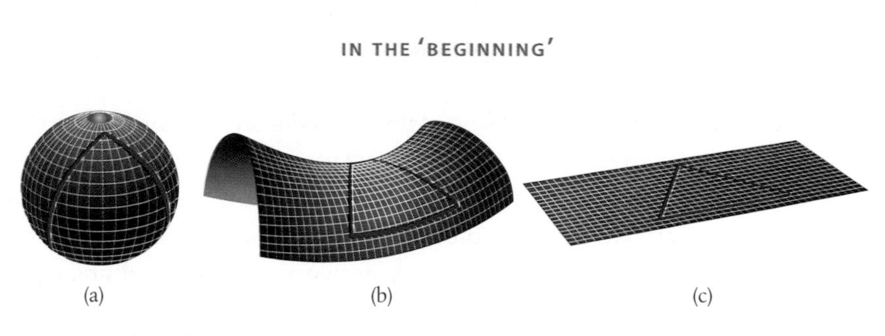

FIGURE 5 The different model universes that Friedmann identified in solutions to Einstein's gravitational field equations produce differently curved spacetimes. The 'closed' universe has a positively curved spacetime as shown in (a). The 'open' universe has a saddle-shaped negatively curved spacetime, (b). A universe with a finely tuned balance between the rate of expansion and the amount of mass-energy it contains will exhibit a flat spacetime, (c).

But why should our universe exhibit such a balance? Scientists get rather uncomfortable when confronted with evidence of such fine-tuning, especially when they can't explain it using physical mechanisms.

This is where inflation comes to the rescue. It really makes no difference precisely how much energy the early universe contained, or whether this was equal to the critical density required to produce a flat spacetime. It doesn't matter what the initial rate of expansion was. It doesn't even matter if inflation is applied to the whole of the early universe or just one small bubble of spacetime within this. No matter what the shape of spacetime prior to inflation, when inflation was done, *flat* spacetime was the only possible result.

Think of it this way. The skin of a deflated balloon may be wrinkled like a prune, curving this way and that. Pump it up and the wrinkles are quickly flattened out. The visible universe may be flat because it comes from a tiny bit of grand-unified universe blown up out of all proportion by inflation.

Inflation theory has many critics, and some have argued that there are ways of arriving at a flat spacetime that do not require inflation, or fine-tuning for that matter. Nevertheless, for the time being at least, it remains an important component of the broadly accepted theory describing the evolution of the universe.

REHEATING

Inflation is thought to have ceased around 10^{-32} seconds after the big bang. The temperature is now several orders of magnitude below the transition temperature.

There are particles in the universe, but these remain largely unfamiliar. The temperature of this early big bang universe is still too high. Although the strong force has separated, this is not the strong force as it is manifested in the universe that we know today. Likewise, the particles that prevail at this time can be thought of as, at best, precursors of the particles that we will eventually come to know. And none of these particles has yet acquired mass.

Any particles produced before inflation would have been spread very thinly as the universe inflates. But, just as the latent heat in supercooled water is eventually released, returning the water to its freezing point, so the energy stored in spacetime is released at the end of the inflationary period. This energy pours into the creation of many more particles, and as the universe fills it reheats to the transition temperature.

For reasons which I hope are quite clear by now, the events as I've described them here are highly speculative and based on assumptions and extrapolations that might—or might not—be justified. I've described a sequence that runs from the Planck Epoch to the Grand Unification Epoch to Cosmic Inflation to Reheating. But this sequence depends on assumptions about the temperature of the universe at the end of inflation (which is itself purely hypothetical). If this temperature were higher—and it could well have been—then inflation and reheating occurs *before* the grand unification symmetry is broken and the strong force is separated. At the present time, we simply have no way of knowing.

Let's nevertheless take stock. The universe is now about 10^{-32} seconds old, blown up by inflation with a temperature of 10^{27} kelvin, filled with a fizzing sea of massless particles that represent precursors to the particles we will come to know. It continues to expand, though at a more leisurely pace, and gravity is already at work. Spacetime has been hammered flat by inflation although, as we will soon see, some tiny wrinkles do remain.

And there's something else. There remains some residual vacuum energy. It's not at all clear where this comes from. This could be vacuum energy somehow left over at the end of inflation. Or maybe this amount of vacuum energy is an intrinsic feature of spacetime itself. But whatever its origin, it acts very much like a cosmological term, gently accelerating the rate of expansion, just as gravity is applying a brake in a kind of cosmic tug-of-war. For the time being, the density of the energy in the universe is high and gravity is winning.

But the vacuum energy—which we will come to know as 'dark energy'—is biding its time.

2

BREAKING THE SYMMETRY

The Origin of Mass

You will no doubt have noticed that I was rather cagey about the nature of the particles that filled the universe in its first 10^{-32} seconds of existence. Actually, I was rather cagey about quite a lot of things. Our ability to understand these very earliest moments in the life of the universe is severely constrained by the inadequacy of our scientific theories when extrapolated to the kinds of conditions that we think must have prevailed at this time.

Don't despair. By the time we reach the end of the next epoch, a trillionth of a second after the big bang, we will find ourselves in somewhat more familiar and comforting territory.

But, before we can continue, it is first necessary to establish a better understanding of the particles that, in our story, will go on eventually to shape the very essence of ourselves and the world around us.

Now, I'm not sure what might flit across your mind when I use a word like 'particle', but when I was younger (and more innocent), I'd have likely imagined a tiny self-contained bit of material substance, the 'stuff' from which everything is made, much as the philosophers of ancient Greece conceived the idea of the 'atom'. Lacking a more vivid imagination, I'd have pictured these 'particles' as tiny solid incompressible indivisible spheres, bouncing off each other like so many microscopic billiard balls.

If this is the kind of thing you imagine, then please be assured that you couldn't be further from the truth.

THE NATURE OF QUANTUM PARTICLES

Physicists in the early decades of the 20th century experienced a massive intellectual shock. They discovered that the microscopic particles that make up the atoms and molecules—the constituents from which everything in our world is formed—behave rather oddly. Under certain circumstances, these particles behave just as we might have imagined in our naivety: as self-contained bits of stuff following clear paths, or trajectories, through space. But there are other circumstances in which these same particles behave like wave disturbances, with peaks and troughs moving much like the ripples that spread out on the surface of a pond where a stone has been thrown.

This is a phenomenon known as *wave–particle duality*. Perhaps you won't be surprised to learn that the developments which led to this understanding were begun by Einstein, in 1905.

Einstein was concerned with the nature of light. Although Newton had conceived of light as consisting of tiny particles (which he had called 'corpuscles'), by 1905 the scientific evidence was pointing very firmly in the direction of a model based on light waves. For example, when forced to squeeze through a narrow aperture or slit in a metal plate, light spreads out, a phenomenon known as diffraction. This is behaviour that's really hard to understand if light is presumed to be composed of particles obeying Newton's laws of motion and moving in straight lines. It's much easier to understand if light is assumed to be composed of waves.

Light also exhibits interference. Shine light on two narrow apertures or slits side by side and it will diffract through both. The wave spreading out beyond each aperture acts as though it comes from a 'secondary' source of light, and the two sets of waves run into each other. Where the peak of one wave meets the peak of the other, the result is constructive interference—the waves mutually reinforce to produce a bigger peak. Where trough meets trough the result is a deeper trough. But where peak meets trough the result is destructive interference: the waves cancel each other out (Figure 6).

The result is a pattern of alternating brightness and darkness called interference fringes, which can be observed using photographic film

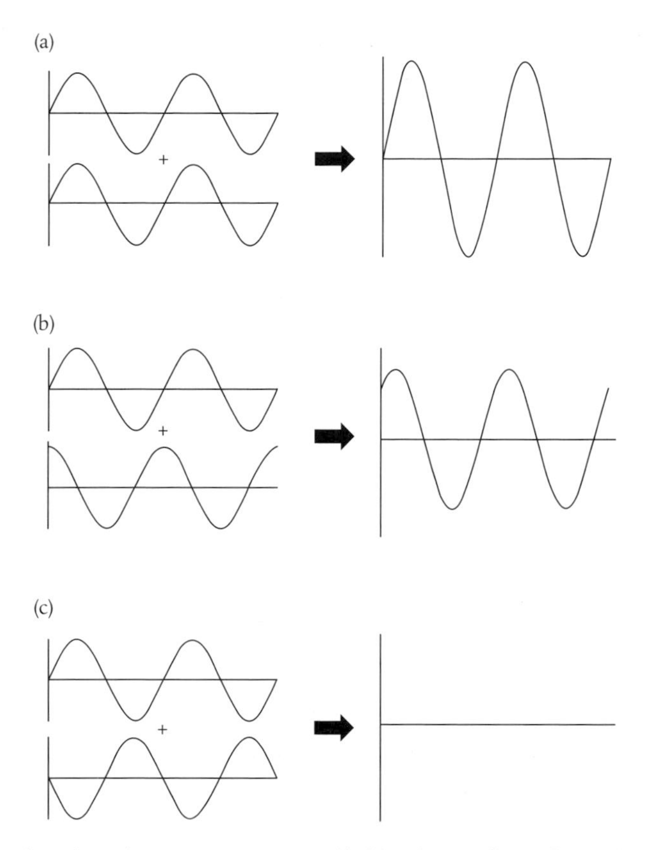

FIGURE 6 When aligned, as in (a), two waves will add and mutually reinforce. This is constructive interference. Moving the waves out of alignment reduces the amplitude, as in (b). If the waves are completely misaligned, the result is destructive interference, (c).

(Figure 7). The bright bands are produced by constructive interference and the dark bands by destructive interference.

But waves are disturbances *in* something. Throwing the stone causes a disturbance in the surface of the water, and it is waves *in the water* that ripple across the pond. What, then, are light waves meant to be disturbances in? Physicists in the 19th and early 20th centuries had argued that these must be waves in a tenuous form of matter called the ether, which was supposed to pervade the entire universe. But no experimental evidence for the ether could be found.

Just as Einstein didn't believe in absolute space and time, so he didn't believe in the existence of the ether. He argued that earlier work in 1900 by German

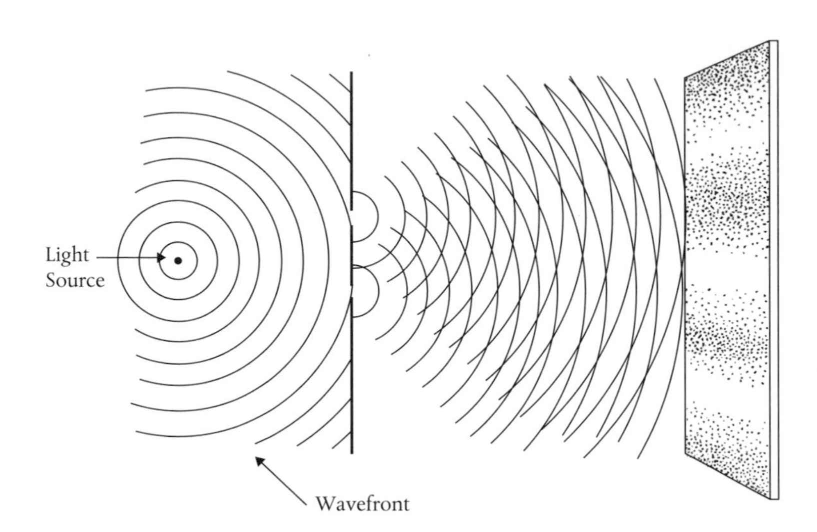

FIGURE 7 When passed through two narrow, closely spaced apertures or slits, light produces a pattern of alternating light and dark fringes. These can be readily explained in terms of a wave theory of light in which overlapping waves interfere constructively (giving rise to a bright fringe) and destructively (dark fringe).

physicist Max Planck hinted at an altogether different interpretation for light. Planck had concluded that light could be modelled successfully *as though* it was absorbed and emitted by matter in discrete units or 'bundles' of energy that he had called *quanta*. Einstein now boldly suggested that this happens because light actually *consists* of quanta.

Einstein hadn't simply reverted to Newton's corpuscles. Nobody was denying all the evidence for wave-like behaviour, such as diffraction and interference. In his paper, Einstein retained the central property of frequency in the description of what he called 'light-quanta', and frequency is a property of waves (it is the number of times a wave oscillates from peak-to-trough and back again in a given unit of time, typically one second).

So, how could quanta or 'particles' of light also be waves? Particles are by definition localized bits of stuff—they are 'here' or 'there'. Waves are delocalized disturbances in a medium, they are 'everywhere', spreading out beyond the disturbance. How could these light-quanta be here, there, *and* everywhere?

These were highly speculative ideas, and physicists did not rush to embrace them. But Einstein used his light-quantum hypothesis to make some predictions for a phenomenon known as the photoelectric effect, and these predictions were all borne out ten years later, in 1915, in experiments performed by American physicist Robert Millikan. In 1921, Einstein was rewarded with the Nobel Prize for Physics for this work, although the light-quantum hypothesis remained rather controversial.

Two years later, American physicist Arthur Compton and Dutch theorist Pieter Debye showed that light could be 'bounced' off electrons, with predictable changes in light frequency. These experiments demonstrate quite unambiguously that light does indeed consist of particles moving in trajectories, like small projectiles. Gradually, the light-quantum became less controversial and more acceptable. The American chemist Gilbert Lewis coined the name 'photon' in 1926.

But the story doesn't end there. In 1923, Prince Louis de Broglie,* the younger son of Victor, fifth duc de Broglie, made a startling proposal:[1]

> After long reflection in solitude and meditation, I suddenly had the idea, during the year 1923, that the discovery made by Einstein in 1905 should be generalized by extending it to all material particles and notably to electrons.

If light waves with a certain characteristic frequency possess associated particle-like properties then, de Broglie reasoned, perhaps particles such as electrons, with a certain characteristic mass, might possess associated wave-like properties.

De Broglie was suggesting that, in certain circumstances, the electron might behave as a wave disturbance. This meant that, in principle, a beam of electrons could be diffracted, just like a beam of light. A beam of electrons passing through two closely spaced slits would exhibit interference. This seems really rather bizarre: how can self-contained particles with mass exhibit interference effects?

It gets worse. Imagine what might happen if we limit the intensity of a beam of electrons so that, on average, only *one* electron at a time passes through the slits. We detect the electrons on the other side of the apparatus using a piece of photographic film. What could we expect to see?

Our instinct might be to assert that each electron *must* pass through one or other of the slits. After all, it's supposed to be an elementary *particle*, an indivisible bit of material substance carrying a unit of negative electrical charge. In our

---

* Pronounced 'de Broy'.

naivety we would judge it impossible for such particles to produce interference effects. And, indeed, when we do the experiment, we find that a single electron passing through the slits does produce one—and only one—spot on the photographic film. There you are. Told you so.

But wait. As we now allow more electrons to pass through the slits—one at a time—we accumulate more and more spots. And lo. The interference pattern is revealed (Figure 8).

The only way to rationalize these kinds of experimental results is to acknowledge that each electron behaves as a wave, passing through both slits. When the resulting wave impinges on the photographic film, it 'collapses', in a seemingly random manner, to produce a single spot. However, the probability that it will produce a spot is much higher where the amplitude of the wave has been reinforced through constructive interference,

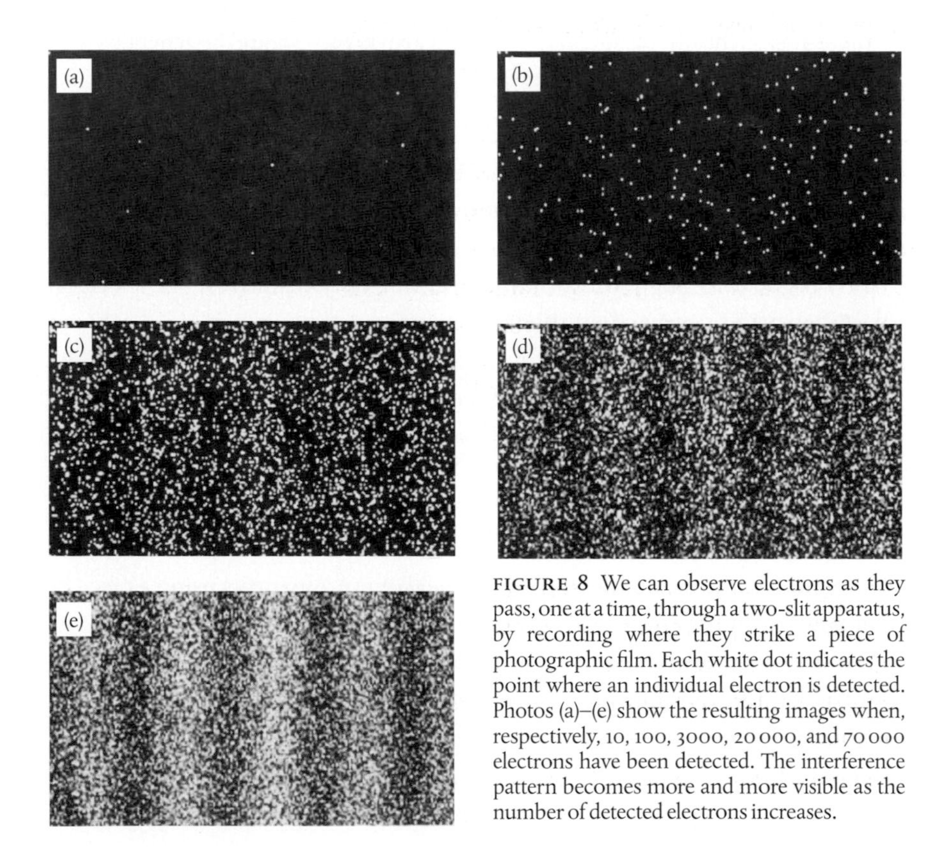

FIGURE 8 We can observe electrons as they pass, one at a time, through a two-slit apparatus, by recording where they strike a piece of photographic film. Each white dot indicates the point where an individual electron is detected. Photos (a)–(e) show the resulting images when, respectively, 10, 100, 3000, 20 000, and 70 000 electrons have been detected. The interference pattern becomes more and more visible as the number of detected electrons increases.

much lower where the amplitude has been cancelled out through destructive interference.

De Broglie's shocking idea was one of the foundations on which quantum theory was constructed in the 1920s and 1930s.

How are we meant to interpret this behaviour? The truth is, we don't really know. We do know that the wave-like behaviour belies a property called *phase* that all types of quantum particle possess. In one sense, the phase of a wave is simply its position in its peak-and-trough cycle. Look back at Figure 6. The waves in (a) are aligned, or 'in phase'. In (b) they're a little (actually, 90°) out of phase. In (c) they're completely (180°) out of phase.

However, in practical terms, the phase of a quantum particle can never be observed directly. Measurements only reveal properties that are affected by the phase, and not the phase itself. Whatever its origin, the property of phase results in the observation of behaviour that we interpret in terms of waves. This helps, but it doesn't really take us any further forward.

Danish theorist Niels Bohr and German theorist Werner Heisenberg argued that particles and waves are merely the shadowy projections of an unfathomable reality into our empirical world of measurement and perception. They claimed that it makes no sense to speculate about what photons or electrons *really are*. Better to focus on how these quantum particles appear—in this kind of experiment they appear as waves, in that kind of experiment they appear as particles. This state of affairs was rather neatly summarized by the playwright Tom Stoppard in a passage in his 1988 play *Hapgood*:[2]

> . . . So, we do it again, exactly the same except now without looking to see which way the bullets [of light] go; and the wave pattern comes back. So we try again while looking, and we get particle pattern. Every time we don't look we get wave pattern. Every time we look to see how we get wave pattern, we get particle pattern. The act of observing determines the reality.

The approach favoured by Bohr and Heisenberg (and Austrian physicist Wolfgang Pauli) is known as the *Copenhagen interpretation*, named for the city where Bohr had established a physics institute where much of the debate about the interpretation of quantum theory took place in the 1920s and early 1930s.

It goes without saying that not everyone is satisfied with this interpretation, and debates have raged for nearly a hundred years.

QUANTUM FIELDS AND FORCES

The debate about interpretation is likely to run and run, but there's no denying the brute empirical fact of wave–particle duality. In contemporary theoretical physics, this aspect of the nature of elementary quantum particles (such as photons and electrons) is captured in a structure called *relativistic quantum field theory*; 'relativistic', because the theory conforms to the demands of Einstein's special theory of relativity.*

There are many different kinds of 'fields' in physics. Any physical property that can be measured over different points in space and time can be represented as a field. For example, if we measured atmospheric temperature and pressure systematically (and instantaneously) at fixed spatial intervals from the ground to the upper reaches of the atmosphere, all around the world, we could use these to define spherical 'fields' of temperature and pressure surrounding the Earth, and then watch as these fields change in time. However, temperature and pressure are not fundamental physical properties: they are secondary properties resulting from the density and energy of motion of molecules of oxygen, nitrogen, and trace atoms of noble gases in the atmosphere.

In contrast, quantum fields are fundamental fields. As far as we know, they don't result from some property even more fundamental.

These fields may appear mysterious, but they are also quite familiar. Sprinkle iron filings on a sheet of paper held above a bar magnet. The iron filings organize themselves along the 'lines of force' of the magnetic field, reflecting the strength of the field and its direction, stretching from north to south poles (Figure 9). The field exists in the 'empty' space around the outside of the bar of magnetic material. In truth, magnetic forces cannot be separated from electrical forces, and this old schoolroom experiment is actually making startlingly visible some aspects of the quantum electromagnetic field.

The nature of the field depends on the nature of the property being measured. The field may be *scalar*, meaning that it has magnitude but acts in no particular direction in space—it doesn't push or pull. It may be *vector*, with both magnitude and direction, like an electromagnetic field or an old-fashioned Newtonian gravitational field. Finally, it may be *tensor*, a more elaborate version of a vector field for situations in which the geometry is more complicated than

* There's no problem joining quantum theory and special relativity. It's when physicists try to unify relativistic quantum field theory with general relativity that things start to go horribly wrong.

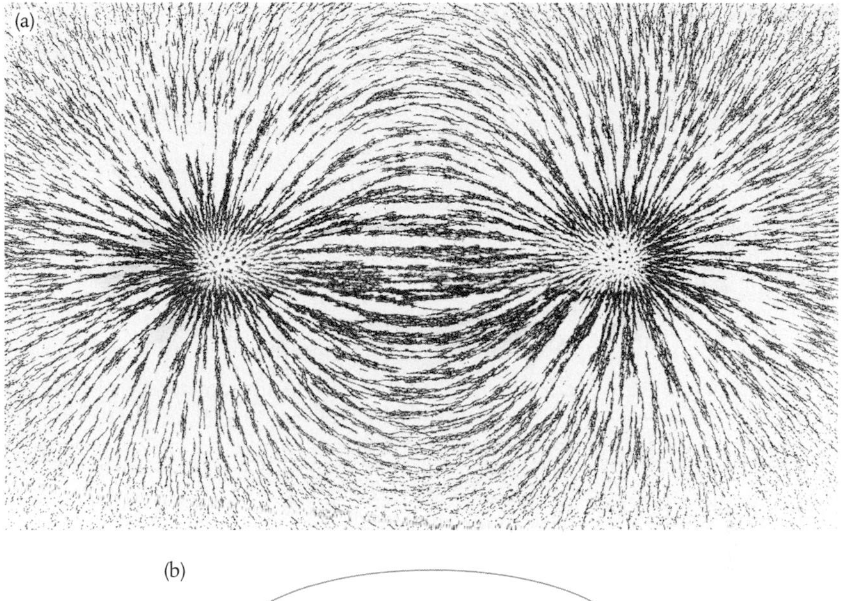

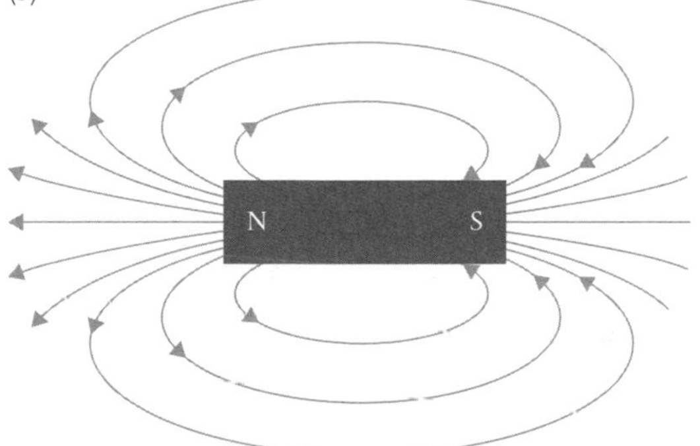

FIGURE 9 (a) Iron filings sprinkled on a sheet of paper held above a bar magnetic reveal the 'lines of force' of the magnetic field stretching between the north and south poles. (b) The pattern of the 'lines of force' is shown schematically here. By convention, the lines of force 'flow' from north to south poles.

everyday Euclidean geometry. Tensor fields crop up in the field equations of Einstein's general theory of relativity.

In quantum field theory, the 'particle' is, in essence, a fundamental field quantum, an elementary fluctuation, disturbance, or vibration of the field. So, for example, the electron is the quantum of the electron field. The photon is the quantum of the electromagnetic field.

It began to dawn on physicists working on early versions of quantum field theory that they had figured out a very different way to understand how forces between quantum particles actually work. Let's imagine that we bounce two electrons off each other (the technical term for this is 'scatter'). We can suppose that, as the two electrons approach each other in space and in time, they feel the mutually repulsive force generated by their negative electrical charges.

But how? We might speculate that each moving electron generates an electromagnetic field and the mutual repulsion is felt in the space where these two fields overlap, much like we feel the repulsion between the north poles of two bar magnets when we try to push them together. But in the quantum domain, fields are also associated with particles, and interacting fields with interacting particles. In 1932, German physicist Hans Bethe and Italian Enrico Fermi suggested that this experience of force is the result of the *exchange* of a photon between the two electrons.

As the two electrons come closer together, they reach some critical distance and exchange a photon. In this way the particles experience the electromagnetic force. The exchanged photon carries momentum from one electron to the other, thereby changing the momentum of both. The result is recoil, with both electrons changing speed and direction and moving apart.

The exchanged photon is a 'virtual' photon, because it is transmitted directly between the two electrons and we don't actually see it pass from one to the other. In fact, there's no telling in which direction the photon actually goes. In diagrams drawn to represent the interaction the passage of a virtual photon is simply illustrated using a squiggly line, with no direction indicated between the electrons (see Figure 10).*

Here was another revelation. The photon was now no longer simply the quantum of the electromagnetic field. It had become the 'carrier' of the electromagnetic force. When we try to push the north poles of two bar magnets together, the resistance we feel is the result of virtual photons passing between them.

SPIN

It will help in what follows to explore the properties of quantum particles a little further. Aside from carrying energy (which is directly related to the wave frequency),[3] photons are also characterized by their intrinsic angular momentum,

* Astute readers will notice that a virtual photon travelling from electron 1 to electron 2 in this diagram must travel backwards in time. This is only one of many ways that quantum theory can mess with your head.

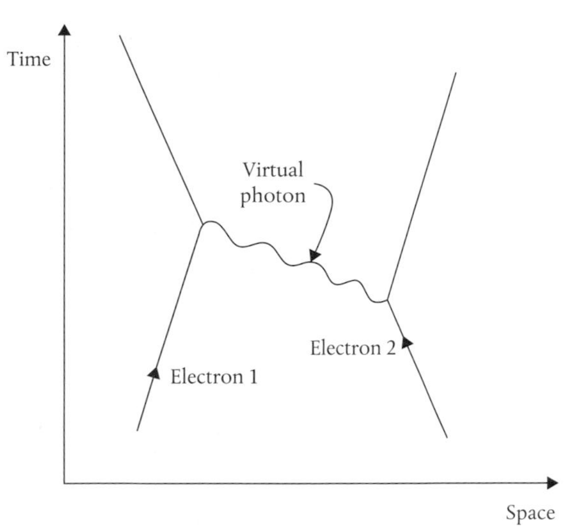

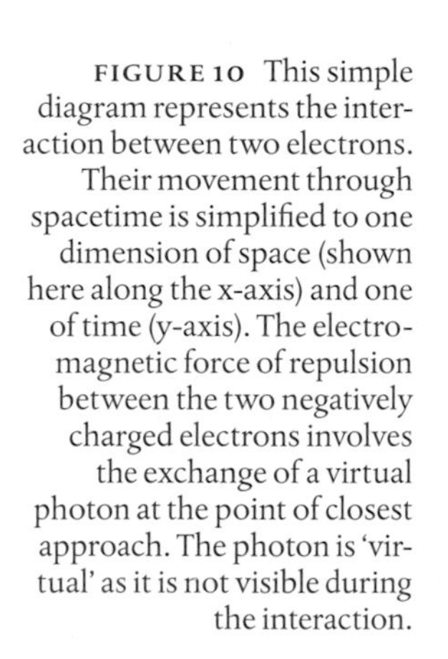
FIGURE 10 This simple diagram represents the interaction between two electrons. Their movement through spacetime is simplified to one dimension of space (shown here along the x-axis) and one of time (y-axis). The electromagnetic force of repulsion between the two negatively charged electrons involves the exchange of a virtual photon at the point of closest approach. The photon is 'virtual' as it is not visible during the interaction.

or what physicists call *spin*. In classical Newtonian mechanics, we associate angular momentum with objects that spin around some central axis. In fact, when we watch a figure-skater spinning ever faster in a tight circle at the end of their routine, we are actually watching the law of conservation of angular momentum in action.

This obviously can't apply to photons. For one thing, photons are massless: they have no central axis they can spin around. They can't be made to spin faster or slower. The term 'spin' is actually quite misleading. It is a hangover from the early days of quantum theory when it was thought that this property could be traced to an intrinsic 'self-rotation', literally quantum particles spinning like tops. This was quickly dismissed as impossible, but the term 'spin' was retained.

It doesn't help to look too closely at the property of spin and ask what a photon is really doing. We do know that the property of spin is manifested as angular momentum. The interactions between photons and matter are governed by the conservation of angular momentum, and many experiments performed over many years have demonstrated this. When we create an intense beam of photons (such as a laser beam) with the spins of the photons aligned, the angular momentum of all the individual photons adds up and the beam imparts a measurable torque, or turning force. Point the beam at a target and the target will be rotated.

Physicists characterize the spin properties of quantum particles according to a *spin quantum number*, which provides a measure of the particles' intrinsic angular momentum. This quantum number can take half-integral ($\frac{1}{2}$, $\frac{3}{2}$, etc.) or integral values (0, 1, 2, etc.). Quantum particles with half-integral spins are called *fermions*, named for Italian physicist Enrico Fermi. If we persist in pushing the spinning top analogy, we find that fermions would have to spin *twice* around their axes in order to get back to where they started (which shows that persisting with classical analogies in the quantum world usually leads only to headaches).

The photon has a spin quantum number of 1, which classifies it as a *boson*, named for Indian physicist Satyendra Nath Bose.

It turns out that the quantum particles that make up matter—quarks and leptons—are all fermions. The particles that carry forces between the matter particles are bosons. Particles associated with scalar quantum fields are bosons with a spin of zero.

I think that's enough to be going on with. Now, where were we?

THE ELECTRO-WEAK EPOCH: 10^{-32} TO 10^{-12} SECONDS

Following the short burst of cosmic inflation, the latent energy stored as vacuum energy in spacetime is released, filling the universe with massless particles and reheating it back to the transition temperature of 10^{27} kelvin. From this point, the universe continues to expand and cool.

We are now in the Electro-weak Epoch. Distorted, high-energy forms of gravity, the strong force, and a combined electro-weak force prevail among quantum fields with particles that are precursors to the particles we will come to know. All these particles are massless. There are no photons. Yet.

This situation persists until the universe is a trillionth of a second old and has cooled to a temperature of about 10^{15} kelvin. At this point, we encounter another phase transition. The weak force separates from electromagnetism and we arrive at the four forces, the particles they act on, and the particles that carry them, pretty much as we know them today.

But, once again, something is different.

SPONTANEOUS SYMMETRY-BREAKING

The other thing I was pretty cagey about in the last chapter was the precise nature of the physical mechanism responsible for triggering the phase transitions we have encountered thus far. This mechanism goes by the rather grand name of 'spontaneous symmetry-breaking'.

Remember that in my simple kitchen experiment with the pan filled with ice, I talked about how raising the temperature forces the ice to melt into liquid water, and the water then vaporizes into steam. In each of these phase transitions, the symmetry is increased: the ice is less symmetrical (it looks different in different directions) compared to the water. The water is less symmetrical compared to the steam. As we ascend through these different phases, from ice to water to steam, the symmetry increases.

When this process is worked backwards, the steam condenses into water and the water freezes to ice. Now, at each phase transition, the symmetry is reduced, or 'broken'. This symmetry-breaking happens spontaneously as the temperature falls.

These are examples of spontaneous symmetry-breaking in everyday life. Here's another that's perhaps a little less 'everyday'. If we had enough patience, we could imagine that we could somehow balance a pencil finely on its tip. We would discover that this is a very symmetric, but very unstable, situation. The vertical pencil looks the same from all directions around it.

But tiny disturbances in our immediate environment, such as small currents of air, are enough to cause the pencil to topple over. When this happens, the pencil topples in a specific, though apparently random, direction. The horizontal pencil now no longer looks the same from all directions. The symmetry is spontaneously broken.

In this example, it is the *environment* in which the pencil stands vertically (specifically, the barely detectable current of air in this environment) that triggers the transition. In much the same way, small inhomogeneities within the environment of the pan provide centres around which water droplets condense from steam, and ice crystals nucleate from water. But, you may well ask, what was it in the environment of the universe that triggered the phase transitions that we think may have occurred in its early history?

Physicists realized that they needed to add something to the background environment of the universe that would help to force a distinction and so break the symmetry of the quantum fields at different moments in time as the universe cooled. Perhaps not surprisingly, they reached for more quantum fields.

But there's a problem. The quantum fields associated with the kinds of particles with which we are familiar necessarily reduce to zero in 'empty' space, defined simply as space with no particles in it.* What was needed was a special type of scalar quantum field that doesn't reduce to zero, a field that stubbornly fixes at a specific magnitude, at every point in spacetime. Such a field is said to have a non-zero vacuum expectation value, which is a kind of fancy average value.

The idea of using a scalar quantum field to break the symmetry was developed in the early 1960s (and *not* for the purposes of describing the early history of the universe). Its origins are rather convoluted but can be traced to the work of Japanese-American theorist Yoichiro Nambu, British theorist Jeffrey Goldstone, and remarks by American physicist Philip Anderson. In 1964, there appeared a series of papers detailing a mechanism published independently by American physicist Robert Brout and Belgian François Englert, English physicist Peter Higgs at Edinburgh University, and also Americans Gerald Guralnik and Carl Hagen with British physicist Tom Kibble at Imperial College, London. From about 1972, the mechanism has been commonly referred to as the *Higgs mechanism* and the scalar fields are referred to as Higgs fields.†

With one exception. The scalar quantum field thought to be responsible for triggering cosmic inflation is given the special name 'inflaton field'.

Until very recently, these were inventions, needed to make the theories work. As we will see shortly, the first experimental evidence for the existence of at least one of these fields has recently become available.

It's quite important for us to understand the nature of these fields and their associated particles as best we can, because the Higgs field that triggered the phase transition at the end of the Electro-weak Epoch is also responsible for the origin of mass in the universe.

THE HIGGS MECHANISM AND THE 'GOD PARTICLE'

At the end of the Electro-weak Epoch, the universe cools to a temperature of about 10^{15} kelvin and another Higgs field becomes fixed with a non-zero vacuum expectation value.‡ From our perspective, this was to prove to be one of the most important events in the entire history of the universe.

---

* This is okay for now, but in the next chapter I'll need to qualify what it means to have a zero field amplitude in empty space.

† Much to the chagrin of the others, it must be said.

‡ This is referred to as the electro-weak Higgs field, which is a bit of a mouthful. In what follows, I will call it simply 'the Higgs field', on the understanding that I'm actually referring specifically to the Higgs field that appeared at the end of the Electro-weak Epoch.

The weak nuclear force and electromagnetism have now separated, but the Higgs field does more than force the distinction between these. The massless particles themselves interact with the field, and the end result is the extraordinary array of quarks, leptons, and force-carrying bosons that we know today.

Precisely what happens when a massless particle interacts with the Higgs field depends on what kind of particle it is. For many, the interaction slows them down. Or, if you prefer, the field resists their acceleration, so they can't be accelerated back to the speed of light. It's as though the Higgs field applies a brake: the particles' acceleration is resisted to an extent that depends on the strength of the interaction. The field drags on them like molasses.

Now, ever since Galileo, we have been inclined to think of an object's resistance to acceleration as the result of its *inertial mass*. Our instinct is therefore to assume that mass is a primary or intrinsic quality of material substance, and we identify inertial mass with the amount of 'stuff' that an object possesses. The more stuff it has, the harder it is to accelerate.

But the Higgs mechanism turns this logic on its head. *The extent to which an otherwise massless particle's acceleration is resisted by the Higgs field is now interpreted as the particle's inertial mass.* Mass has become a secondary quality. It is the *result* of an interaction, rather than something that is intrinsic to matter.

Let's see how this works for the specific case of the force carriers of the weak force and electromagnetism. This was, in fact, the first formal application of the Higgs mechanism to a problem in particle physics, published by Steven Weinberg in 1967 and subsequently by Pakistani theorist Abdus Salam in 1968, the latter building on earlier work by himself and British theorist John Ward.

It had long been understood that the weak force and electromagnetism were somehow related, and that the difference in the strengths of these forces must be due to the fact that, in contrast to the massless photon, the carriers of the weak force—the W and Z particles—are for some reason 'heavy'. The trouble was, the version of quantum field theory available at the time predicted that they should all be massless, just like the photon.

The Higgs mechanism furnished a solution. Prior to breaking the symmetry, the electro-weak force is carried by four massless particles which, for the sake of simplicity, we will call the W^+, W^o, W^-, and B^o. Interactions with the electro-weak Higgs field cause the W^+ and W^- particles to gain mass. The W^o and B^o particles of the electro-weak force mix together to produce a massive Z^o particle and the massless photon. We associate the massive W^+, W^-, and Z^o particles with the weak force, and the massless photon with electromagnetism.

It is this interaction with the background Higgs field that causes the forces to split and go their separate ways.

In 1967, Weinberg estimated that the W particles would each have a mass of about 85 times that of the proton, and the Z^0 would be slightly heavier, with a mass of about 96 times the proton mass. The W particles were discovered at CERN in Geneva in 1982/3, with masses 85 times that of the proton, just as Weinberg had predicted. The discovery of the Z^0 was announced later in 1983, with a mass about 101 times the mass of a proton.

Remember, every quantum field has an associated field particle. In 1964, Higgs had referred to the possibility of the existence of what would later become known as a 'Higgs boson', the elementary particle of the Higgs field. In his 1967 paper, Weinberg had found it necessary to introduce a Higgs field with four components. Three of these give mass to the W^+, W^-, and Z^0 particles. The fourth appears as a physical particle—a Higgs boson with a spin of zero. Another way to think about this is to imagine that the four field components represent four Higgs bosons. The W^+, W^-, and Z^0 particles acquire mass by 'absorbing' three of these, leaving one behind.

On 4 July 2012, scientists at CERN's Large Hadron Collider declared that they had discovered a new particle 'consistent with' the Higgs boson. This was found to have a mass around 133 times that of a proton and interacts with other standard model particles in precisely the way expected of the Higgs. Analysis of the data gathered through 2011 and 2012 was assessed in a further press release dated 14 March 2013 which concluded: '. . . we are dealing with a Higgs boson though we still have a long way to go to know what kind of Higgs boson it is.'[4]

In his popular book *The God Particle: If the Universe is the Answer, What is the Question?* first published in 1993, American physicist Leon Lederman (writing with Dick Teresi) explained why he'd chosen this title:[5]

> This boson is so central to the state of physics today, so crucial to our final understanding of the structure of matter, yet so elusive, that I have given it a nickname: the God Particle. Why God Particle? Two reasons. One, the publisher wouldn't let us call it the Goddamn Particle, though that might be a more appropriate title, given its villainous nature and the expense it is causing. And two, there is a connection, of sorts, to another book, a much older one . . .

Lederman went on to quote a passage from the Book of Genesis.

The name 'God particle' has stuck. Not many practising physicists like it, not least because it rather overstates the importance of the Higgs boson. Of course, the Higgs boson is of considerable interest, but it is the Higgs *field* that is fascinating. Scalar quantum fields were introduced into quantum field theory as devices designed to break the symmetry and create mass. To 'see' the Higgs boson, it has been necessary to build a particle collider capable of achieving the kinds of energies needed to disturb or vibrate the Higgs field. These are energies that would have prevailed towards the end of the Electro-weak Epoch, a mere trillionth of a second after the big bang.

Now we have evidence that at least one kind of Higgs field exists. The 2013 Nobel Prize in Physics was awarded to Higgs and Englert in recognition of their achievement, 49 years previously.*

THE STANDARD MODEL OF PARTICLE PHYSICS

If the new particle discovered at CERN in 2012 can eventually be proved to be the characteristic field particle of the electro-weak Higgs field, then this will complete the theoretical structure known as the standard model of particle physics. I have already introduced the particles and the forces that the standard model describes, but as these represent the building blocks for everything that follows, it's worth our while to take a closer look at these now.

The standard model particles and forces are summarized in Figure 11. The quarks and leptons form three 'generations', distinguished by the particle masses. As we've seen, this means that they're distinguished by the extent to which the particles in each generation interact with the Higgs field.

The leptons include the first-generation electron (usually given the symbol e^-) and the electron-neutrino (ν_e). The electron is, of course, very familiar. It carries a negative electrical charge of −1 and has a small mass, about 0.00054 times that of a proton.[6]

The electron-neutrino will likely be less familiar. Neutrino is Italian for 'small neutral one'. It is a neutral particle, and puts in an appearance in beta radioactivity, in which a neutron inside an unstable atomic nucleus breaks down eventually to form a proton, which is left behind in the nucleus, and a high-speed electron and electron-neutrino, which are ejected. For some time, neutrinos were thought to be massless particles, but they are now believed to have very

* Sadly, Robert Brout died in 2011 after a long illness, and the Nobel Prize is not awarded posthumously.

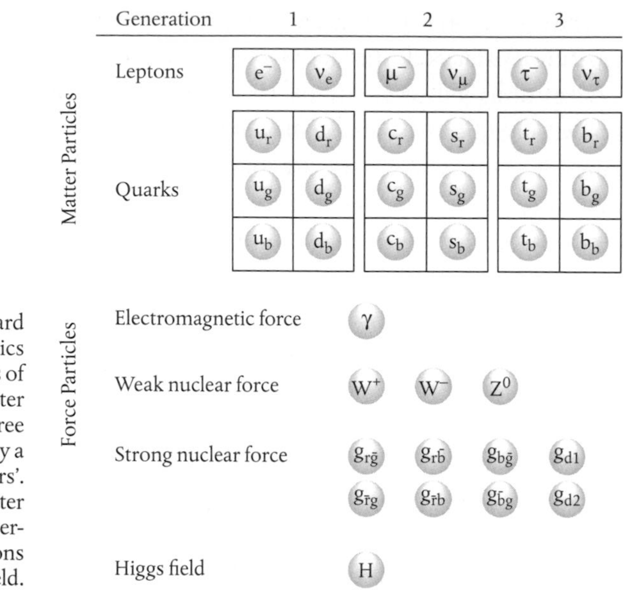

FIGURE 11 The standard model of particle physics describes the interactions of three generations of matter particles through three kinds of force, mediated by a collection of 'force carriers'. The masses of the matter and force particles are determined by their interactions with the Higgs field.

small masses, so small that these have been hard to measure accurately, and only upper limits are quoted.

The second-generation leptons are the muon (μ^-) and muon-neutrino (ν_μ). The muon also carries an electrical charge of −1 and has a mass of about 0.113 times the proton mass. It is a heavier version of the electron. The third-generation tau (τ^-) is even heavier, weighing in at 1.89 times the proton mass. It is accompanied by the tau-neutrino (ν_τ).

There's more. In August 1932, American physicist Carl Anderson discovered a particle that to all intents and purposes behaves just like an electron, but with a *positive* electrical charge.* He called it the *positron*, the first in a long line of discoveries of *anti-matter* particles, particles with the same properties of mass and spin, but with opposite electrical charge. All particles in the standard model have anti-matter counterparts.† For simplicity, these are not shown in Figure 11.

* When English theorist Paul Dirac developed the first relativistic version of a quantum theory of the electron in 1927, he found that it produced twice as many solutions as he thought he needed. In addition to two solutions corresponding to the different spin orientations of the electron, he also found two 'negative energy' solutions. These were quite puzzling, until he finally accepted in 1931 that they must correspond to 'a new kind of particle, unknown to experimental physics, having the same mass and opposite charge to the electron'. Paul Dirac, *Proceedings of the Royal Society*, **A133** (1931), pp. 60–72.

† Neutral particles are their own anti-particles.

The quarks also form three generations. The first generation includes the up quark (u), with an electrical charge +⅔, and the down quark (d) with a charge of −⅓. Now, the idea of fractional electrical charges might be new to you and, in truth, no single fractionally charged quark has ever been observed directly. But please be assured that there are solid experimental grounds (albeit indirect) for believing that quarks exist and have the properties we think they have. 'Up' and 'down' represent the different 'flavours' of quarks, and each flavour comes in three 'colours': red, green, and blue (subscript r, g, and b).

The up and down quarks are quite light, though they are heavier than the electron, weighing in at 0.0018 to 0.0033 times the proton mass (u) and 0.0044 to 0.0061 times the proton mass (d). Only mass ranges are quoted, as it's so far been impossible to derive accurate mass data from particle physics experiments.

If quarks are a little less familiar, then protons and neutrons are probably particles with which you might feel more comfortable. In fact, a proton consists of two up quarks and a down quark (uud), with a total electrical charge of +1. A neutron consists of an up quark and two down quarks (udd), with a charge of zero. You might be puzzled by the simple fact that the masses of two up quarks and a down quark don't add up to give the proton mass. In fact, there seems to be quite a lot of mass 'missing' (from the information above we would estimate that the combination uud should have a mass range of 0.008 to 0.013 times the proton mass, implying that about 99% of the proton has gone AWOL). Don't fret—the answer is coming.

This same pattern is repeated in the second- and third-generation quarks. The charm quark (c) carries a charge +⅔ and has a mass about 1.37 times the proton mass. The strange quark (s) has a charge −⅓ and mass of about 0.11 times that of the proton. Similarly, the top quark (t) has charge +⅔ and mass 184 times the proton mass and the bottom quark (b) has charge −⅓ and mass 4.47 times the proton mass.

All the quarks and leptons have a spin of ½ (they are fermions). All have anti-matter counterparts.

The force of electromagnetism acts on all electrically charged particles and is carried by the massless photon (γ). The weak force acts on both quarks and leptons and, as I've already explained, is carried by the 'heavy photons', the W^+, W^-, and Z^0 particles.* In fact, in beta radioactivity a down quark

* I should own up. Quantum particles also have a property called 'parity', which can be traced back to properties in the particles' wave descriptions. Parity has no counterpart in our everyday world

inside a neutron in an unstable atomic nucleus transforms into an up quark, turning the neutron into a proton. The down quark emits a W^- particle, which carries away the balancing electrical charge. The W^- quickly decays into an electron and what I can now more correctly describe as an electron anti-neutrino.

The strong force acts only on quarks, as these are the only matter particles that possess 'colour charge'. It is the strong force that binds quarks together inside protons and neutrons. But this force is somewhat counter-intuitive.

When we imagine a force acting between two particles, we probably tend to think of examples such as gravity or electromagnetism, in which the force grows stronger as the particles move closer together (Figure 12). But the strong force doesn't behave in this way. It is as if the quarks are fastened to either end of a piece of strong elastic or a coiled spring. When the quarks are close together, the elastic or the spring is loose, and there is little or no force between them. But as we try to pull the quarks apart we begin to stretch the elastic or the spring. The force increases the harder we pull.

This is illustrated in Figure 12(b), in terms of the interaction between a quark and an anti-quark. Such quark–anti-quark particles are called *mesons*. For

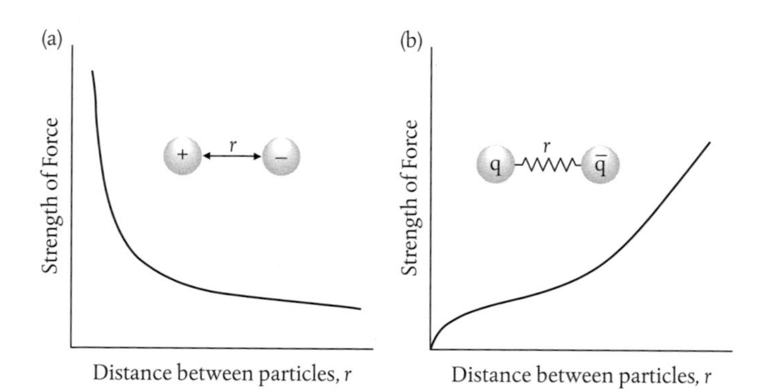

FIGURE 12 (a) The electromagnetic force between two electrically charged particles increases in strength as the particles move closer together. But the colour force that binds quarks together behaves rather differently. (b) In the limit of zero separation between a quark and an anti-quark (for example), the force falls to zero. The force increases as the quarks are separated.

of direct experience, but it is related to, and governs, angular momentum in particle interactions. Particles can be thought to have 'left-handed' or 'right-handed' parity. Curiously, only left-handed particles and right-handed anti-particles are affected by the weak force. Nobody really knows why.

example, the particle formed from an up quark (charge $+\frac{2}{3}$) and an anti-down quark (charge $+\frac{1}{3}$) is a meson called the positive pion (π^+), with a charge of +1.

The strong force is carried by eight different massless gluons (g), which pass colour charge from one quark to the next, binding them together. And here lies the answer to the mystery of the proton's mass. In principle, the energy of a single, isolated quark is infinite, which is why individual quarks have never been observed. Inside the proton the colour charge of each quark is masked by combining it with two other quarks of different colour, such that the net colour charge is zero. For example, a red up quark, a green up quark and blue down quark combine to give a 'white' or 'colourless' proton.

However, within the confines of the proton, it is not possible to mask the exposed colour charges completely. This would require that the quarks be piled directly on top of one another, occupying the same location in space and time. But quarks are fermions, and their half-integral spins prevent them from being 'condensed' in this way.

Nature settles for a compromise. Inside a proton the colour charges are exposed and the energy—manifested in the associated gluons that pass between them—increases. The increase is manageable, but it is also substantial. The energy of the gluons inside the proton builds up and, although the gluons are themselves massless, through $m = E/c^2$ their energy accounts for the other 99% of the proton's mass.

All the force carriers—the photon, W and Z particles, and the gluons—have a spin of 1 (they are bosons). The final particle in the standard model is, of course, the electro-weak Higgs boson, with spin zero.

As it stands, the standard model of particle physics is a great triumph of the human intellect. But it is also really rather unsatisfying. It consists of 61 elementary particles. Count them: There are three generations of matter particles each consisting of two leptons and two flavours of quark which each come in three different colours (making 24 in total), the anti-particles of all these (making 48), 12 force particles—a photon, W^+, W^-, and Z^0 particles and eight gluons (making 60) and a Higgs boson (61).

And, although we now have good grounds to believe that each particle with mass gains this from interacting with the Higgs field, we have no understanding of what governs the strengths of these interactions. We therefore have no way of predicting the particle masses, and there appears to be no rhyme or reason in the mass values themselves.

DARK MATTER

And this is still not the last of our problems. To explain the large-scale structure of the universe as it appears to us today, we need at least one further kind of particle, one that is not included in the standard model.

I'm going to ask you to suspend your disbelief for the moment and hold all your questions until Chapter 4. At this stage all I want to acknowledge is that studies of the motions of galaxies and clusters of galaxies have led to the overwhelming conclusion that there must be another kind of matter that is otherwise invisible and so far undetectable. In a great failure of imagination it is called 'dark matter'. And, as it must account for more than eighty per cent of all the matter (visible and invisible) in the universe today, it is, quite frankly, an embarrassment.

It is believed that a small proportion of dark matter exists in the form of exotic astronomical objects composed of ordinary matter which emit little or no radiation (and which therefore appear 'dark'). These are called Massive Astrophysical Compact Halo Objects, or MACHOs.

However, the vast majority of the dark matter is thought to consist of currently unknown particles, called Weakly Interacting Massive Particles, or WIMPs.* These particles have many of the properties of neutrinos, but are required to be far more massive and therefore move through spacetime much more slowly.

THE QUARK EPOCH: 10^{-12} TO 10^{-6} SECONDS

So, we now have all the building blocks of the visible (and invisible) universe. Spacetime itself is imbued with vacuum energy, which we call dark energy, which will make itself felt in about nine billion years' time. The four forces of nature have taken forms that we will later come to recognize. The universe contains a large proportion of dark matter particles which are susceptible to gravity and the weak force and little else. There's nothing more to be said about these for the time being because, well, we simply don't know anything about them.

We turn our attention instead to the visible matter that we will in any case come to be rather more concerned about—the quarks, leptons, and force particles of the standard model.

---

* Of course, these acronyms are not coincidental. WIMP was coined first, apparently inspiring the subsequent development of MACHO.

The temperature of the universe at the end of the Electro-weak Epoch is still a tremendously high 10^{15} kelvin. Despite the fact that the energy of an isolated colour charge is theoretically infinite, there's still too much energy in the universe for quarks to sit comfortably bound to anti-quarks (in mesons), and in larger particles such as protons and neutrons (which are collectively called *baryons*). In fact, mesons (from the Greek *mésos*, meaning 'middle') and baryons (from the Greek *barys*, meaning 'heavy') are sub-categories of a class of particle called *hadrons* (from the Greek *hadros*, meaning 'thick' or 'heavy').*

Instead, the contents of the universe are dominated by a fizzing plasma of quarks, gluons, leptons, neutrinos, and photons. There's enough energy around for all three generations of leptons and quarks and their anti-particles to form, and the gluons run madly back-and-forth between the quarks in an attempt to shield the exposed colour charges. If larger particles do form, they're ripped apart in a fraction of an instant.

This is the Quark Epoch.

Not much is known about the quark–gluon plasma, but recent experiments at CERN's Large Hadron Collider are designed to create such a plasma and study its properties. Although much of the focus of the effort at CERN has been on the search for the Higgs boson, when the collider hasn't been smashing protons together in pursuit of the Higgs, it's been used to smash lead nuclei together. Lead nuclei? There are four naturally occurring isotopes† of lead (symbol Pb). One of these, ^{206}Pb, contains 82 protons and 124 neutrons. So, smashing two lead nuclei together at the kinds of energies that prevailed during the Quark Epoch really is going to create an awful mess.

And this is what the ALICE detector at CERN is used to study.‡ The protons and neutrons inside the colliding lead nuclei 'melt', liberating the quarks and gluons from inside them and, for a fleeting instant, re-creating the quark–gluon plasma at temperatures around 5.5×10^{12} kelvin. This quickly cools and forms a veritable blizzard of exotic material particles and 'jets', energetic sprays of quarks and gluons that condense to form mesons and other particles.

One early observation from these kinds of studies, at CERN and elsewhere, is that the quark–gluon plasma behaves more like an ultra-hot, low-viscosity *liquid* than a gas.

---

* Which is why the Large Hadron Collider is so named.

† Isotopes have the same number of protons in the nucleus but different numbers of neutrons.

‡ ALICE stands for A Large Ion Collider Experiment.

But the universe is continuing to expand and cool, and about a microsecond (10^{-6} seconds) after the big bang the universe has cooled to a temperature around 3×10^{12} kelvin, cold enough for the quarks to become 'confined' within mesons and baryons. Once trapped inside these larger particles, there is now insufficient energy to pull the quarks apart.

THE HADRON EPOCH: 10^{-6} SECONDS TO 1 SECOND

The matter in the universe is now dominated by hadrons and their anti-matter counterparts. Among the matter particles are the familiar proton and neutron, and the less familiar pions, charged π^+ (formed from up and anti-down quarks), π^- (formed from down and anti-up quarks), and neutral π^0 (a mixture of up-anti-up and down-anti-down). The more exotic second- and third-generation quarks have disappeared, either through decay into more stable particles or through 'annihilation' reactions, which occur when matter and anti-matter particles collide.

Also present in the mix are electrons, muons, their associated neutrinos and their anti-matter counterparts, and photons. The third-generation taus have annihilated. The neutrinos are very light particles that travel at virtually the speed of light. Once they cease interacting with other particles they 'disengage', ignoring most of the rest of the matter and anti-matter and simply expanding along with spacetime itself.

There is now no exposed colour charge—the quarks and gluons are safely tucked away inside the hadrons. In a sense, the strong force has become 'internalized'; it is now manifested only inside larger particles. The forces at play in the resulting plasma of electrically charged particles are the weak force and (predominantly) electromagnetism.

The temperature continues to fall. As it falls below about 10^{12} kelvin, hadrons and anti-hadrons annihilate. Pairs of π^+ and π^- particles annihilate and the π^0 particles decay into photons. Protons and anti-protons annihilate to form sprays of mesons, which in turn decay into photons. Protons annihilate anti-neutrons, and anti-protons annihilate neutrons.

That's a lot of annihilation. In truth, these kinds of matter–anti-matter annihilation reactions have been going on all the time, but at higher temperatures there's enough energy in the photons that are produced for these to spontaneously re-create the matter–anti-matter pairs, so there's no net loss. Now, as the temperature falls, there's no coming back.

Simple logic suggests that this should proceed quite equitably, with matter and anti-matter disappearing in equal quantities, transforming eventually into light. However, the universe that we observe today appears to contain only matter. It's not impossible that great chunks of anti-matter exist in space—perhaps in distant clusters of anti-matter galaxies. It would actually be quite impossible to tell the difference between matter stars and anti-matter stars in such galaxies based on measurements that we can make from Earth or satellite-borne instruments.

But, although at the moment we can't prove it either way, the idea that there might be great chunks of anti-matter somewhere 'out there' feels all a bit unnatural. This leaves us having to accept that, when all the hadron annihilation reactions are done, what is left is a small residual amount of matter. This might be random chance. If there is some subtle physical mechanism which determines that matter could be expected to dominate, we do not yet know what this mechanism is.

THE LEPTON EPOCH: 1 TO 10 SECONDS

Most of the hadrons and anti-hadrons have now been wiped out, leaving a small residual number of hadrons. The matter in the universe is now dominated by leptons.

As the universe continues to expand and the temperature falls below about 5×10^9 kelvin, electrons and positrons start to annihilate each other and their respective neutrinos become disengaged. The photons continue to dance back and forth between matter particles and so remain at the temperature of the hot plasma, but the neutrinos now stand apart, cooling along with the general expansion of spacetime.

The disappearance of electron–positron pairs and the disengagement of the neutrinos means that reactions converting protons to neutrons and neutrons to protons no longer occur. The numbers of protons and neutrons in the universe now 'freeze' in the ratio of about five to one.

The annihilation reactions continue until the universe is about ten seconds old. The temperature has now fallen to about 10^9 kelvin, and there is now insufficient energy for new electron–positron pairs to form. Once again, for reasons we don't understand, matter triumphs over anti-matter, leaving a small residual number of leptons (primarily electrons) whose negative charge balances the net positive charge of the remaining hadrons.

Let's take stock once more. The universe is ten seconds old. The mass-energy of the universe is dominated by dark energy, the energy of 'empty' spacetime. The mysterious dark matter particles account for a further sizeable chunk. Visible matter makes a small contribution to the total mass-energy, consisting of small residual numbers of hadrons (mostly protons and neutrons in the ratio five to one), with balancing numbers of leptons (mostly electrons), and photons. The last now dominate the visible matter and radiation in the universe, as we enter the Photon Epoch.

3

THE LAST SCATTERING SURFACE

The Origin of Light

The universe will never again be as simple, in terms of its contents, as it is now. It is a universe dominated by light, with photons outnumbering baryons (protons and neutrons) by a factor of about 1.6 billion.

But if we could peek into this universe, we wouldn't be able to see very much. Electrically charged protons and electrons still run free, in a dense, charged plasma. The photons run back and forth between them, their interactions particularly with free electrons creating an impenetrable 'fog' in the mother of all electrical storms. Welcome to the Photon Epoch, which will last from ten seconds to 380 000 years after the big bang, and which is the subject of this chapter.

Although photons rule, and dark matter and disengaged neutrinos stand as mute witnesses to the unfolding events, watching and waiting, our attention is drawn nevertheless to the baryons themselves. They may be outnumbered and out-gunned, but they are the primordial building blocks of visible matter, and they're just starting to get their act together.

PRIMORDIAL NUCLEOSYNTHESIS

About a hundred seconds or so after the big bang, the temperature of the universe has cooled to just below one billion (10^9) kelvin. Weak-force reactions responsible for dictating the balance between the numbers of protons and neutrons in the universe have all but ceased, 'freezing' the ratio. But the neutron is inherently unstable. One of the two down quarks inside it is prone to the weak force interaction which turns it into an up quark, thereby transforming the neutron into a proton. In other words, free neutrons are radioactive.

A free neutron has a half-life of about 610 seconds (a little over ten minutes), meaning that within this time we can expect that half the initial number of neutrons will have decayed into protons (or, alternatively, there's a 50% chance that each neutron will have decayed).[1] The universe hasn't existed quite this long, but free neutrons have been around from the beginning of the Hadron Epoch, a millionth of a second after the big bang, so within the time that has now elapsed the number of neutrons will have declined by a further 25%. This increases the proton-to-neutron ratio to about seven to one.

Protons and neutrons sometimes go by the name 'nucleons', meaning the constituents of atomic nuclei. At this temperature, nuclear reactions between the protons and neutrons now start to produce heavier combinations of these nucleons. They have flirted with each other before, coming together in combinations that have instantly broken up again. Now they come together and persist long enough to become involved in further reactions. This signals the onset of something called primordial nucleosynthesis.

The most important nuclear reactions are pictured in Figure 13. Here the proton is shown as a red ball, the neutron as a blue ball. As the proton is also the nucleus of the hydrogen atom, I've labelled it ^{1}H, where the 'H' represents the chemical element hydrogen, the superscript-1 indicates that it contains just one nucleon—in this case one proton. Although I'm not showing it, don't forget that the hydrogen nucleus also carries a unit of positive electrical charge (^{1}H$^+$).

The starting point for nucleosynthesis is the reaction between a single proton and a single neutron. The proton picks up a neutron, turning into ^{2}H. This is still a hydrogen nucleus, as its chemical identity is determined by the number of protons in it, which is still only one. This 'heavy' isotope of hydrogen is called deuterium.

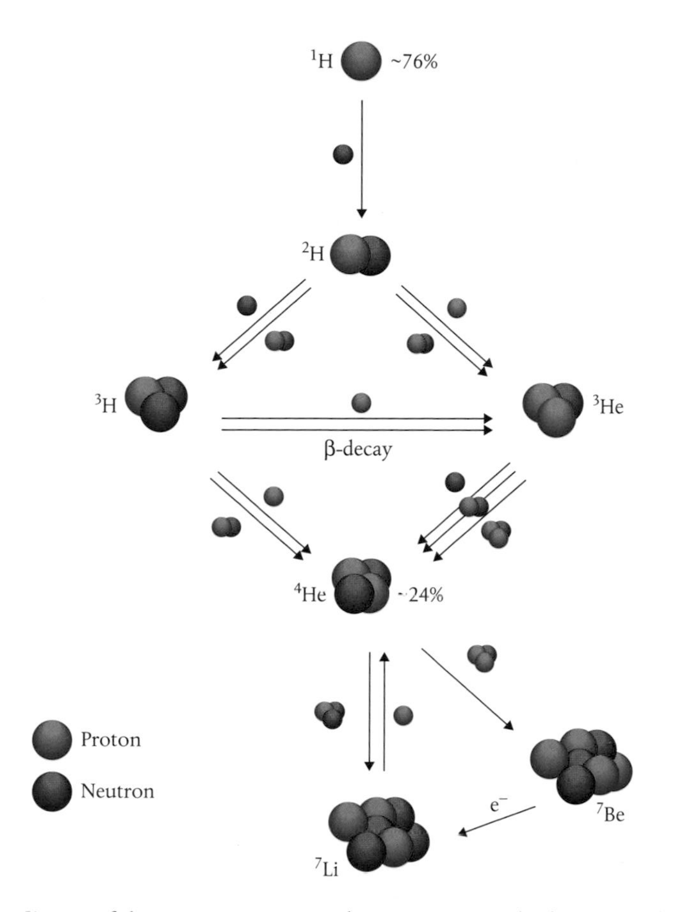

FIGURE 13 Sixteen of the most important nuclear reactions involved in primordial nucleosynthesis. These are shown schematically in terms of the reactions of protons (red balls) and neutrons (blue balls). For simplicity, not all the products of the reactions are shown.

The proton and neutron in the deuterium nucleus are quite happy to cling to each other, and the energy of the combination is lower than the energy of a free proton and neutron by an amount called the binding energy. This excess energy is carried away from the ^{2}H nucleus by a photon (not shown in Figure 13).

This begs a question. What force now binds the proton and neutron together in the deuterium nucleus? The answer used to be: the strong force, of course! But we now know that the strong (colour) force binds quarks together *inside* the proton and the neutron. So, what's going on?

There are at least two ways of thinking about this. We can think of the force holding the proton and neutron together as a kind of 'leakage' of the colour force beyond the boundaries of each nucleon. The result is not as strong as the colour force itself, but it is nevertheless strong enough to hold the nucleons together.

Alternatively, we can think of this as a new derivative or secondary force (i.e. one that is not 'fundamental') which acts between protons and neutrons and, as we will soon see, between protons and protons—overcoming the force of electrostatic repulsion between like charges—and neutrons and neutrons. By definition, such a derived or secondary force must have associated force particles. These are, in fact, the pions, π^+, π^-, and the neutral π^0. If a proton (which, remember, consists of two up quarks and a down quark, uud) and neutron (udd) exchange positive and negative pions (consisting of up–anti-down and down–anti-up quarks), then the proton turns into a neutron as the neutron turns into proton. Nothing outwardly changes, but the exchange helps to reduce the energy of the combination, so binding the particles together.

A glance at the top of Figure 13 tells us that building heavier nuclei is not possible without first making deuterium nuclei. This means that the speed or rate at which this first reaction proceeds will broadly determine the rate of the entire chain of reactions that follows. This is known as the *deuterium bottleneck*. Once we have built up a sufficient number of deuterium nuclei, then precisely what happens next will simply depend on the relative rates of the nuclear reactions that follow.

At this stage I can't help but be reminded of some maths classes I attended early in my school career. To learn about calculus, we would be set problems that required us to solve differential equations involving processes happening at different rates. These always seemed to involve filling a bath in which we'd forgotten to put the plug in the plughole.* Water from the tap flows into the bath at one rate, and flows out through the plughole at another rate. The question was, how long will it take to fill the bath?

Solving the differential equation for primordial nucleosynthesis is in principle no different, just more complicated. Figure 13 shows 16 different nuclear reactions. It's as though the water from the bath in my school maths problem

* I attended these classes in the late 1960s, but our schoolbooks had clearly been written in the 1950s. Several problems would begin with 'A man jumps onto the running board of a moving car...'

drains into another bath, which in turn drains into another bath at a different rate, and so on and on until the water trickles right to the bottom of the chain.

To solve the equation we need to measure the rates of all the different nuclear reactions involved. What we get coming out depends on these reaction rates and on the initial conditions that prevailed at the onset of nucleosynthesis, such as the temperature, the ratio of photons to baryons, and the ratio of protons to neutrons.

All this information is available from studies of nuclear reactions and observations of the universe today (we'll come on to consider the latter in what follows). We also know the temperature, which must have been cool enough to allow deuterium nuclei to form in sufficient quantities to get us through the deuterium bottleneck.

So let's continue the story. A deuterium (^{2}H) nucleus may now react further, picking up another neutron to form a nucleus of tritium (^{3}H)—another, even heavier isotope of hydrogen—plus a photon (again, not shown in Figure 13). Or it may react with another deuterium nucleus to form tritium plus a free proton. Or it may react with a neutron or another deuterium nucleus to form ^{3}He, consisting of two protons and a neutron, which is an unstable isotope of helium.

Like the free neutron, the ^{3}H nucleus is susceptible to the weak force. A neutron inside the ^{3}H nucleus (one proton, two neutrons) may transform into a proton, turning it into a ^{3}He nucleus (two protons, one neutron). This is shown in Figure 13 as a reaction labelled 'β-decay', since this is the process involved in beta radioactivity.

You get the picture. Both ^{3}H and ^{3}He may undergo a variety of reactions to form ^{4}He, the nucleus of the most common isotope of helium. This combination of two protons and two neutrons happens to be particularly stable. Although it is susceptible to further reactions (with ^{3}He to produce an unstable isotope of beryllium, ^{7}Be, or with ^{3}H to produce an isotope of lithium, ^{7}Li) the rate at which the ^{4}He 'bath' drains is a lot slower than the rate at which it fills. The end result is that an appreciable number of ^{4}He nuclei accumulate.

Now, not all the nuclear reactions thought to be involved in primordial nucleosynthesis are shown in Figure 13, but the most important ones are. So why does the chain appear to end at ^{7}Li? After all, we know that today there are heavier elements in the universe, such as carbon, oxygen and nitrogen, because these are what we're made of.

The answer is that only certain combinations of protons and neutrons have lasting stability, determined by the quantum properties of their constituent quarks. The next element in the periodic table is beryllium, but both ^{7}Be and ^{8}Be are unstable (the most common isotope of beryllium is ^{9}Be, consisting of four protons and five neutrons). This is not so much a gap as a yawning chasm: one that can't be bridged under the conditions that prevailed in the first few hundred seconds after the big bang. We will see in the next chapter how the chasm is bridged in the interiors of stars, which is where most of the elements heavier than helium are synthesized.

So, what do we get? Given the starting conditions and the measured rates of all the nuclear reactions involved, the differential equation can be solved numerically (on a computer) to give the end result. Although ^{2}H is a critical intermediate in this chain of reactions, it tends to be consumed almost as rapidly as it is produced, with the end result that the number of ^{2}H nuclei left at the end is very small (the ratio of deuterium to hydrogen nuclei is about 0.00003). Much the same is true for ^{3}He (the ratio of ^{3}He to hydrogen nuclei is estimated to be less than 0.00007).

Because it is so stable, the number of ^{4}He nuclei builds up and, when nucleosynthesis is done, these account for an impressive 24% of the total by mass. The number of ^{7}Li nuclei remaining at the end is also very small (the ratio of ^{7}Li to hydrogen nuclei is about 5×10^{-10}), which means that the balance by mass of 76% is accounted for by unreacted ^{1}H nuclei (protons).

All the neutrons in the universe have now been incorporated into ^{4}He nuclei, which prevents any further loss through radioactive processes. The number of ^{4}He nuclei that result is therefore determined precisely by the number of neutrons that were available at the onset of nucleosynthesis. As there are two protons and two neutrons in each ^{4}He nucleus, then the proportion of ^{4}He by mass is expected to be simply twice the proportion of neutrons in the initial mix. A seven-to-one proton–neutron ratio at the start implies a neutron percentage of about 12%. Twice 12% is 24%.[2]

Nucleosynthesis changes the nature of the charged plasma, from one of protons, neutrons, electrons, and photons to one of positively charged protons (^{1}H), positively charged ^{4}He nuclei, electrons, and photons. And, although they have had no part to play in these events, let's not forget the disengaged neutrinos and the dark matter particles.

The universe is still a charged plasma, and the electrical storms continue to rage.

The amount of hydrogen and helium in today's universe is something that we can measure. For sure, the composition will have changed in the 13.8 billion years that have elapsed. As we will see in the next chapter, both hydrogen and helium are consumed in the interiors of stars and converted into heavier elements. Stars 'cook' hydrogen nuclei, making more helium in the process. But there are some interstellar gas clouds that have compositions not much different from that of the primordial universe, from which it is possible to infer rather precisely the results of this early nucleosynthesis. These inferences are entirely consistent with the 76% hydrogen, 24% helium mixture that was produced just a few hundred seconds or so after the big bang.

ALPHER, BETHE, GAMOW

One of the first nuclear physicists to take the idea of a big bang origin of the universe seriously was the Ukrainian-born émigré George Gamow. In the 1930s, Lemaître had speculated that the big bang had involved the radioactive disintegration of some kind of 'primeval atom', containing all the protons and neutrons in the universe.

Though picturesque, this model could not explain the relative abundances of hydrogen, helium, and the sprinkling of other atoms in the universe. Such a primeval atom would indeed be unstable and would decay rather rapidly, but the decay would most likely involve nuclear *fission* reactions, in which the nuclei split up into smaller fragments rather than fall apart completely to form individual protons and neutrons. To Gamow, it seemed rather more logical to start with an early universe consisting of primordial neutrons (with protons formed from weak-force decays) and electrons and apply the principles of nuclear physics to work out what would most likely happen next.

With support from his American postgraduate student Ralph Alpher, Gamow did precisely this. Despite their erroneous starting point (quarks hadn't been discovered at this stage and we now know that protons and neutrons were formed in near-equal numbers in the Hadron Epoch), Alpher and Gamow nevertheless successfully predicted the relative abundance of hydrogen and helium that resulted from primordial nucleosynthesis.

On submitting a paper describing their calculations to a noted scientific journal, Gamow chose to add the name of fellow émigré physicist Hans Bethe to the list of authors. Bethe had not been involved in the work but Gamow, author of the successful *Mr Tompkins* series of popular science books, had a reputation as a prankster.* The possibilities afforded by a paper authored by Alpher, Bethe,

---

* *Mr Tompkins in Wonderland* was first published in 1940. This was followed in 1944 by *Mr Tompkins Explores the Atom*.

and Gamow had captured his imagination. Inevitably, it became known as the alpha-beta-gamma paper.

The paper was published in 1948, on April Fool's Day. Gamow had originally marked the paper to indicate Bethe as an author *in absentia*, but the journal editor removed this note. Bethe (who, as it turned out, was asked to review the manuscript) didn't mind. 'I felt at the time that it was rather a nice joke, and that the paper had a chance to be correct, so that I did not mind my name being added to it.'[3]

However, Alpher was unimpressed. The paper was a summary of his doctoral dissertation. Both Gamow and Bethe were established physicists with international reputations. Anyone reading the paper would likely conclude that these more esteemed physicists had done all the work.

RECOMBINATION

The universe now experiences an unprecedented period of relative calm. A short and rather frantic beginning, with forces and then elementary particles emerging out of the initial burst of inflated space, time, and energy, gives way to 380 000 years without much change in its overall composition.

Of course, the universe continues to expand and cool through this period, the temperature eventually falling to about 3000 kelvin. This triggers what will prove from our perspective to be another very significant moment in the evolution of the universe. At this temperature, the free helium nuclei (^4He) have already combined with free electrons to produce neutral helium atoms. Now, at 3000 kelvin, the free protons (^1H) combine with the free electrons to form neutral hydrogen atoms.

The photons that, up until this point, had flitted between and careened off the charged particles that formed the plasma, are now released. They become disengaged from matter in the same way that the neutrinos had become disengaged thousands of years earlier.

The subtitle I've chosen for this chapter is possibly a little misleading. Photons have existed in the universe from almost the very beginning, of course, so this moment of recombination* can hardly be said to signal the origin of

* Yes, I know. This is the first time in the history of the universe that charged particles have combined together to form neutral atoms, so calling it 're'-combination doesn't really seem appropriate. However, this is the name by which this moment has come to be known and, as scientists are generally quite conservative folks, the name has stuck.

the elementary particles of light. But the photons now released from their ties to matter will persist for all eternity, streaming freely through the universe in much the same way that light streams towards us from the Sun. This might not be the origin of light particles, but it is the origin of the phenomenon of light as we experience it.

And it is the oldest light in the universe.

The universe now becomes transparent, as it floods with the light released by recombination. Some of this light is visible, although there is obviously nobody around to see it. It spans a broad range of frequencies or wavelengths, peaking in the infrared. The universe is literally bathed in a warm glow, an 'afterglow' of creation.

The light ceases to have any further role to play, and simply cools as it is carried along for the ride by the expansion of spacetime. But locked within its properties are clues to the circumstances of its origin. Like dark matter particles and neutrinos, this light stands as mute witness to more turbulent times in the early history of the universe. But, unlike dark matter particles and neutrinos, the testimony of this witness is one that we will one day be able to hear.

THE COSMIC BACKGROUND RADIATION

Alpher became rather frustrated because the model for primordial nucleosynthesis that he had developed with Gamow didn't predict the production of any elements heavier than helium. In the meantime, Gamow had forged ahead on other aspects of early post-big-bang physics. In the summer of 1948 he sent Alpher the manuscript of a paper he had recently submitted to the British journal *Nature*. The paper was concerned with the densities of matter and radiation at the moment of recombination.

Gamow was working through the summer months at the US atomic weapons research laboratory at Los Alamos. When Alpher and his colleague Robert Herman* realized that Gamow's calculations were seriously wrong, they hastily informed him in a telegram. Gamow judged that it was now too late to withdraw his paper, and instead urged Alpher and Herman to submit a note correcting his error for publication 'back-to-back' in the same journal.

* Gamow joked that he'd tried to persuade Herman to change his name to Delter.

Alpher and Herman decided to use this opportunity to go a little further. They argued in this short note that the radiation released after recombination would have persisted to the present day, forming a kind of cosmic 'background', pervading the entire universe and filling all of space. It could be expected to have the characteristics of so-called 'black-body' radiation.

Heat any object to a high temperature and it will gain energy and emit light. We say that the object is 'red hot' or 'white hot'. Increasing the temperature of the object increases the intensity of the light it emits and shifts it to a higher range of frequencies (shorter wavelengths). As it gets hotter, the object glows first red, then orange-yellow, then bright yellow, then brilliant white.

A 'black body' is a theoretical, completely non-reflecting (i.e. totally black) object that absorbs and emits light radiation without favouring any particular range of wavelengths. The intensity of radiation emitted by a black body is then directly related to the amount of energy in it when it is in thermal equilibrium with its surroundings. It was the study of black-body radiation that in 1900 had led Planck to discover what Einstein would later call 'light-quanta', or photons.

Alpher and Herman argued that if, as expected, the plasma of atomic nuclei, electrons, and photons maintained a thermal equilibrium until the moment of recombination, then the radiation released at this moment would have a distribution of wavelengths characteristic of a black body with a temperature of 3000 kelvin (Figure 14). Once disengaged from matter, this cosmic background radiation would have cooled as the universe expanded. Alpher and Herman estimated that it would today have an average temperature just 5 degrees above absolute zero, or 5 kelvin. Although they didn't say so, black-body radiation with a temperature of 5 kelvin would have a distribution of wavelengths peaking in the microwave and infrared region.*

Now this was an unexpected, out-and-out *prediction*, one that only a cosmology based on a hot big bang could make. The evidence that our universe began in the fireball of a big bang should be all around us in the form of invisible, but nevertheless detectable, microwave radiation.

But, it seems, the prediction was largely ignored. Cosmology was not yet a serious science, disputes were raging about astronomical observations that implied that the universe was—absurdly—*younger* than the Earth, and there

* The infrared region of the electromagnetic spectrum extends from about 700 nanometres to 1 millimetre (or 0.1 centimetres). The microwave region includes wavelengths from 1 millimetre up to 1 metre. A small, home microwave oven uses wavelengths typically of the order of 122 millimetres.

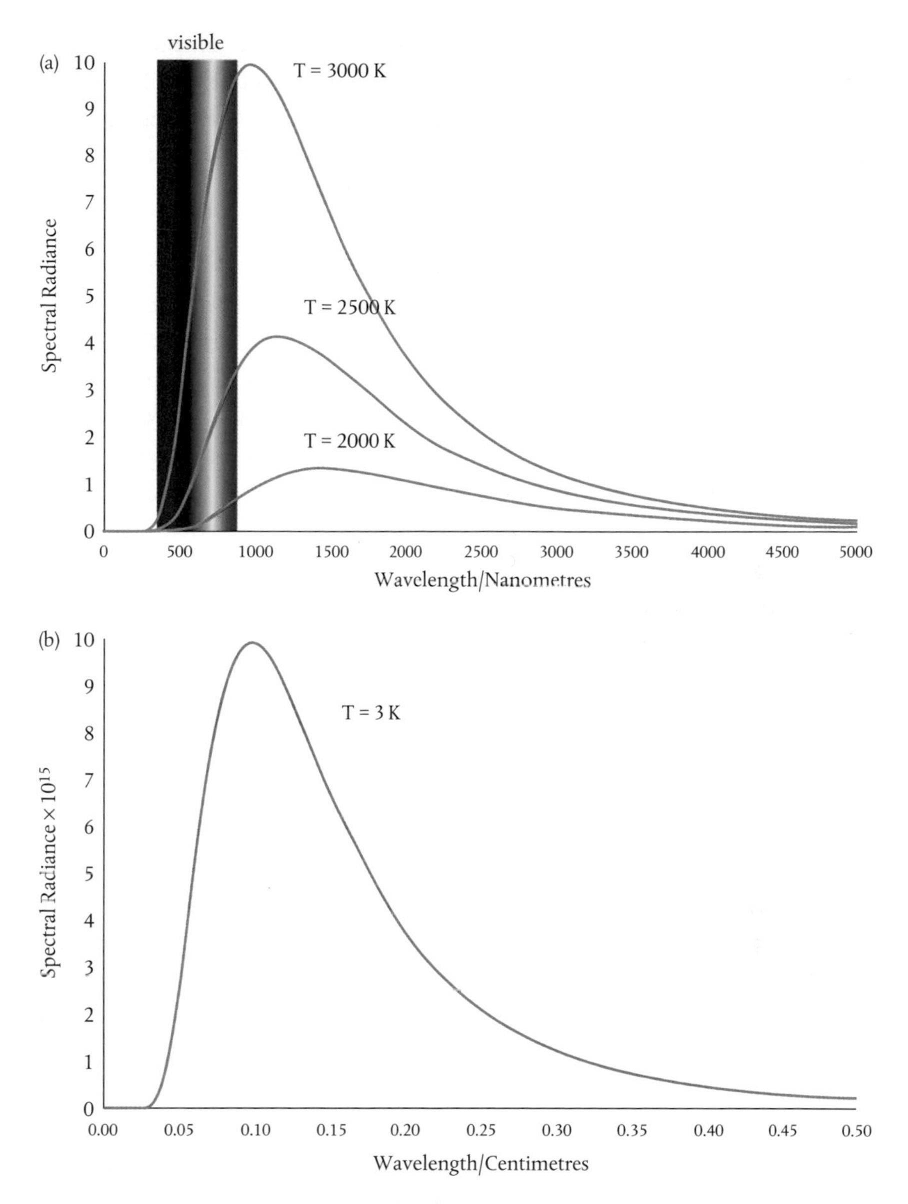

FIGURE 14 A perfect 'black body' emits light with a characteristic pattern of intensity, or spectral radiance, over a distribution of wavelengths determined by the temperature of the body. (a) shows three distributions at temperatures of 3000, 2500, and 2000 kelvin. Note how the spectral radiance sharply declines and the distribution moves to longer wavelengths as the temperature falls. (b) shows the same distribution for a temperature of 3 kelvin, but with the spectral radiance magnified by a factor of 10^{15} compared to (a) and the wavelength axis changed from nanometers to centimetres.

were other theories that did not require a big bang. Alpher and Herman also later wondered if Gamow's reputation for playfulness might have been a factor. In his reflections on the situation in his popular book *The First Three Minutes*, Steven Weinberg famously noted:[4]

> This is often the way it is in physics—our mistake is not that we take our theories too seriously, but that we do not take them seriously enough.

Whatever the reason, even though the technology that was needed to detect the cosmic background radiation was arguably available in the 1950s, there was no great rush to begin a systematic search for it.

Consequently, when in the summer of 1964 Princeton University physicist Robert Dicke suggested: 'Wouldn't it be fun if someone looked for this radiation?' Alpher and Herman's earlier prediction had been largely forgotten. Dicke had independently re-discovered the possibility of cosmic background radiation and assigned the task of working out how to detect it to two young radio astronomers, Peter Roll and David Wilkinson. Turning to Jim Peebles, a young theorist from Manitoba, he said: 'Why don't you go and think about the theoretical implications?'[5]

Peebles went home and thought about it. He re-invented a version of the big bang model that Gamow, Alpher, and Herman had developed and used it to predict a cosmic background radiation with a temperature of about 10 kelvin. But when he submitted this work for publication it was rejected by the journal editor, on the basis that Alpher, Herman, and Gamow had already covered this ground some years before. This was news to Peebles.

Just 30 miles away, at the Bell Laboratories' Holmdel research facility in New Jersey, radio astronomers Arno Penzias and Robert Wilson were puzzling over the source of some rather annoying microwave interference that appeared to be coming from all directions in the sky. Despite their best efforts, this interference just wouldn't go away.

They were using the 20-foot horn antenna that had been built to demonstrate the feasibility of satellite-based communications (Figure 15). However, the Telstar 1 satellite, launched with great fanfare in July 1962, had prematurely gone out of service in November that year, largely due to the deleterious effects of atmospheric nuclear weapons tests. A workaround brought the satellite temporarily back to life in January 1963 but its transistors were finally overwhelmed a month later, and it shut down. This had freed up time on the Holmdel antenna for some radio astronomy.

FIGURE 15 The 20-foot microwave horn antenna installed at the Bell Telephone Laboratories' Holmdel research facility. This is now designated as a US National Historical Landmark.

Penzias and Wilson were looking for microwave radiation that was thought to be emitted by the glowing cloud of gas surrounding our own Milky Way galaxy. They began by pointing the antenna at the sky in a direction at right angles to where they expected to find the radiation, intending to establish a 'baseline' against which they could eventually measure their signal. To their great surprise, they found a substantial signal, a baseline so large it would actually overwhelm the signal they were hoping to study.

From their measurements at a single wavelength of 7.35 centimetres, they estimated a black-body temperature for this signal of 3.5 kelvin, plus or minus 1 kelvin. Such a measurement was not proof that this was black-body radiation; it was rather an estimate based on the assumption that they were sampling from a black-body spectrum.*

* A glance at Figure 14(b) reveals that 7.35 centimetres lies far to the long-wavelength end of the 3 kelvin black-body spectrum.

Like all good experimentalists, they set about the task of identifying the source of this unwelcome 'noise' so that they could eliminate it. The radiation did not seem to depend on direction—the signal was coming uniformly from all directions in the sky, with a uniform temperature. The signal didn't change with the time of day or season. They ruled out microwave pollution from nearby Manhattan. They evicted a pair of pigeons that were roosting inside the horn and cleared their droppings from its surface. Alas, these were homing pigeons and, when they inevitably returned, the astronomers had to adopt a rather more permanent method of removal.

The pigeons died in vain, as the source of microwaves persisted.

No nearer to solving their problem, in December 1964, Penzias attended a meeting of the American Astrophysical Society in Montreal. He mentioned these puzzling observations to a colleague and was eventually advised that Jim Peebles at Princeton had recently done some work that suggested the existence of cosmic background radiation with a temperature of 10 kelvin. Penzias acquired a copy of Peebles' unpublished manuscript. Sensing a possible connection, he picked up the phone and called Bob Dicke, Peebles' research advisor.

Dicke, Peebles, Wilkinson, and Roll had gathered together in Dicke's office for a lunchtime meeting to discuss the design of their own microwave antenna, to be constructed on top of the Palmer Physical Laboratory at Princeton. The meeting was interrupted by the phone call from Penzias. Dicke listened intently and muttered a few remarks before putting the phone down. 'Well boys,' he said, 'We've been scooped!'[6]

Dicke, Roll, and Wilkinson piled into a car and drove the short distance to Holmdel. Penzias and Wilson and the Princeton group published companion papers announcing the discovery of the cosmic background radiation in May 1965.

The oldest light in the universe, witness to events that had occurred at some of the very earliest moments of creation, was ready to testify.

ANISOTROPY AND THE ORIGIN OF STRUCTURE

Roll and Wilkinson soon followed up with their own measurement, at a wavelength of 3.2 centimetres. They estimated that the radiation has a temperature between 2.5 and 3 kelvin. Several further measurements at different microwave wavelengths were subsequently reported, all confirming a black-body temperature between 2.7 and 3 kelvin.

But these measurements were all to the long-wavelength side of the expected black-body spectrum, shown in Figure 14(b). There was a simple reason for this. The Earth's atmosphere is fairly transparent to radiation longer than about 0.3 centimetres. However, radiation of shorter wavelengths tends to be absorbed by the atmosphere. Radiation with wavelengths around 0.1 centimetres and shorter is actually infrared (heat) radiation. Unfortunately, this is the wavelength at which black-body radiation with a temperature of 3 kelvin is expected to peak.

The only way to be sure that this was really cosmic background radiation was to make measurements up above the atmosphere. In 1977, just as Weinberg was making some final corrections to the proofs of *The First Three Minutes*, he received a newsletter announcing the formation of a scientific team to investigate the possibility of mapping the cosmic background radiation using satellite-borne instruments. If the satellite was ever built, it would be called the Cosmic Background Explorer (COBE).

As the scientific community waited patiently for closure on this question, a couple of really puzzling implications started to emerge. In applying general relativity to the entire universe in 1917, Einstein had assumed that the universe is uniform in all directions. And, indeed, the cosmic background radiation appears to have a temperature that is uniform in all directions. The technical term is *isotropic*.

But how could this be? As the universe expands up to the moment of recombination, the distances between different parts of the universe inevitably increase. Eventually, these distances become so great that different parts of the universe are no longer in causal contact with one another. This means that light simply takes too long to travel from one part to another and, as nothing can travel faster than light, these different parts of the universe can no longer hold any physical influence over each other.

Here's an everyday example. We know that if we place a hot object in contact with a cool object, the temperatures of both objects will equilibrate and become uniform. Such uniformity in temperature implies a dynamic exchange of energy, with energy flowing from the hot object to the cool one, which happens because they are in causal contact.

But the last moment that the cosmic background radiation could have been influenced by such a dynamic exchange of energy was at the moment of recombination. After that moment, the radiation was freed from matter and would interact with it very rarely, if at all. The trouble was, the 380 000-year-old universe was now already too big. The finite speed of light would not allow all its different parts to remain in contact.

If different parts of the universe had become disconnected, as expected, then there was really no good reason why the temperature of the cosmic background radiation should be so isotropic. This is called the *horizon problem*.

A potential answer was first devised in December 1979 and we have already considered it in Chapter 1. It is called cosmic inflation. At the onset of inflation, every part of the universe *was* in causal contact with every other. Equilibrium prevailed. By the time inflation was done, that uniformity had been imprinted on a scaled-up universe that was now much, much larger. It didn't matter that parts of the universe were now no longer in contact with each other, as every part was guaranteed to have the same post-inflation conditions, and simple physics would ensure that each would evolve in the same way.

Cosmic inflation resolves both the flatness *and* horizon problems.

But *still* something nagged. Okay, if we look up at a night sky we can accept that, in a very coarse-grained sense, the universe looks much the same in different directions, as Einstein had assumed. We see stars, distant galaxies and empty space. However, a child, unfazed by the scientific authority of this observation, will simply point out that the pattern of points of light over here on the left looks very different compared with the pattern of points of light over there on the right. In a fine-grained sense, we have to own up to the simple fact that the universe has a *structure*.

And, indeed, detailed studies of patterns of galaxies published towards the end of the 1980s began to establish the true extent of this structure. Galaxies are not randomly but uniformly distributed across the sky. They have formed along 'strings' and 'walls', surrounding great 'voids'. For sure, it is a haphazard pattern, but it is a pattern nonetheless. We will take a closer look at this structure in the next chapter.

Now, if we accept the argument based on cosmic inflation and conclude that the cosmic background radiation is perfectly isotropic, this implies that the distribution of matter at the moment of recombination must also have been isotropic.

Let's just pause for a moment to reflect on this. Imagine a simple two-dimensional universe filled with a uniform (isotropic) distribution of matter in the form of tiny billiard balls (Figure 16(a)). Because the balls are all the same distance from each other (and we assume our universe has no edges), gravity

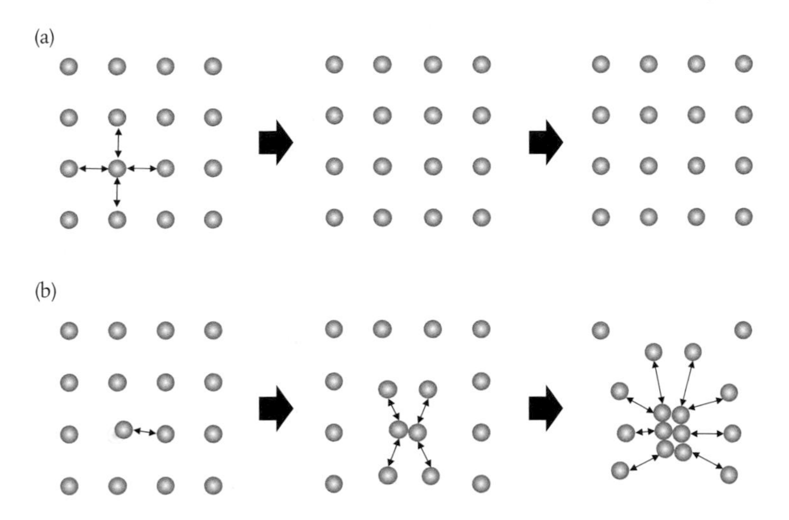

FIGURE 16 (a) In a perfectly uniform or isotropic distribution of matter, gravity acts equally in all directions and the matter remains frozen in place, even though spacetime itself might be expanding. In (b) we introduce a small anisotropy by displacing one of the balls slightly. Now the displaced ball feels a slightly stronger gravitational attraction to its nearest neighbour. Over time, the matter clumps together, forming a structure.

acts equally on all of them. The matter in such a universe is frozen in place. Even if the spacetime in such a universe is expanding, the balls will spread out but will remain equally distant from one another. No structure forms.

Now let's displace one of the balls just a little, creating a small *anisotropy* (Figure 16(b)). The ball that is displaced now feels a slightly stronger gravitational pull from its nearest neighbour. As these balls are drawn together by the force of gravity, they exert a stronger gravitational pull on some of the other balls that surround them. The process snowballs and eventually the balls clump together, forming a structure.

There was no alternative but to conclude that, despite appearances to the contrary, the universe *must* have been anisotropic and, by implication, the temperature of the cosmic background radiation must be anisotropic, too.

But by how much? It was possible to work backwards from the structure that we see in the present-day universe. Assuming the existence of sufficient dark matter to provide the gravitational instabilities, it was estimated that an anisotropy as small as a few parts in 100 000 would be enough.

The nature of the questions changed. What could have caused an anisotropy in the distribution of matter in the universe of the order of a few parts in

100 000? And, if this anisotropy is real, can we see it reflected in small differences in the temperature of the cosmic background radiation in different directions across the sky?

HEISENBERG'S UNCERTAINTY PRINCIPLE

To answer the first question, we need to go back once more to our understanding of the quantum nature of elementary particles. What are the consequences of dealing with entities that are both particles *and* waves?

Suppose we were somehow able to localize a quantum wave–particle in a specific region of space so that we could precisely measure its position. In the wave description, this is in principle possible by combining together a large number of waves of different wavelengths in what is known as a superposition, such that they add up to produce a resultant wave which has a large amplitude in one location in space and is small everywhere else. This is called a 'wavepacket' (Figure 17). Such wavepackets are inherently unstable, but that's not the point. We can see how creating such a superposition would allow us to measure an instantaneous position.

Now what about the measurement of a particle-like property, such as momentum? That's a bit of a problem. We localized the wave by combining

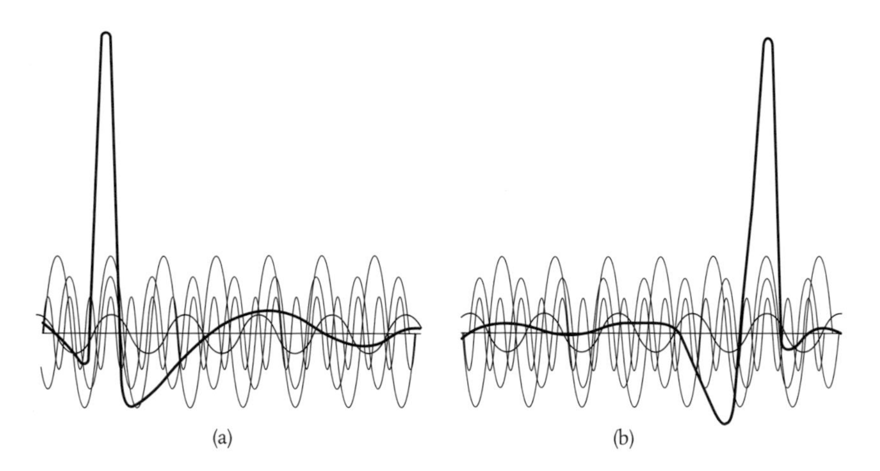

FIGURE 17 Although waves are by definition delocalized, we can nevertheless add together lots of waves with different frequencies to create a superposition which produces a strong peak in one specific location in space, (a). This is called a 'wavepacket'. As the collection of waves moves, the peak moves along with it, giving the impression of a particle-like trajectory through space, (b).

lots of waves with different wavelengths. This means that we have a spread of wavelengths in the superposition. But in 1923 de Broglie figured out that the wavelength of a quantum wave–particle is inversely proportional to its momentum.[7] The spread of wavelengths therefore means there's a spread of particle momenta.

In other words, we can measure the position of a quantum wave–particle with arbitrary precision, but only at the cost of uncertainty in its momentum.

The converse is also true. If we have a quantum wave–particle described by a wave with a single wavelength, this implies that we can measure the momentum with arbitrary precision using the relation devised by de Broglie. But then we can't localize the particle—it remains spread out in space. We can measure the momentum of a quantum wave–particle with arbitrary precision, but only at the cost of uncertainty in its position.

This is Heisenberg's famous uncertainty principle, which he discovered in 1927.[8]

The uncertainty principle is not limited to position and momentum. It applies to other pairs of physical properties, called *conjugate* properties, such as energy and the rate of change of energy with time.

The consequences are quite profound. Suppose we now create a perfect vacuum, completely insulated from the external world. We might be tempted to argue that there is 'nothing' at all in this vacuum, but what does this imply? It implies that the energy of an electromagnetic field (or any other field) in the vacuum is zero. It also implies that the rate of change of the energy of this field is zero, too. But the uncertainty principle denies that we can know precisely the energy of an electromagnetic field and its rate of change. They *can't* both be exactly zero.

What happens is that the vacuum suffers a bad case of the jitters. It experiences fluctuations of the electromagnetic field which average out to zero, both in terms of energy and its rate of change, but which nevertheless can be non-zero at individual points in space and time.

As we now know, fluctuations in a quantum field are equivalent to quantum particles. The vacuum fluctuations of the electromagnetic field can be thought of as *virtual photons*; 'virtual' again not because they are not 'real', but because they are not directly perceived.

Sheer fantasy? No. Take two small metal plates and place them side by side a few millionths of a metre apart in a vacuum, insulated from any external electromagnetic fields. There is no force between these plates, aside from an utterly insignificant gravitational attraction between them which, for the purposes of this experiment, we can safely ignore.

And yet, although there can be no force between them, the plates are actually pushed very slightly together. What's happening? The narrow space between the plates limits the number of virtual photons that can persist there. The density of virtual photons between the plates is then lower than the density of virtual photons elsewhere. The end result is that the plates experience a kind of virtual radiation pressure, the higher density of virtual photons on the outsides of the plates pushes them closer together. This is the Casimir effect, first identified by the Dutch physicist Hendrik Casimir in 1948.

And here's the answer. Quantum fluctuations in the inflaton field that pervaded the universe just 10^{-35} seconds after its birth became imprinted on the large-scale structure of the universe by cosmic inflation. Quantum fluctuations are responsible for an anisotropy of a few parts in 100 000.

COBE, WMAP, AND PLANCK

The horizon of the universe at the moment of recombination represents a 'last scattering surface'. Think of this surface as a kind of three-dimensional mirror. The image reflected in the mirror shows us what the universe was like 380 000 years after the big bang. The image also carries an imprint of the quantum fluctuations that prevailed just 10^{-35} seconds after the big bang, magnified to universe-sized proportions by cosmic inflation, like some giant thumbprint left at a cosmic crime scene. And, because the light released by recombination has been left largely unaffected by the subsequent evolution of the universe (aside from cooling and shifting its range of wavelengths), the image persists even though the mirror is long gone.

Now, to the second question. Can we look upon this image? Can we see the fluctuations in the cosmic background radiation?

We need to be careful. As it turns out, there are several sources of anisotropy in the cosmic background radiation and we need to distinguish between these. There are three so-called *primary* anisotropies. From our perspective, the most important of these arises from primordial quantum fluctuations reflected in small differences in the distribution of matter (and, subsequently, radiation temperature).

The second results from a competition between gravity and radiation pressure in the fluid-like plasma up to and at the moment of recombination. As gravity tries to pull the atomic nuclei in the plasma together, radiation pressure tries to push them apart and this competition results in so-called acoustic oscillations: sound waves bouncing back and forth through the fluid.* The sound waves produce regions of slightly higher matter density (compressions) and lower matter density (rarefactions), overlaying the primordial quantum fluctuations. These differences in matter density are translated into small differences in radiation temperature.

The third source of primary anisotropy results from gravitational fluctuations, both at the moment of recombination and subsequently, which can cause small shifts in the radiation frequency. In addition, primordial gravitational fluctuations (so-called gravitational 'waves'), thought to be produced during the burst of cosmic inflation, are believed to result in anisotropy in the so-called B-mode polarization of the cosmic background radiation.†

It is possible to build a theoretical model to describe all of these sources. It turns out that the model is sensitively dependent on a number of parameters, such as the densities of baryonic matter, dark matter, and dark energy. Changing these parameters even by a small amount changes the predicted patterns of primary temperature variations in the cosmic background radiation.

There are also expected to be a couple of important *secondary* sources of anisotropy in the radiation. The first is due to the simple fact that our Milky Way galaxy is moving. This motion causes a small 'dipole variation': the background radiation appearing slightly hotter as we move towards it, slightly cooler as we move away. This variation must be measured and subtracted from the temperature data if the underlying primary fluctuations are to be revealed.

The second is due to the scattering of cosmic background photons by electrons in intergalactic space within clusters of galaxies along the 'line of sight' of our observations. This effect changes not only the temperature, slightly, but

---

* Those happy to indulge a poetic turn of mind might like to imagine that the universe is *singing*.

† These are extremely difficult to detect and there was no end of fuss in March 2014, when scientists working on the second generation of an instrument called the Background Imaging of Cosmic Extragalactic Polarization (BICEP2) announced that they had discovered such fluctuations in the B-mode polarization signature. If this was true, it would represent powerful evidence in favour of inflationary big bang cosmology. Unfortunately, the announcement was premature and was subsequently retracted. The conclusion of further collaborative work involving BICEP2 and scientists working on the European Space Agency's Planck satellite (see later in this chapter) suggests that the results are strongly influenced by polarization caused by interstellar dust grains.

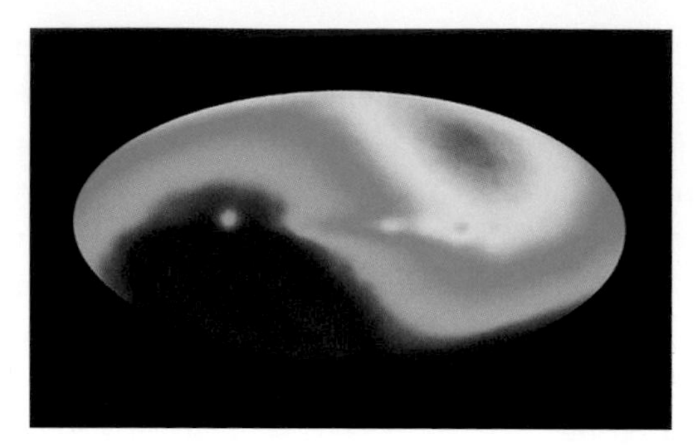

FIGURE 18 The motion of the Milky Way galaxy through the cosmic background radiation causes a small shift in measured temperatures, the radiation appearing a little hotter as we move towards it, a little cooler as we move away. This 'dipole variation' is shown in this false-colour, all-sky temperature map. Hotter regions (with a temperature a little over 0.003 kelvin higher than average) appear red, cooler regions (0.003 kelvin lower than average) appear blue. In this picture, the plane of the Milky Way galaxy lies horizontally along the centre.

also the shape of the black-body spectrum. Fortunately, the density of electrons in intergalactic space is pretty small and the anisotropy is not pronounced. Its characteristic dependence on photon frequency makes it easy to distinguish from the primary anisotropy.

In 1977, a group from the University of California at Berkeley detected and mapped the dipole variation using instruments on board a Lockheed U-2 reconnaissance aircraft flying at altitudes of about 70 000 feet.* A more recent all-sky map of the dipole variation is shown in Figure 18. From analysis of this picture, it is possible to deduce that the Local Group of galaxies, of which the Milky Way forms part, is moving roughly in the direction of the Virgo Cluster at a speed of about 600 kilometres per second.

The Berkeley group could find no evidence for any of the predicted primary anisotropy, however, and it became clear that only satellite-borne instruments would offer sufficient sensitivity.

The launch of the COBE satellite was originally planned for 1988 as part of a Space Shuttle mission, but the shuttles were grounded following the Challenger disaster on 28 January 1986. COBE was eventually placed into Sun-synchronous orbit on 18 November 1989.

---

* Also known as the U-2 'spy plane', this aircraft was operated by the US Central Intelligence Agency during the cold war and features in many cold war 'incidents'. CIA pilot Gary Powers was flying a U-2 over Soviet territory when he was shot down in 1960. Major Rudolf Anderson, Jr. was flying a U-2 when he was shot down in 1962 during the Cuban missile crisis. It also inspired the name of the Irish rock band U2.

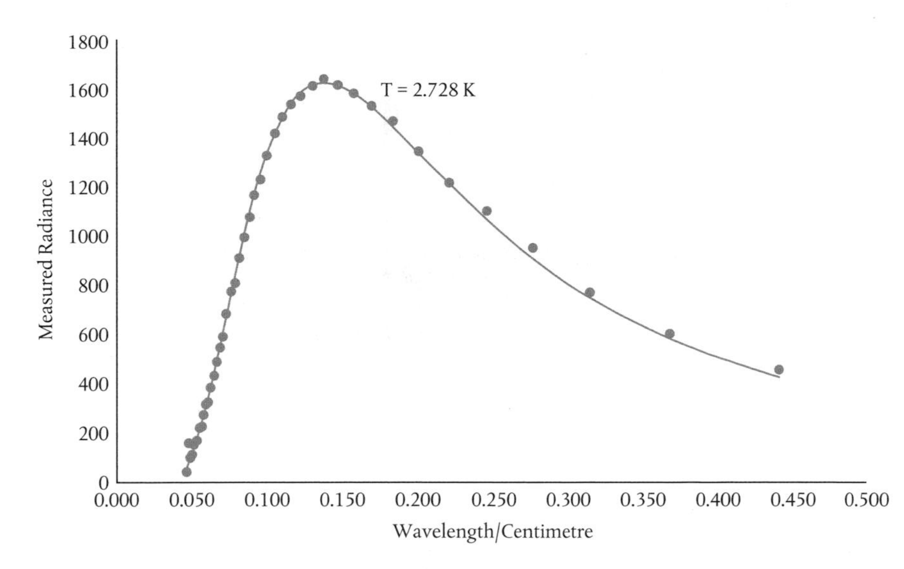

FIGURE 19 Measurements of the spectral radiance of the cosmic background radiation as a function of wavelength were reported by the COBE team in 1990. The spectrum is that of a black body with a temperature of 2.728 kelvin. The points represent the experimental measurements and the continuous line is the prediction based on a black-body spectrum. Adapted from D. J. Fixsen, E. S. Cheng, J. M. Gales, J. C. Mather, R. A. Shafer, and E. L. Wright, astro-ph/9605054, 10 May 1996.

On 13 January 1990, John Mather, a scientist at the NASA Goddard Spaceflight Center and member of the COBE team, delivered a short, ten-minute presentation on COBE's progress at a meeting of the American Astronomical Society in Crystal City, Virginia. He explained that things were going well, but that it would take a year or more for the instrument on board the satellite to complete its all-sky map of the cosmic background radiation. However, he did have one early result to announce. Based on just nine minutes of measurement time, it was now possible to confirm that the spectrum of cosmic infrared and microwave radiation was indeed that of a black body, with a temperature of 2.736 kelvin, revised in subsequent analyses to 2.728 kelvin (Figure 19).

The COBE team completed its work in early 1992. The dipole variation (which is shown in Figure 18) was measured and subtracted from the all-sky temperature map. The result was announced on 23 April 1992. The residual temperature map did indeed reveal the primary anisotropies, dominated by quantum fluctuations from the very beginning of the universe (Figure 20). The announcement was reported worldwide and ran on the front page of the *New York Times*.

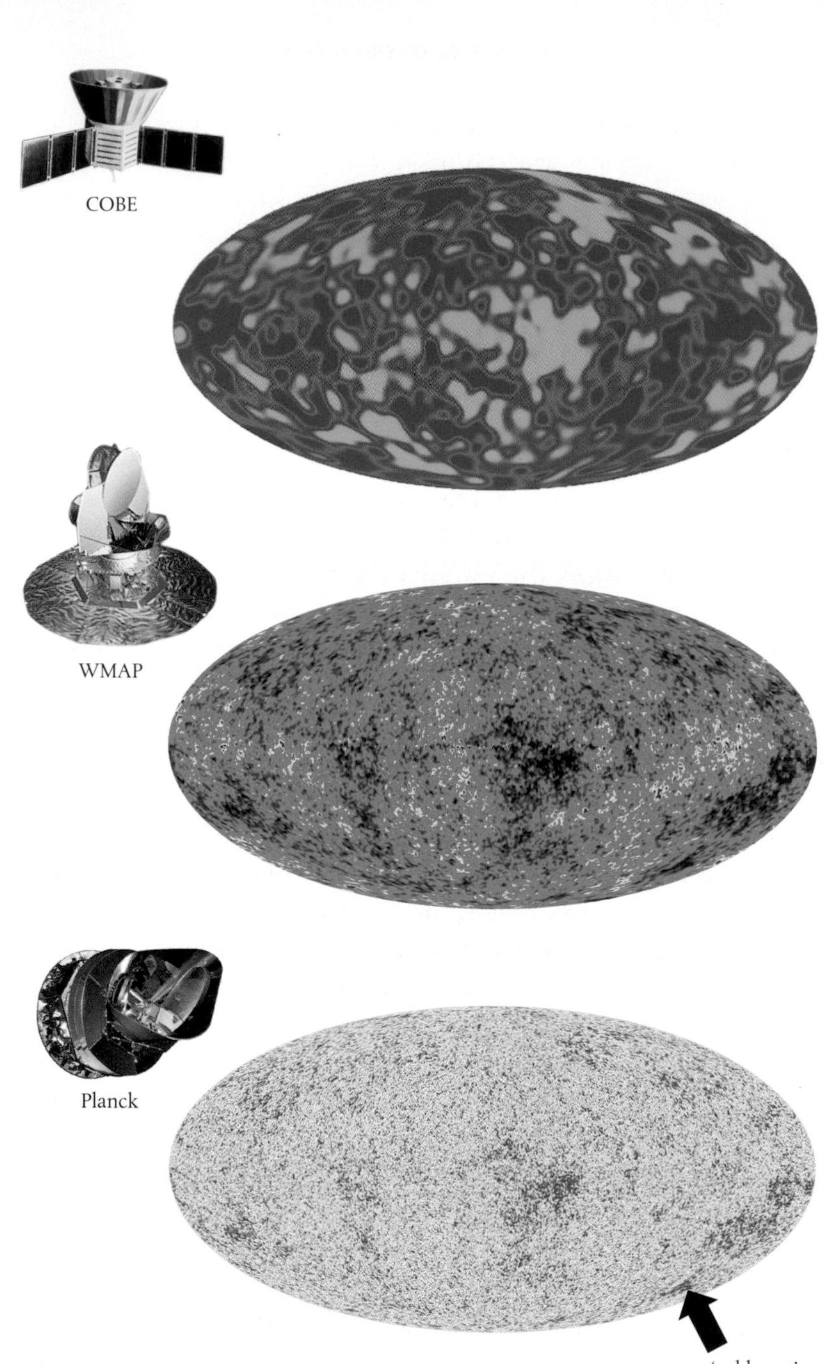

FIGURE 20 The detailed, all-sky map of temperature variations in the cosmic background radiation derived from data obtained from the COBE, WMAP (9-year results), and Planck satellites. The temperature variations are of the order of ±200 millionths of a degree and are shown as false-colour differences, with red indicating higher temperatures and blue indicating cooler temperatures. The angular resolution of this map has increased dramatically with successive missions.

It's difficult to look on this map without a creeping sense of awe. Throughout much of human history, debates about the origin and evolution of the universe have been religious or philosophical debates. Only in the last century has it become possible to devise plausible scientific explanations for the structure of the universe. At the time of the announcement, George Smoot, the COBE team member with overall responsibility for the analysis of the temperature variations, became lost for words. 'Well, if you're religious,' he explained to the assembled audience, 'it's like seeing the face of God.'[9]

These tiny temperature variations have since been measured in even more exquisite detail by the Wilkinson Microwave Anisotropy Probe* (WMAP), which was launched on 30 June 2001. This was 45 times more sensitive than COBE. It was originally intended that WMAP would provide observations of the cosmic background radiation for two years, but mission extensions were subsequently granted in 2002, 2004, 2006, and 2008. WMAP produced four data releases, in February 2003, March 2006, February 2008, and January 2010. Figure 20 shows the all-sky map derived from the nine-year results reported in 2010.

On 14 May 2009, the European Space Agency (ESA), with considerable mission-enabling support from NASA, launched the Planck satellite into orbit about 1.5 million kilometres from Earth, on the other side of the Earth from the Sun. Its instrumentation provided 2.5 times the angular resolution that was available from WMAP. ESA scientists released the most recent, and most detailed, all-sky temperature map of the cosmic background radiation on 21 March 2013. This is also shown in Figure 20. Note the improvement in angular resolution that has been achieved by successive satellite missions.

In all of these maps the dipole variation and interference from our own Milky Way galaxy has been carefully subtracted out. Different temperatures are shown as different false colours, with the differences magnified by a factor of 100 000 to make them more clearly visible; red represents higher temperature and blue representa lower temperature. The total temperature variation from red to blue is just two hundred *millionths* of a degree.

Here, then, is the answer. These small temperature variations betray small variations in the density of matter (light and dark) at the moment of recombination. This is all the variation needed to set in motion a chain of events that will lead to the structure that we see in the universe today. The hot spots betray a

---

* Named for David Wilkinson, formerly of Dicke's research group at Princeton, a member of the COBE team and leader of the design team for WMAP. He died in 2002 after a long battle with cancer.

higher density of matter which will be the seeds for future star and galaxy formation. Cold spots indicate a lower density of matter which will become voids.

THE Λ-CDM MODEL

So, this is how we know. For sure, the cosmic background radiation is not the only piece of observational evidence we have for the inflationary big bang model of creation, but it is spectacularly conclusive evidence.

Consensus has gathered around a version of inflationary big bang cosmology called variously the 'concordance' model, the 'standard model of big bang cosmology', or the Λ-CDM model, where Λ stands for the cosmological constant first introduced in 1917 by Einstein, and CDM stands for 'cold dark matter'. The version of events that I have told in these opening chapters is the version broadly described by this model.

The Λ-CDM model is based on six parameters. Three of these are related to the density of dark energy, which depends on the size of the cosmological constant, the density of cold dark matter, and the density of baryonic matter, of neutral hydrogen and helium atoms.

The agreement between theory and observation is quite remarkable. Figure 21 shows the 'power spectrum' derived from the squares of the

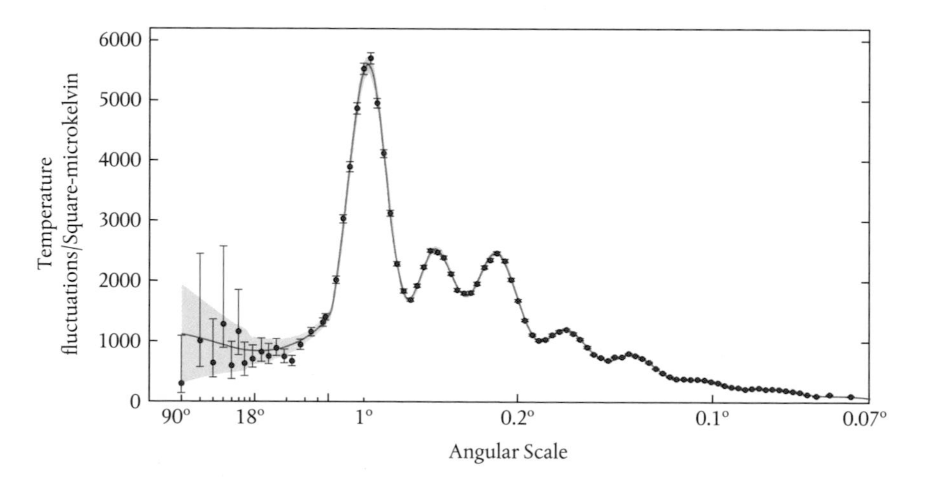

FIGURE 21 The variation of temperature fluctuations (measured in square microkelvin) with angular scale across the sky, derived from measurements of the cosmic background radiation by the Planck satellite. The Planck data are illustrated by the points with associated error bars and the best-fit Λ-CDM model prediction is shown as the continuous line.

temperature differences in the cosmic background radiation as mapped by the Planck satellite. Taking the squares of the differences means that cold regions, where the temperature dips below the average, are not differentiated from hot regions and the focus is instead on the absolute *magnitudes* of the variations across the sky.

This is a complex plot, but it is enough to know that the oscillations in this spectrum are a direct result of the sound waves that bounced back and forth in the plasma up to the moment of recombination. The positions of these oscillations and their damping with angular scale depend sensitively on the model parameters. But the characteristic pattern of a single large peak, followed by two peaks of nearly equal height, followed by damped oscillations in a long tail, can be described quite adequately by simple hydrodynamic models which treat the plasma as a fluid.

The best-fit curve is based on a model in which the universe began 13.8 billion years ago and in which dark energy accounts for about 68.3% of the total mass-energy, with dark matter accounting for a further 26.8%. Ordinary baryonic matter—or what we used to think of as 'the universe' not so very long ago—accounts for just 4.9%.

This visible matter is carried along on a body of dark matter much like the foam on the top of your morning cappuccino. The push derived from the gentle acceleration of the expansion of spacetime due to dark energy and the pull of the gravity of mostly dark matter will now go on to shape the evolution of the universe beyond the Photon Epoch.

4

SETTING THE FIRMAMENT ALIGHT

The Origin of Stars and Galaxies

Now begins the slow but inexorable process of building structure from the small anisotropies in the distribution of matter across the young universe. Spacetime continues to expand everywhere, but in those regions where the density of matter is a little higher, the expansion is slowed. The pattern of anisotropy in the distribution of matter is translated into a pattern of growing contrasts in the expansion rates in different regions of the universe.

There's a simple logic to this kind of cosmic tug-of-war. As I mentioned in Chapter 1, there are three different kinds of solution to Einstein's gravitational field equations. These model universes evolve in different ways depending on the relationship between the amount of mass-energy in the universe and its rate of expansion. A universe with a sufficiently high density of mass energy will eventually slow down the rate of expansion. The universe will turn around and collapse back in on itself. Such a universe is closed. A universe with insufficient mass-energy to turn the expansion around will expand forever. It is open.

We use much the same kind of logic when thinking about the trajectories of objects hurled through the air. A cannonball shot from the castle battlements describes a parabolic trajectory. It loses speed as it moves through the air before it runs out of steam and succumbs to the force of Earth's gravity, at which point it falls back to the ground in a graceful arc. We could say that its trajectory is closed. A very powerful cannon could in theory shoot the cannonball at a speed greater than about 11.2 kilometres per second, which is the Earth's escape velocity. The cannonball would then break free of Earth's gravity and escape into space. Its trajectory would be open.

Similar principles apply when we consider individual regions of spacetime *within* the universe. When the different rates of expansion between regions of high matter density and 'empty' spacetime grows to a factor of about ten, the high matter region stops expanding. It falls back in on itself, reaching a kind of equilibrium. Structures start to form. The time required for this turn-around to happen simply depends on the initial rate of expansion and the amount of matter in the region. Different parts of the universe therefore turn around at different rates and in different times.

Remember that this matter consists of about 85% dark matter and about 15% baryonic gas, mostly neutral hydrogen and helium atoms. In Chapter 2, I asked you to suspend your demand for evidence for the existence of dark matter. I can now reassure you that this evidence, such as it is, will be forthcoming towards the end of this chapter. We just need to form some galaxies first.

And here's the curious thing. When it comes to dark matter, we remain almost completely in the dark.* We know next to nothing about it. Yet it is fundamental to the physics of galaxy formation, according to a mechanism first devised by astrophysicists Simon White and Martin Rees in 1978, and which is now broadly accepted more or less in its original form.[1]

How, you might ask, can we construct a mechanism based on a substance that we can't directly detect? In truth, scientists do this all the time. From the evidence which I will provide shortly, we can deduce enough about how dark matter behaves even though we have no real idea what it is.

This is much more than educated guesswork. It's perhaps worth noting that calculations based on the baryonic matter we find in the universe today demonstrate quite unequivocally that if this were really all the matter that there were, then *no* parts of the universe would have condensed to form the structures of stars and galaxies that we see today. Given the extent of the anisotropy revealed in the cosmic background radiation and the sizes of the resulting gravitational instabilities, there simply isn't enough baryonic matter to overcome the

* Pun intended.

expansion of spacetime. Dark matter (whatever it is) is essential to explain the large-scale structure of the visible universe.

DARK MATTER HALOS

In a region of spacetime with a slightly higher than average density of matter (the slightly hotter regions in the cosmic background radiation shown in Figure 20), the excess dark matter gathers to form a *halo*. The earliest halos begin to form between 10 and 20 million years after the big bang, but they're not yet large enough to exert much gravitational influence over the heavy baryonic matter. They break out across the universe like an invisible rash.

These dark matter halos are mobile. They are attracted to one another by their gravity. They merge with other halos in their vicinity to form larger halos in a 'merger tree' (Figure 22). After a further 100 to 150 million years, the halos become large enough—exerting sufficient gravitational pull—to begin to trap baryonic matter, which starts to gather within the halos. The baryonic matter cools, condenses, and becomes concentrated.

The formation of dark matter halos is a relatively slow process. The cosmic background radiation has cooled along with the expansion of spacetime that has taken place since the moment of recombination, its black-body spectrum declining in intensity and shifting towards the invisible infrared region. The most ancient light in the universe has gone out. The baryonic matter now

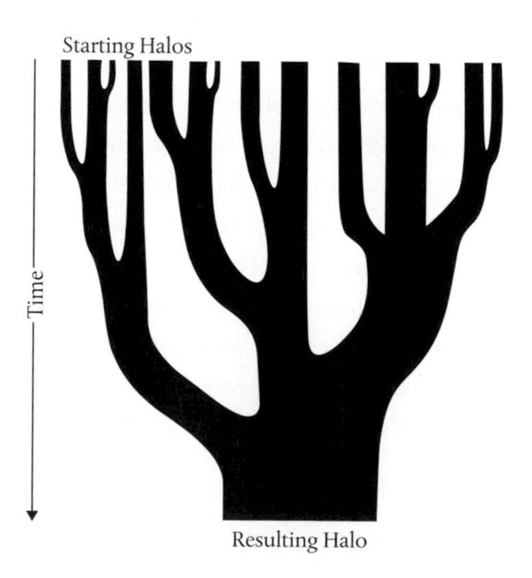

FIGURE 22 As dark matter halos form, they are attracted to other halos in their vicinity. The halos merge and coalesce, their history tracing the outlines of a 'merger tree' as illustrated here. As time passes, a single, large halo is formed. Adapted from Cedric Lacey and Shaun Cole, *Monthly Notices of the Royal Astronomical Society*, **262** (1993), p. 636.

becoming trapped within the largest dark matter halos has yet to condense to form stars. The universe is a bleakly dark place.

Welcome to the Dark Ages, which stretch from 380 000 years to about 300 million years after the big bang.[2]

THE FIRST STARS

The baryonic matter caught in the gravity of a dark matter halo may now condense to form a protogalaxy of protostars. Although the mechanism by which the first stars are formed still remains the subject of some scientific debate, most of the physical principles are broadly understood.

Baryonic matter, consisting of neutral hydrogen and helium atoms formed through primordial nucleosynthesis and recombination, gathers together within a dark matter halo. Once again, the future evolution of this matter is determined by the tug-of-war between the kinetic energy of the atoms in the gas (or the gas temperature) and the force of gravity. Too much kinetic energy and the gas cloud will persist. But a cloud with a low kinetic energy has no defence against gravity, which will slowly pull the matter together and concentrate it.

What this tells us is that there can be no stars if the trapped cloud of hydrogen and helium atoms is too hot, as the atoms in a hot gas move faster and have a higher average kinetic energy. How cool do the atoms need to be? British physicist James Jeans worked out the principles in 1902, and relatively simple relationships can be deduced between critical mass, density (the number of atoms per unit volume), and temperature.

These principles are illustrated in Figure 23, which shows temperature–density curves for clouds of hydrogen atoms with three different masses, 100 times the mass of the Sun (which is about 1.99×10^{30} kilogrammes, given the special symbol $M_\odot$), 500 $M_\odot$ and 1000 $M_\odot$. For a given mass and density, a cloud with a temperature above the curve will remain a cloud. As the temperature falls below the curve, the cloud will collapse to form a star.

In this early universe, the temperature of the cosmic background radiation is still relatively high (between a few hundred kelvin and 30 kelvin, depending on the age of the universe at the time the cloud is contracting) and there are few physical mechanisms available that can serve to cool the gas and reduce its kinetic energy. One possible mechanism involves radiative cooling by molecular hydrogen, H_2, formed in the cloud by joining together individual hydrogen atoms.

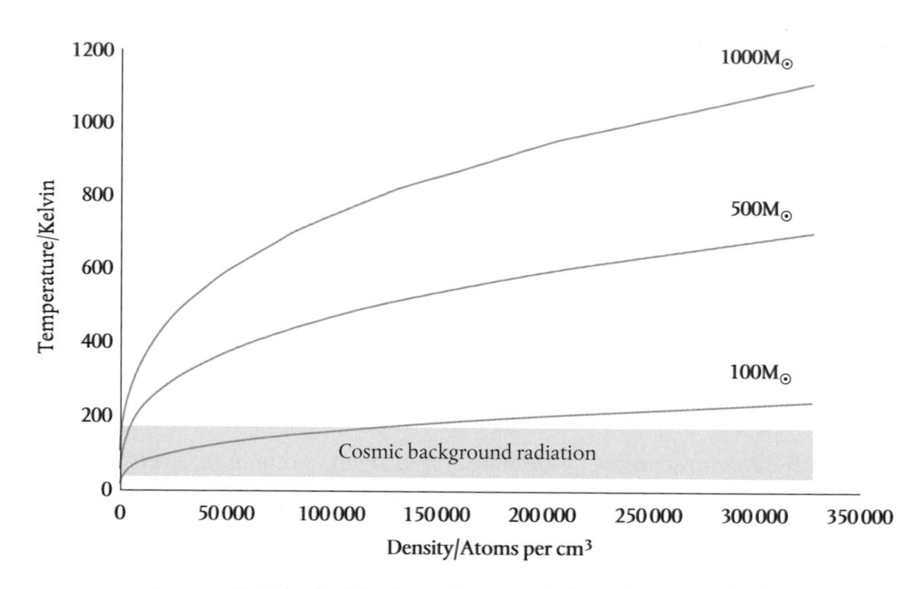

FIGURE 23 In 1902, British physicist James Jeans worked out the principles that determine whether or not a gas cloud will collapse to form a star. What will happen depends on the interplay between the cloud's mass, density, and temperature, as illustrated in these curves calculated for masses of hydrogen atoms of 100, 500, and 1000 $M_\odot$, where $M_\odot$ stands for the mass of the Sun. For a given mass and density, clouds with temperatures above the curve will not collapse, those with temperatures below the curve will collapse. The grey band indicates the temperature range of the cosmic background radiation 30–400 million years after the big bang. Under the conditions that prevailed in the early universe, the first stars are predicted to have had very large masses.

Where does the H_2 come from? There are various mechanisms by which molecular hydrogen can be formed from hydrogen atoms. The most important involves the reaction of neutral hydrogen atoms with stray electrons, which we can write: $^1H + e^- \rightarrow {}^1H^- + \gamma$, where γ represents a photon. The negatively charged hydrogen ion $^1H^-$ then goes on to react with another neutral hydrogen atom to form molecular hydrogen: $^1H^- + {}^1H \rightarrow {}^1H_2 + e^-$.

Anyone who has stood in the open on a cloudless night has experienced radiative cooling, as infrared (heat) radiation from the surface of the skin escapes into the cooler air. As hydrogen and helium atoms collide with hydrogen molecules in the gas cloud they may transfer some of their kinetic energy into the energy of molecular vibrations and rotations.

These molecular motions are complicated quantum phenomena, but we can picture them quite simply, as two hydrogen atoms rocking back and forth as though tethered together by a spring (vibrations), and as motions in which the molecule tumbles and twirls through space like a cheerleader's baton (rotations). Energy transferred into these motions is radiated away into the cooler space surrounding the cloud in the form of microwaves and infrared light. This is not a particularly efficient process, but it does allow the gas to cool to the temperature of the cosmic background radiation, by which time gravity has taken over.

Figure 23 shows that, for modest densities of 50 000 atoms per cubic centimetre and less, cloud temperatures of a few hundred kelvin favour the formation of more massive stars.* The cloud will fragment, as smaller regions within it break ranks and cross the threshold, collapsing to produce a variety of protostars with different masses.

In some recent computer simulations, the baryonic matter trapped in a dark matter halo with a mass of 10^5–10^6 $M_\odot$ produced a distribution of protostars with masses ranging from 10 $M_\odot$ to 1000 $M_\odot$, with the majority concentrated around a few tens to a few hundreds of solar masses.[3]

STELLAR NUCLEOSYNTHESIS

Let's now fix on a specific protostar within this distribution. As the gas is further compressed by gravity the temperature starts to rise, which means that the kinetic energy of the atoms increases. But it's now too late for them to escape gravity's grip. The hydrogen and helium atoms are once more stripped of their electrons and, as the temperature and pressure rise further, a series of nuclear fusion reactions is initiated.

But this is not a repeat of the sequence of reactions that was involved in primordial nucleosynthesis. Recall that a few hundred seconds after the big bang, protons and neutrons in a ratio of about seven to one combined to produce a mixture of about 76% hydrogen nuclei (protons) and 24% helium nuclei by mass. All the neutrons were locked away inside the helium nuclei.

Now, trapped inside a protostar with temperatures climbing in excess of 10^7 kelvin, there are no free neutrons to react with. Instead, two hydrogen nuclei are squeezed together with sufficient force to make them combine. How? There is no stable atomic nucleus consisting of just two protons. One of the protons must undergo a weak force decay, turning into a neutron and

* For comparison, the density of liquid water is about 3×10^{22} molecules per cubic centimetre.

a W^+ particle which carries away the positive electrical charge. The W^+ particle quickly decays into a positron (e^+) and an electron neutrino (ν_e). This is the opposite of beta-decay, in which a neutron transforms into a proton with the emission of a W^- particle (which decays into an electron and an electron anti-neutrino).

This is a rare event. The coming together of two protons like this is *very* uncomfortable, and only the conditions inside the collapsing protostar can make it happen. It's just as well. If this reaction were any easier the hydrogen inside the Sun's core would have already been exhausted long ago, making life on Earth rather difficult. As it happens, when it was formed there was enough hydrogen in the Sun's core to last ten billion years.

The upshot of all this is that two hydrogen nuclei (^1H) combine to produce a deuterium nucleus, ^2H. We know from our earlier encounter with primordial nucleosynthesis that the deuterium nucleus will react with another hydrogen nucleus to produce an unstable isotope of helium, ^3He, emitting a photon in the process.

There are two possible fates for the ^3He nucleus. It may react with another ^3He to form the stable ^4He nucleus, spitting out two protons in the process. In the Sun, this reaction happens about 85% of the time. Or the ^3He may react with a ^4He nucleus* to produce ^7Be, with the emission of another photon. This reaction happens about 15% of the time. These reactions are summarized in Figure 24, which shows that the sequence—known as the proton-proton (p-p) chain—has three branches.†

In the first branch, called ppI, a quick count-up reveals the net result. Four hydrogen nuclei are fused together to produce one helium nucleus, with two of the protons turning into neutrons. Two energetic photons are produced, one for each of the ^3He nuclei required. The two positrons produced in the weak force decays are quickly annihilated by two free electrons, producing four more energetic photons.

If we now carefully add up all the masses of the nucleons involved in this branch of the p-p chain we discover a small discrepancy, called the *mass defect*. About 0.7% of the mass of four hydrogen nuclei has gone 'missing', converted into about 26 million electronvolts of radiation energy according to Einstein's formula $E = mc^2$.

* Remember, the gas inside the protostar contains 24% helium by mass, so there are plenty of ^4He nuclei around.

† There is believed to be a fourth branch—called ppIV—which is much rarer than the three illustrated in Figure 24.

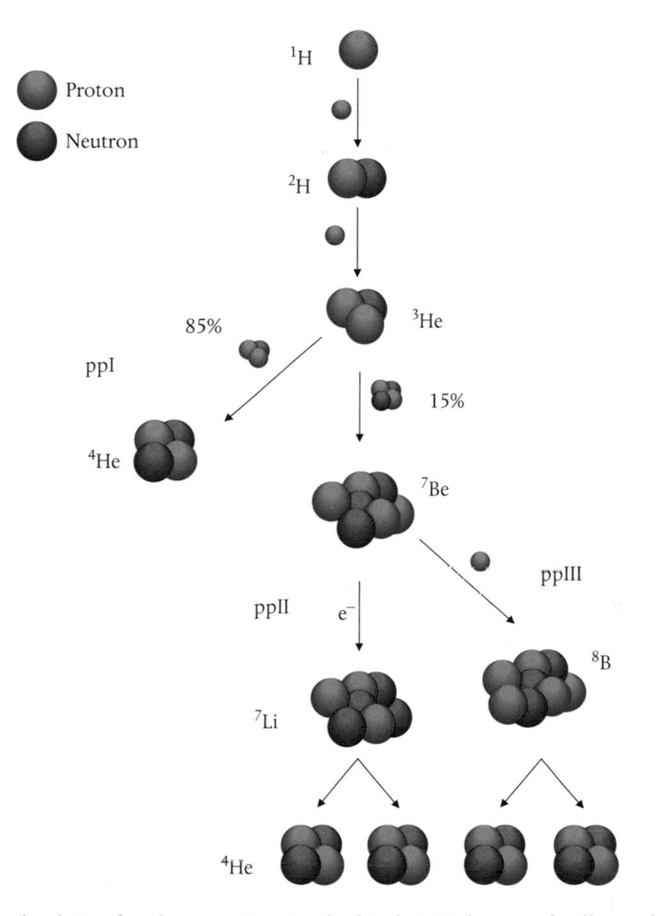

FIGURE 24 The chain of nuclear reactions involved in the initial stages of stellar nucleosynthesis is known as the proton-proton (p-p) chain.

Now, a 100-watt light bulb burns energy at a rate of about 600 billion billion electronvolts per second, so 26 million electronvolts might not sound too impressive. But wait. This is the energy generated from the fusion of just *four* hydrogen nuclei. It is estimated that in the Sun's core about 3.7×10^{38} hydrogen nuclei are reacting *every second*, releasing energy equivalent to about 4 million billion billion 100-watt light bulbs. Now that's more like it.

As the dark matter halos continue to merge and grow, the baryonic matter caught within them accumulates. Now the hydrogen and helium atoms cool quite efficiently without the aid of collisions with molecular hydrogen. The pace picks up, as hundreds of billions of nuclear furnaces are lit, and a galaxy is formed. This happens across the universe, tracking the ribbon of anisotropies in the primordial distribution of matter.

The Dark Ages have ended, and there is light in the universe once more.

THE SHORT BUT SPECTACULAR LIVES OF POPULATION III STARS

But the stars that are now forming are not the stars we see in the night sky today. These first stars are formed from the baryonic matter that's available to them, some time around 300 to 550 million years after the big bang. This matter consists of primordial hydrogen, helium, and very, very small amounts of heavier elements such as lithium and beryllium.

The relative abundance of elements heavier than hydrogen and helium is referred to by astronomers as *metallicity*. Of course, there are many heavier elements that are not metals (such as carbon, oxygen, and nitrogen) but for the purposes of developing theories of stellar evolution astronomers tend to lump these together and considered them collectively as 'metals'.

So, these first generation stars are by definition of very low metallicity. They are called—rather confusingly—Population III stars. The reason for this will become apparent shortly. Although searches are under way for Population III stars that are still visible today, none has been observed so far. We therefore have no empirical evidence that these stars ever existed, although there are some fairly strong logical and theoretical grounds for thinking that they might have been an important stepping stone to the kinds of stars we know today.

If they did form, then (as we have seen) these first stars would likely have been very massive, up to several hundred $M_\odot$. Despite the rarity of the first reaction in the proton-proton chain, the high temperatures and pressures at the cores of such stars would have caused them to burn through their hydrogen fuel very quickly. They would have been much larger and much more luminous than the stars we see today, glowing with predominantly ultraviolet light.

What happens when all the hydrogen inside the core of such a star is exhausted? The three branches of the proton-proton chain illustrated in Figure 24 all result in the formation of stable ^{4}He nuclei and so, as the ^{1}H nuclei are burned, ^{4}He nuclei accumulate. The pressure of the radiation produced in this process helps to hold the star up against further gravitational contraction. But when all the ^{1}H nuclei are gone, the radiation ceases. The core contracts, driving up the temperature and pressure.

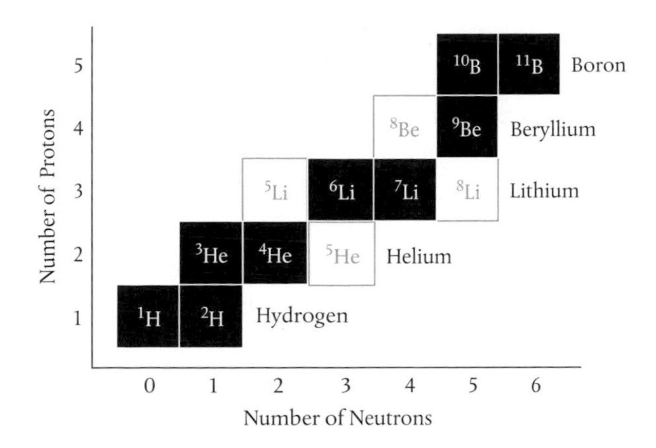

FIGURE 25 Adding single protons and neutrons to ^{1}H allows the building of a sequence of stable elements (black squares) up to ^{7}Li. But here we hit a bottleneck: there are no stable nuclei with eight nucleons. Adding a proton to ^{7}Li produces an unstable ^{8}Be nucleus, which decays very rapidly.

As the temperature in the core rises above 10^8 kelvin, the helium nuclei fuse together, releasing radiation once more. The temperature in the core starts to run away, increasing the rate of the helium fusion reactions and causing a dramatic increase in the luminosity of the star, called the *helium flash*.

It is at this point that we must confront the challenge posed by the absence of stable atomic nuclei containing eight nucleons (Figure 25). This was the gap that couldn't be bridged under the conditions which prevailed during primordial nucleosynthesis, a few hundred seconds after the big bang, which is why the chain of reactions terminated with lithium and beryllium isotopes. But the conditions in the interior of a Population III star are thought to be rather different.

What happens now is that two ^{4}He nuclei combine to produce an ^{8}Be nucleus, with the emission of an energetic photon. The ^{8}Be nucleus contains four protons and four neutrons, and is highly unstable. Under normal circumstances, it will very quickly fall apart, back into two helium nuclei (the half-life of ^{8}Be is about 7×10^{-17} seconds). But these are not normal circumstances. In the core of a helium-burning star, there is now a chance that an ^{8}Be nucleus will encounter another ^{4}He nucleus before it can break up, fusing with it to form ^{12}C, the nucleus of a carbon atom.*

* These two reactions are collectively referred to as the 'triple-alpha' process. The $^{4}He^{2+}$ nucleus is also an alpha-particle, emitted from unstable nuclei susceptible to alpha radioactivity. The two reactions then involve the fusion of three alpha particles into a stable carbon nucleus.

For sure, the chance is slim. But it is enhanced by virtue of the fact that the ^{12}C nucleus possesses an excited state with an energy that corresponds closely to that of the combined energies of the separate ^{8}Be and ^{4}He nuclei. This excited state provides a 'resonance', giving the nuclei a path by which they can come together, bridging the gap before the moment is lost.

Think of it this way. The ^{8}Be bath has *two* plugholes. One leaks very quickly away back to the bath of ^{2}He nuclei. This would otherwise be the end of the matter, but the existence of the resonance pulls another plug, one which spills even more rapidly into the bath of ^{12}C.

There is simply no other way that heavier elements can form inside a star. And yet we know that carbon and heavier elements are relatively abundant in the universe today. In 1954, the British physicist Fred Hoyle concluded that there *must* be a mechanism which allows ^{12}C to form before the unstable ^{8}Be breaks up. He went on to predict the existence of a resonant state of the carbon nucleus which aids the formation of ^{12}C. This state is now sometimes referred to as the 'Hoyle state' and was discovered, with an energy very close to Hoyle's prediction, in 1957.

Once formed, a ^{12}C nucleus may now react with another ^{4}He nucleus to form ^{16}O, the nucleus of an oxygen atom. This opens up a new nucleosynthesis route involving nuclei of carbon, nitrogen (formed indirectly), and oxygen as catalysts in the further conversion of hydrogen into helium, and the conversion of carbon into nitrogen and oxygen (Figure 26).

When the helium is exhausted there are further nuclear reaction chains that can be ignited, involving carbon burning to form neon, sodium, and magnesium; neon burning to form oxygen and magnesium; oxygen burning to form silicon; and silicon burning to form a range of heavier elements, such as sulphur, calcium, and iron.

Fusing iron nuclei together costs more energy than can be released and, as iron accumulates in the core, the battle with gravity is finally lost. The result is a catastrophic collapse, the core contracts at speeds of 1000 kilometres per second. All the hard work building heavy nuclei through nucleosynthesis is now undone, as intolerable levels of radiation in the core disintegrate the iron nuclei into helium nuclei, which further disintegrate into protons and neutrons. The protons undergo weak force decay, forming a core of almost pure neutrons with the density of an atomic nucleus.

These stars burned brightly. They lived briefly but spectacularly, with an average lifetime of just a few million years. Now they die.

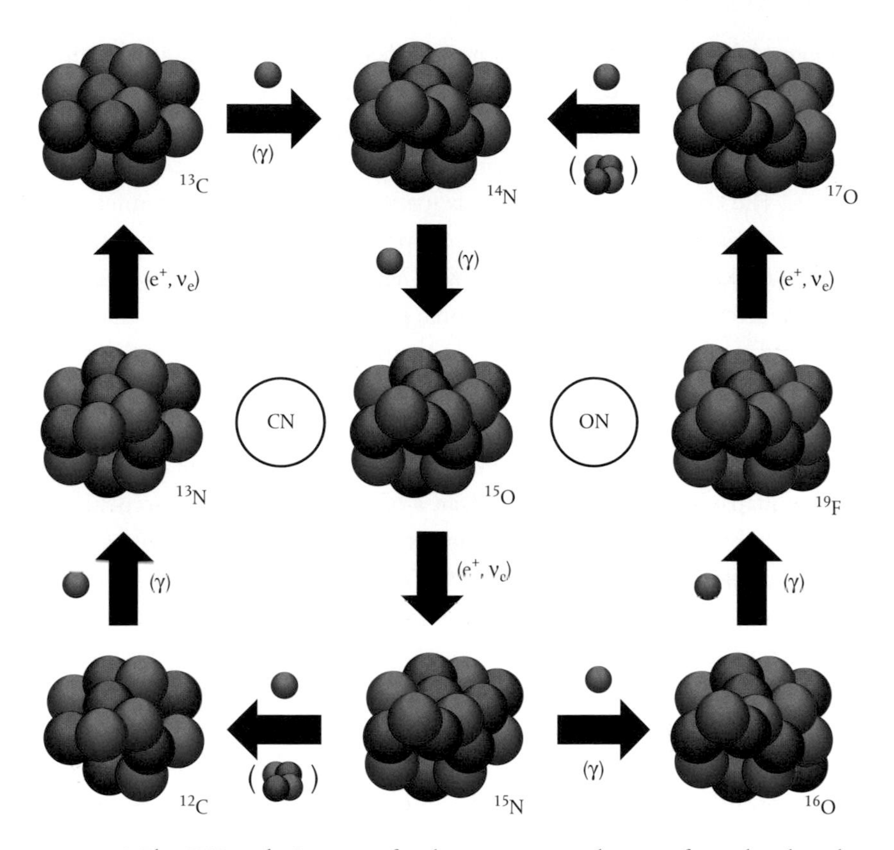

FIGURE 26 The CNO cycle. Isotopes of carbon, nitrogen, and oxygen form closed catalytic loops. Starting from the bottom left, a ^{12}C nucleus picks up a free proton to form ^{13}N, an isotope of nitrogen, emitting a high-energy photon (shown in brackets). This undergoes a weak-force decay, emitting a W^+ particle which transforms into a positron (e^+) and an electron neutrino (ν_e) to form ^{13}C. This picks up another proton to form ^{14}N, and another proton to form ^{15}O, an isotope of oxygen, which undergoes a weak force decay to ^{15}N. The CN cycle closes with a reaction involving the fusion of another proton and the emission of a ^{4}He nucleus (an alpha particle), bringing us back to ^{12}C. In this cycle, four protons have once again been combined to produce a helium nucleus. The ON cycle shown on the right does much the same job.

THE FATES OF POPULATION III STARS

What happens next depends on the initial mass of the star. In the computer simulations I mentioned above, first-generation Population III stars with initial masses between about 8 $M_{\odot}$ and 25 $M_{\odot}$ undergo a *core-collapse supernova*. The strong nuclear force defends the neutron core against further collapse and the core rebounds, sending a shockwave through the in-falling material which blows it apart in a spectacular explosion. For a brief moment, the star shines brighter than an entire galaxy. The energy released in the explosion is sufficient to reassemble heavier nuclei, to iron and beyond.

What's left is a frantically spinning *neutron star,* a ball of mostly neutrons with a mass of 1–3 $M_{\odot}$ and a radius of just 5–10 kilometres. Let's just put that in perspective. The London orbital motorway (called the M25) has a circumference of 188 kilometres and forms a rough circle around the city with a radius of about 30 kilometres. We could fit about half a dozen neutron stars within this boundary.

The neutron star consists of neutrons, protons and electrons with at least 200 neutrons for every proton, which ensures that the neutrons themselves don't decay away through weak-force reactions.

For heavier Population III stars, with initial masses 25–80 $M_{\odot}$, even the strong force holds no defence against gravity, and the remnant left following a core-collapse supernova is a *black hole,* a body so massive and so dense that not even light can escape its surface. In fact, its dimensions are defined by the radius of its event horizon, also known as the Schwarzchild radius (named for German physicist Karl Schwarzchild), which defines the threshold beyond which any in-falling object isn't coming back.

As some of the initial mass of the star is lost during the supernova, the mass of the black hole remnant will inevitably be lower. A remnant with a mass lower than about 3 $M_{\odot}$ will happily persist as a neutron star. A 5 $M_{\odot}$ black hole has a Schwarzchild radius of about 15 kilometres. A 30 $M_{\odot}$ black hole has a Schwarzchild radius of about 90 kilometres.

Population III stars of intermediate initial mass of 80–240 $M_{\odot}$ may experience a *pair-instability supernova,* in which energetic photons in the star's core produce electron–positron pairs. Energy that was directed into radiation pressure holding the star up against gravity is now diverted into pair-production. The radiation pressure falls rapidly, and the core collapses. As the core over-pressurizes, runaway thermonuclear reactions, involving the burning of oxygen and silicon, are triggered and the star is completely torn apart, leaving no remnant.

The largest stars, with initial masses greater than 240 $M_{\odot}$, suffer the ultimate fate. They collapse directly to form black holes.

About 70% of the nuclear furnaces that were lit at the end of the Dark Ages now flash brilliantly, each shining brighter than the whole galaxy for a short time, before going out.

But their job is done. The black holes that are left behind may now merge, producing supermassive black holes that are thought to exist at the centres of most of the galaxies that we observe today. Clouds of gas, now sprinkled liberally with heavy elements formed by nucleosynthesis inside the first stars and by the explosive energy released during the supernovae, gather together once again in the interstellar space enclosed within a dark matter halo. Shockwaves from nearby supernova explosions may hasten the process.

The cycle begins again. Now the presence of heavy elements in the clouds further aids the cooling of the gas, allowing a range of stars with smaller masses to form. As the clouds collapse, the p-p chain is initiated once more. In heavier stars the production of helium nuclei is supported by the CNO cycle because, this time, nuclei of carbon, nitrogen, and oxygen already exist within the gas cloud.

The resulting second-generation Population II stars will burn less brightly, but they will be more stable and will live longer.

POPULATION II STARS: THE HERTZSPRUNG–RUSSELL DIAGRAM

If they formed, the large stars of Population III now give way to more modest Population II stars. The galaxies of stars now taking shape will one day be recognizable. As we will see, the formation of stars and galaxies is not an event—it doesn't happen just once. It is rather a process, continuing to this day.

The older stars in our own Milky Way galaxy are difficult to age accurately but one of the oldest is a Population II star called HD 140283 (nicknamed the 'Methuselah star'), which lies about 1800 trillion kilometres from Earth.[4] So we're now dealing with stars that are visible in our night sky and which we can study.

These stars evolve in ways that are broadly predictable. As hydrogen burning gets under way, the radiation released in the protostar's interior exerts a pressure that resists the self-gravity of the gas which is causing it to contract. It's not all plain sailing, however. Just as pockets of hot air rise in the Earth's atmosphere, so masses of hot gas in the interior of the protostar will rise up into its cooler outer layers. Physical processes taking place in the outer layers, involving ionization of the gas and the production of bremsstrahlung ('braking radiation') by atomic nuclei and electrons prevent all the radiation from escaping the surface. In other words, the outer layers are *opaque*, trapping some of the radiation inside the protostar.

As the hot masses of gas cool they sink back down to the interior, forming convection currents. The protostar churns, its luminosity and surface temperature fluctuating wildly. Steep temperature differentials established between the core and the outer layers encourage even stronger convection.

But the physical processes responsible for the protostar's opacity depend *inversely* on temperature—cooler stars are more opaque than hotter ones. So, as the protostar continues to contract under its own gravity the temperatures of both its core and the outer layers increase, causing the opacity to fall.[5] This relieves the pressure somewhat; the temperature differentials become less marked, and the convection currents slow down and eventually cease. At this point the protostar enters *hydrostatic equilibrium*, and becomes a fully fledged star. The gravitational pull of the matter in the star is now balanced by the radiation pressure generated by the hydrogen fusion reactions in its core.

We can follow what happens next by reference to one of the most famous diagrams in astrophysics—the Hertzsprung–Russell diagram, named for Danish astronomer Ejnar Hertzsprung and American astronomer Henry Norris Russell. This is a map of star luminosity versus effective surface temperature. A star's luminosity is determined by its measured brightness and distance from Earth. Its 'effective' surface temperature is the temperature that a perfect black body with the same luminosity and radius would be expected to have.

The diagram is illustrated in Figure 27. Most of the stars that we observe fall within a diagonal band of luminosity versus effective surface temperature which is known as the *main sequence*. The higher the surface temperature, the brighter the star shines and the bluer its colour.

This diagram is really a snapshot of the physical states of many different stars at this particular moment in time and vantage point. But if we could watch these same stars evolve over the next few million years, we would observe that their positions on the chart change with time. A star joins the main sequence at a position that depends on its initial mass, size, age, and evolutionary history. As it burns through its hydrogen fuel it brightens and ascends the main sequence.

Larger stars (with masses of a few $M_{\odot}$) will deplete their fuel reserves more quickly, the catalytic CNO cycle dominating the production of helium nuclei. A star the size of our own Sun will spend about 10 billion years on the main sequence, with the p-p chain providing the most important mechanism for helium production.

When a star has exhausted the hydrogen in its core, hydrogen atoms in the star's outer layers start to burn, and it swells to become a *red giant*. Helium burning in the core may then commence, depending on the size of the star. This

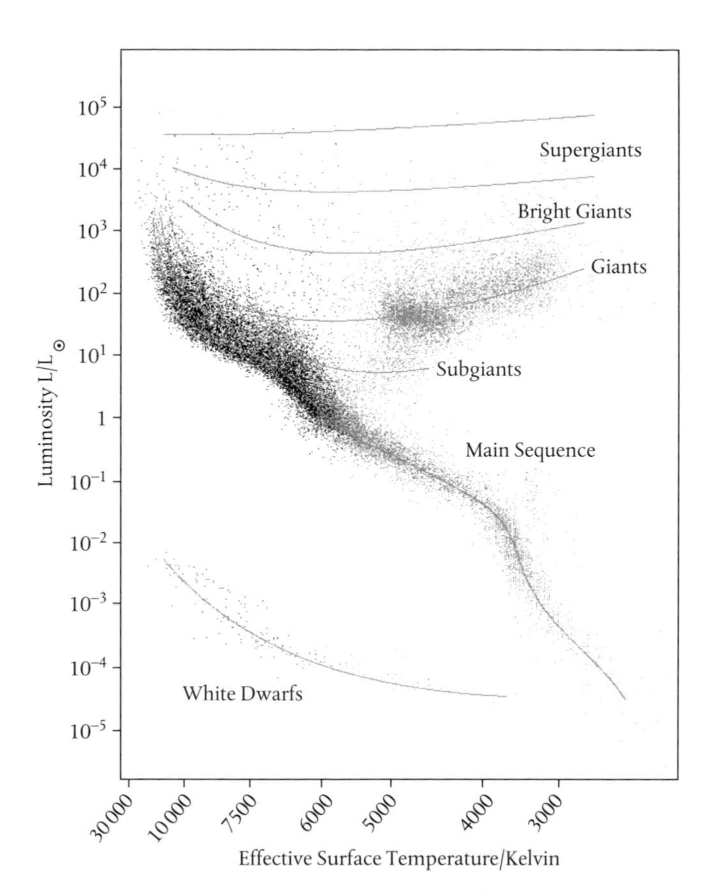

FIGURE 27 The Hertzsprung–Russell diagram maps the luminosity of stars (L, measured relative to the luminosity of the Sun, $L_{\odot}$) against their effective surface temperature. Most of the stars observed fall into a diagonal band called the main sequence. Stars in the main sequence are burning through their reserves of hydrogen. As the hydrogen is exhausted, the stars leave the main sequence to form red giants. Further evolution may result in the loss of mass through thermal pulsing, leaving White Dwarfs that no longer shine but glow through radiative cooling.

creates substantial temperature differentials once again, and hydrodynamic stability is lost. A star of intermediate mass, between 3 $M_{\odot}$ and 8 $M_{\odot}$, may experience cataclysmic thermal pulses (sometimes called a *superwind*) which blow off most of the outer hydrogen envelope, reducing the mass of the star to less than 1.4 $M_{\odot}$.

The story continues pretty much as we have already seen for the hypothetical Population III stars, but the second-generation Population II stars have a distribution biased towards much smaller masses. A star of mass less than or equal to

$0.5\ M_\odot$ isn't large enough to trigger helium-burning reactions, and when the p-p chain reactions involving hydrogen cease, then the star is left with a cooling core of inert helium.

Stars of intermediate initial mass, between $0.5\ M_\odot$ and $8\ M_\odot$, trigger helium burning but, because they lose so much mass during their red giant phase, they are not large enough to trigger carbon burning. When the helium in the core is exhausted, they are left with a cooling core of carbon and oxygen. Similarly, slightly larger stars with initial masses of 8–11 $M_\odot$ may trigger carbon burning to produce neon and magnesium but the loss of mass means that they are unable to progress further. They are left with cores of oxygen, neon, and magnesium.

As there are now no longer any nuclear reactions occurring in the cores of these stars, they no longer shine. They do glow, however, as a result of radiative cooling, much as a cinder snatched from the fire will glow for a time. Strictly speaking, these are no longer stars, and they are collectively known as *white dwarfs*.

Stars that retain a mass greater than about $11\ M_\odot$ will burn their way through all the elements, with silicon burning producing a core of iron. We know what happens next.

ATOMIC ORBITALS AND ENERGY LEVELS

In what remains of this chapter, our attention will shift from the stars to the galaxies that accommodate them. But, before we go on, I need to take you on a short diversion to look at some of the properties and behaviour of the atoms that fuel these stellar furnaces.

Let's wind back briefly to the moment of recombination. Although positively charged $^1H^+$ and $^4He^{2+}$ nuclei would seem to be just crying out for electrons to combine with, this is the first time in the history of the universe that they have come together. And, as they do, something really rather beautiful happens.

Although it is perfectly true to say that a hydrogen nucleus 'captures' an electron to form a neutral hydrogen atom in a process we could write as $^1H^+ + e^- \rightarrow {}^1H + \gamma$, it is both simplistic and rather unromantic. We might be tempted to visualize the resulting atom in terms of the familiar planetary model, with the captured electron orbiting the hydrogen nucleus at some fixed distance, much as the Earth orbits the Sun.

Actually, this is both conceptually wrong and, in my opinion, a lot less visually appealing than what we now understand to be true. An electron moving in the electrostatic (or so-called Coulomb) field generated by the charge of the central proton is expected to radiate energy in the form of light. This is the basis for all television and radio broadcasting. Any energy of orbital motion that the electron possesses would quickly radiate away, and the electron would be sucked down on top of the nucleus. Disaster.

The reason this doesn't happen is that these particles are not the tiny incompressible billiard balls of naïve imagination. They are quantum wave–particles, and they behave altogether differently.

De Broglie wrote up his ideas about the properties of electron wave–particles in a series of papers published in 1923. He assembled these into a doctoral dissertation which he submitted to the University of Paris. In this thesis he drew analogies with music. Musical notes produced by string or wind instruments result from so-called *standing waves*, vibrational patterns which 'fit' between the ends of the stopped length of the string or the length of the pipe. Such standing wave patterns will persist provided that they have zero height (or amplitude) at each of the fixed ends. Examples of some simple standing wave patterns are shown in Figure 28.

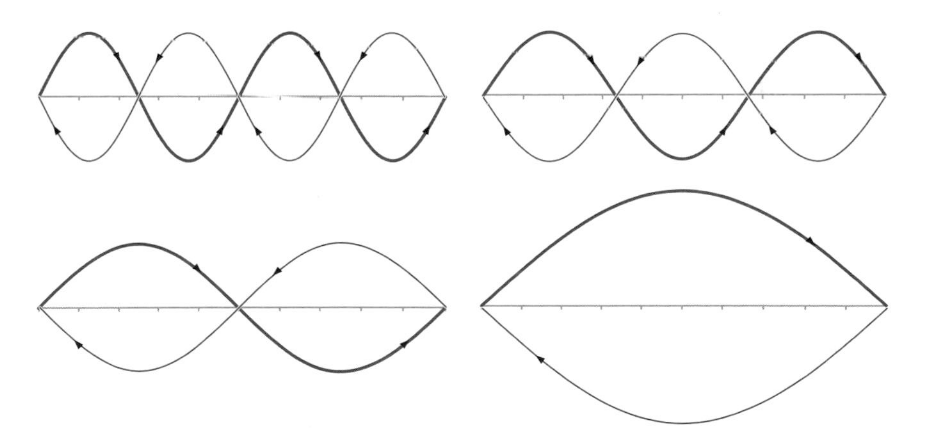

FIGURE 28 Examples of standing wave patterns. These are characterized by having zero height or amplitude at each end of the stopped string or length of pipe. The arrows give some sense of the direction in which the waves 'travel' although, once established, the waves appear to stand still.

What if, de Broglie now reasoned, the electron wave is required to produce a standing wave pattern as it orbits the nucleus of an atom? Put simply, by the time an electron wave has performed one complete orbit around the nucleus and returned to the starting point, the wave amplitude and its phase (its position in its peak–trough cycle) must be the same as at the starting point. If this condition is not satisfied, the wave will not 'join up' with itself; it will interfere destructively and no standing wave will be produced.

It was a bold suggestion, one that seemed vaguely mad. De Broglie's thesis advisor sent a copy of his dissertation to Einstein, and asked his opinion. Einstein wrote back offering encouragement: 'He [de Broglie] has lifted a corner of the great veil'.[6] Such an endorsement was enough, and de Broglie was duly awarded his doctorate in November 1924.

It fell to Austrian physicist Erwin Schrödinger to work out what this really meant. A few days before Christmas 1925, Schrödinger left Zurich for a short vacation in the Swiss Alps. His marriage was in trouble, and he chose to invite an old girlfriend from Vienna to join him, leaving his wife Anny at home. He also took with him his notes on de Broglie's thesis.

We do not know who the girlfriend was or what influence she might have had on him, but when he returned on 8 January 1926, Schrödinger had discovered a version of what we have come to know as *quantum mechanics*.

In essence, Schrödinger had found the three-dimensional standing wave patterns for an electron 'orbiting' the nucleus in a hydrogen atom. Because only certain wave patterns 'fit', and each pattern is associated with a specific energy, then an electron in a hydrogen atom can occupy one of a limited set of discrete *energy levels*. The electron can move between these energy levels by—you guessed it—absorbing or emitting photons in instantaneous transitions called quantum jumps.

Today we call these standing wave patterns *atomic orbitals*. Some of the lowest-energy orbitals of the electron in a hydrogen atom are shown in Figure 29, together with the corresponding energy levels.

The energy levels of the orbitals are fixed, so transitions from one to another are only possible by absorbing or emitting photons of precisely the right energy needed to bridge the gap. For photon energy, read light frequency or wavelength (or, indeed, colour). For example, an electron in a 3d orbital may emit a photon with a wavelength of 656.28 nanometres (billionths of a metre) and fall to the

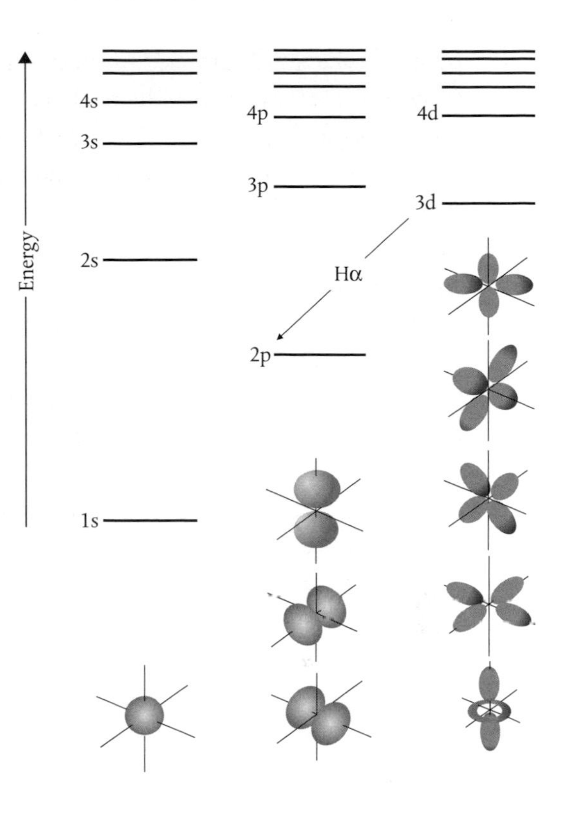

FIGURE 29 A selection of the lowest-energy hydrogen atomic orbitals. Each orbital represents a three-dimensional 'standing wave' pattern for the electron orbiting the nucleus. The lowest atomic energy levels are also shown. An electron in the 3d orbital will lose energy, falling to the 2p orbital, emitting red light with a wavelength of 656 nanometres. This produces a characteristic emission line called the Hα line.

2p orbital.* This gives rise to a characteristic emission line, known as the Hα (H-alpha) line, in the middle of the red part of the visible spectrum. It is this emission that lends a reddish hue to many true-colour pictures of nebulae, such as the Orion nebula (Figure 30(a)).

The result is an absorption or emission *spectrum*. Light emitted by the Sun spans a broad range of frequencies, or wavelengths. Most of this light reaches us unaffected, and if we spread out the different frequencies by passing the light through a prism, we see the familiar rainbow spectrum of colours. But as we look more closely, we see that the rainbow pattern is crossed by a series of dark

* The terms s, p, d, . . . derive from terminology introduced during the early years of atomic spectroscopy, as a shorthand for 'sharp', 'principal', 'diffuse', etc. They were applied first to help characterize the spectral lines and were then used to label the individual atomic orbitals involved.

(a)

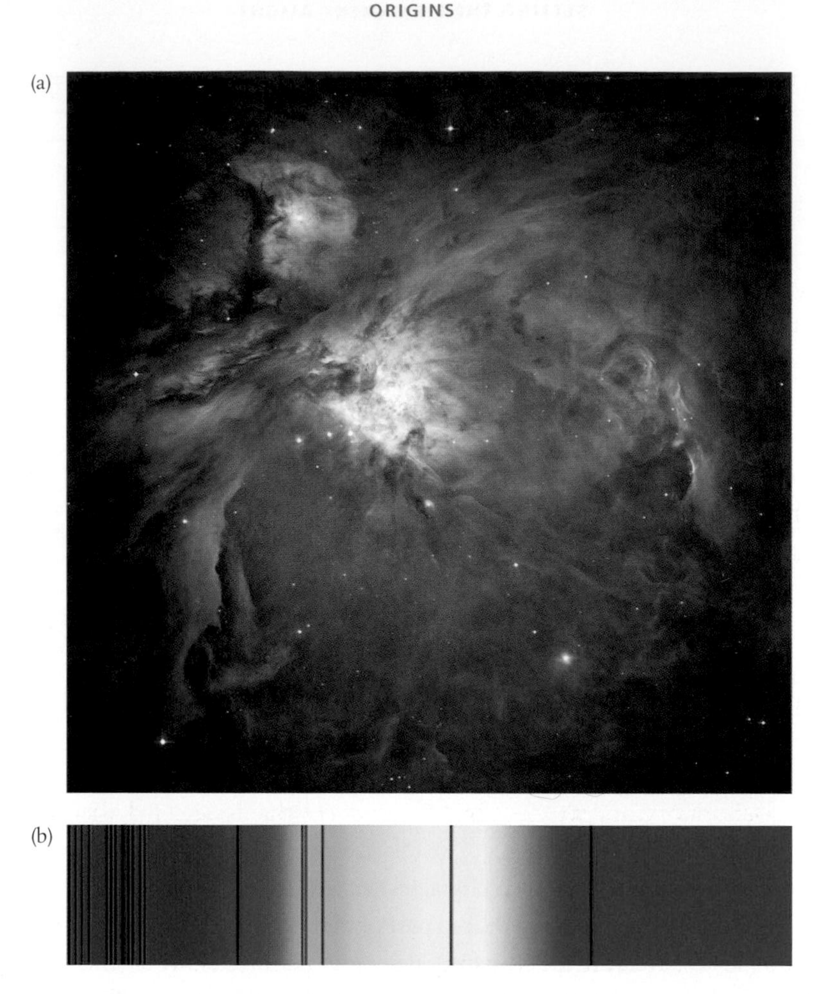

(b)

FIGURE 30 (a) Hα emission from hydrogen atoms lends a reddish hue to many true-colour photographs of nebulae, such as the Orion nebula. (b) If we pass sunlight through a prism, we will get the familiar rainbow pattern of colours in the visible spectrum. However, if we look closely, we may notice that this spectrum is crossed by a series of dark lines. These represent wavelengths of light that have been absorbed by atoms—including hydrogen atoms—in the outer layers of the Sun. This picture shows lines characteristic of hydrogen, including the Hα line in the red part of the spectrum.

lines (Figure 30(b)). At certain very discrete frequencies, the light drops sharply in intensity. These are frequencies that are absorbed by hydrogen atoms present in the Sun's outer layers. The lines correspond to the differences in energy between different energy levels. Figure 30(b) shows a series of hydrogen atomic absorption lines that cross the visible spectrum, including the Hα line.

REDSHIFT, STANDARD CANDLES, AND HUBBLE'S LAW

This business with atomic absorption and emission spectra turns out to be fundamentally important. It provides us with a tool for determining the compositions of distant stars. It also provides a means for us to discover how fast the distant stars and galaxies around us are moving and how far away they are.

In 1912, the universe was thought to consist of just the few hundred billion stars in our own Milky Way galaxy, dotted here and there with rather diffuse objects, called *nebulae*. American astronomer Vesto Slipher at the Lowell Observatory in Flagstaff, Arizona, used the Doppler effect to investigate the speeds of these nebulae.

The technique works like this. When we receive a wave signal (light or sound) from a moving object, we find that as the object approaches, the waves become bunched and their pitch (frequency) is detected to be higher than the frequency that is actually emitted. As the object moves away, the waves become spread or stretched out, shifting the pitch to lower frequencies. The effect is familiar to anyone who has listened to the siren of an ambulance or a police car as it speeds past.

If we know the frequency that is emitted by the source, measured for example in a laboratory, and we measure the frequency that is detected, then we can use the difference to calculate the speed at which the source is moving, towards or away from us.

If light is emitted by atoms in a nebula that is moving relative to our viewpoint on Earth, then the spectral frequencies will be shifted by an amount that depends on its speed. Figure 31 shows what happens. The spectral lines from a distant object moving away from us at high speed are shifted towards the red (b) compared to those same frequencies emitted by the Sun (a).

In his observations, Slipher actually used two lines from the spectrum of atomic calcium. Towards the end of 1912 he discovered that light from the Andromeda nebula is blue-shifted (higher frequencies or shorter wavelengths), suggesting that it is moving at a speed of about 300 kilometres per second *towards* us. However, as he gathered more data on other nebulae over the next five years, he found that most—21 of 25 nebulae (84%)—are red-shifted (lower frequencies or longer wavelengths), suggesting that they are all moving away, with speeds up to 1100 kilometres per second.

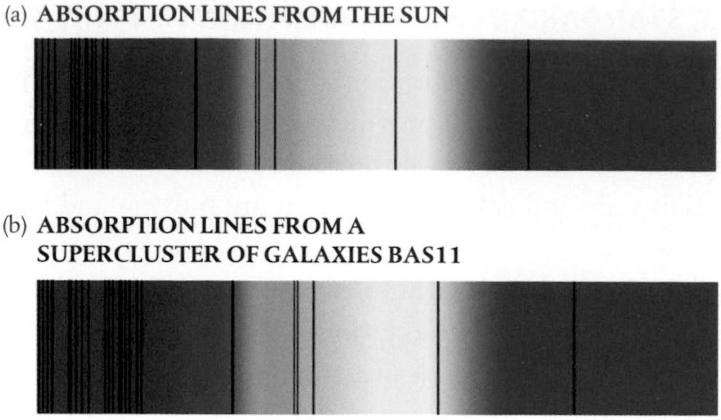

(a) **ABSORPTION LINES FROM THE SUN**

(b) **ABSORPTION LINES FROM A SUPERCLUSTER OF GALAXIES BAS11**

FIGURE 31 Atomic absorption lines crossing the spectrum of light from the Sun, (a), are shifted towards the red in light from a distant object which is moving away from us, (b). From the extent of the redshift, it is possible to work out the speed of the object.

This posed quite a conundrum. If the nebulae were all meant to be contained within the Milky Way, why would most of them be moving away from us, at such vast speeds? Arguments developed that the nebulae aren't diffuse objects sitting in our own galaxy—they are complete galaxies (or 'island universes') in their own right. This point was famously debated in April 1920 by American astronomers Harlow Shapley (who argued that the nebulae exist within the Milky Way) and Heber Curtis (who argued that they are galaxies lying far outside our own). Opinion is divided on who won the debate.

The matter was resolved in December 1924. American astronomer Edwin Hubble discovered that within some of the nebulae are a few stars with characteristics similar to those of Delta Cephei, in the constellation Cepheus. These stars, called *Cepheid variables*, are yellow supergiants which undergo regular pulsations in luminosity, reflecting periodic changes in their opacity. As a Cepheid variable star contracts, it heats up, becomes less opaque and expands.* Radiation is released and the star emits a bright pulse of light. But then it cools and becomes more opaque, and the radiation stops. The star contracts and the cycle repeats.

Earlier astronomical studies had established a relationship between pulsation rate and luminosity in the Cepheids, and Hubble realized that these could

* Remember for our current discussion that a star's opacity is related inversely to temperature—cooler stars are more opaque, hotter stars less opaque.

provide 'standard candles'.* By carefully measuring their pulsation rate and *apparent* luminosity, he could calculate just how far away they must be.

Using Cepheid variables that he had found in the Andromeda nebula, he estimated that this must be just short of a million light-years distant. A light-year is the distance that light will travel in a year, about 9500 billion kilometres. Hubble's estimate put the Andromeda nebula at 9.5 billion billion kilometres from Earth, well outside the Milky Way.† Overnight, the Andromeda nebula became the *Andromeda galaxy*.

Having established the distances to some nearby galaxies, Hubble now used Slipher's redshift measurements to map speed and distance, and arrived at a remarkable conclusion. The galaxies are receding from us at speeds that are directly proportional to their distances. Working together with his assistant Milton Humason, Hubble subsequently extended the data set to include more galaxies and confirmed the relationship between speed and distance, which is known as Hubble's law.[7]

There could be only one explanation. The universe is expanding.

At first sight, the fact that most of the galaxies are receding from us appears to place us in an especially privileged position at the centre of the universe. But remember that in an expanding universe it is spacetime that is doing the expanding, with every point in spacetime moving further away from every other point. The standard analogy is to think of the three-dimensional universe in terms of the two dimensions of the skin of a balloon. If we cover the deflated balloon with evenly spaced dots, then as the balloon is inflated the dots all move away from each other. And the further away they are, the faster they appear to be moving. This is illustrated (with galaxies instead of dots) in Figure 32.[8]

There were plenty of problems still. If the relationship between galaxy speed and distance had remained fixed since the big bang, then taking the inverse of the Hubble constant—the constant of proportionality between them—provides an estimate for the 'Hubble age' of the universe. Hubble's original work suggested that the universe is only two billion years old, rather embarrassingly *younger* than the estimated age of the Earth at the time. But these

* Think of it this way. Suppose we have a candle that is calibrated to emit light with a specific luminosity or brightness. If I now place this some distance away, the candle will appear dimmer. We can then use measures of this apparent brightness, combined with the calibration, to calculate just how far away it is.

† The diameter of the Milky Way is estimated to be between 100 000 to 120 000 light-years.

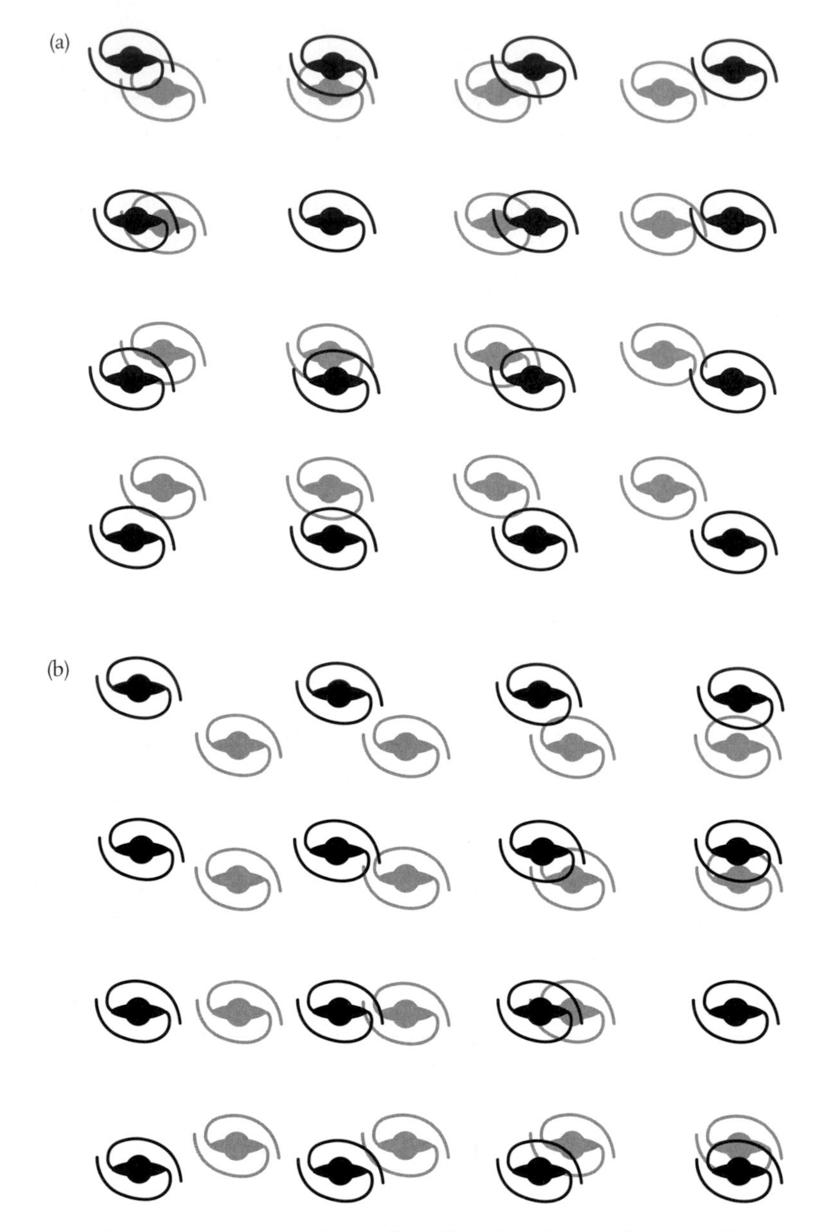

FIGURE 32 Imagine a universe consisting of a uniform distribution of galaxies. The pattern in grey represents this universe at some moment in time. The pattern in black is the distribution a short time later, after the universe has expanded. Let's now look at this from two different vantage points. From the perspective of the galaxy in row 2, column 2, shown in (a), all the galaxies around it have moved away, with more distant galaxies appearing to have moved the furthest. In (b), we fix on the galaxy in row 3, column 4. The result is the same—all the galaxies have moved away, with more distant galaxies moving the furthest. So the fact that we see most galaxies receding from us doesn't mean we're at the centre of the universe.

problems were eventually resolved in the 1950s, most notably by the American astronomer Allan Sandage. And we now know that the expansion of the universe has not been uniform—it has changed over time, which means that the Hubble constant isn't actually constant.

Today, the value of the Hubble constant is determined by data from the Planck mission to be 67.80 ±0.77 kilometres per second per megaparsec.* A galaxy receding with a speed of 1000 kilometres per second has a distance of 1000 divided by 67.8, or 14.7 megaparsecs (48 million light-years or 450 billion billion kilometres).

REDSHIFT SURVEYS

Aside from the local pull of gravity from neighbouring galaxies and clusters of galaxies, there is a sense in which galaxies that are carried along by the expansion of spacetime aren't really moving at all. We think of an object moving with a certain speed in relation to a fixed set of coordinates. But, in this case, it is the coordinates themselves that are moving. On cosmological scales, the redshift of light from distant objects represents a kind of 'dilution' of its frequency, as the light is stretched by the expansion of the space between the object and ourselves, the observers.

Consequently, the redshift of distant astronomical objects such as galaxies isn't really due to the Doppler effect, whose physics applies for objects moving in fixed spatial frames (though it is often mistakenly identified as a Doppler shift). It is, instead, a *cosmological* redshift resulting from the objects standing still (or moving relatively slowly due to the local gravitational pull of nearby objects) in an expanding spacetime.†

The redshift of an object, such as a galaxy, is expressed in terms of the parameter z, which can be calculated as the difference between the observed versus actual wavelengths of spectral lines (as measured in a laboratory on Earth) divided by the actual wavelength. For example, if we find that the red Hα line emitted by hydrogen atoms in a distant galaxy is redshifted to 658.47

---

* A megaparsec is equivalent to 3.26 million light-years, or 30.9 billion billion kilometres.

† There is therefore no limit to the speeds with which objects can recede, and recession speeds greater than the speed of light are in principle possible. This sounds like nonsense, but we must remember that special relativity imposes a speed limit on objects moving *through* space, and measurements depend on the state of motion of the observer. A galaxy receding at a speed greater than light is not physically moving at this speed, so special relativity is not violated. But since emitting light, the spacetime between the galaxy and ourselves has stretched, redshifting the light to an extent *equivalent* to a faster-than-light recession speed.

nanometres, a difference of 2.19 nanometres, then z is simply 2.19 nanometres divided by 656.28 nanometres, or 0.00333. Positive values of z indicate a redshift, negative values indicate a blueshift.

For galaxies moving at speeds much slower than light, z can also be approximated as the ratio of the speed of the galaxy to the speed of light.[9] Conversely, if z is known from a wavelength measurement, the speed of the galaxy can be determined simply by multiplying z by the speed of light. A redshift z of 0.00333 implies a speed of about 1000 kilometres per second.

In an extraordinarily simple set of algebraic manipulations, we can now use Hubble's law to estimate the distance of the galaxy. A speed of 1000 kilometres per second corresponds (as we saw above) to a distance of 48 million light-years.[10]

Through much of Chapters 1–3, I talked about the cosmic background radiation cooling as the universe expands, once again without being very explicit about why this should happen. We can now see quite clearly why this happens—the cosmic background radiation experiences a cosmological redshift. If we look back at Figure 14(a) we see that at a temperature of 3000 kelvin, the black-body curve peaks at around 1000 nanometres, in the infrared region of the electromagnetic spectrum. Figure 14(b) shows the same curve for a temperature of 3 kelvin, which peaks around 0.1 centimetres. In this case the 'actual' or starting wavelength is so much smaller than the observed wavelength as to be negligible when calculating the difference, so we can estimate z simply as 0.1 centimetres divided by 1000 nanometres, giving $z = 1000$. It's no coincidence that this corresponds to the ratio of the temperatures, 3000 kelvin divided by 3 kelvin.

In fact, the temperature of the cosmic background radiation is measured to be a little lower than 3 kelvin. If we use the temperature of 2.728 kelvin shown in Figure 18 we estimate a redshift z of 1100, which is close to the accepted figure derived from the Planck mission data.

Armed with techniques to determine the distance and speeds of galaxies, astronomers have been busily mapping the structure in the universe. The first galactic redshift survey, conducted by the Harvard-Smithsonian Center for Astrophysics (CfA), was begun in 1977, and initial data collection was completed in 1982. A second survey was begun in 1985 and continued for ten years. This was followed by the Two-degree-Field (2dF) Galaxy Redshift Survey conducted by the Anglo-Australian Observatory (now the Australian Astronomical Observatory) near Sydney, between 1997 and 2002. Results were reported in June 2003. The Sloan Digital Sky Survey, named for the Alfred P. Sloan Foundation (an important source of funding), began collecting data at the Apache Point Observatory in New Mexico in 2000. It has produced a series of data releases, the most recent in July 2013.

These surveys have revealed the true nature of the structure in the visible universe, already mentioned in Chapter 3. For example, Figure 33 shows some of the results from the 2dF survey, in which a couple of galaxy superclusters, which form some of the largest known structures in the universe, are picked out.

The Shapley concentration in the constellation of Centaurus is named for Harlow Shapley and sits about 650 million light-years away with a redshift z of 0.046, indicating that it is moving away from us at about 13 800 kilometres per second. The Horologium–Reticulum supercluster, in the constellations Horologium and Eridanus, measures 550 million light-years in length and contains visible mass of $10^{17}\ M_{\odot}$. It lies about 700 million light-years from us with a redshift of 0.063 (18 900 kilometres per second).

The Pisces–Cetus supercluster complex consists of many galaxy superclusters stretching across a billion light-years. The nearest of the clusters that form the complex is about 540 million light-years away, with a redshift z of 0.039 (11 700 kilometres per second). The furthest cluster lies about 970 million light-years

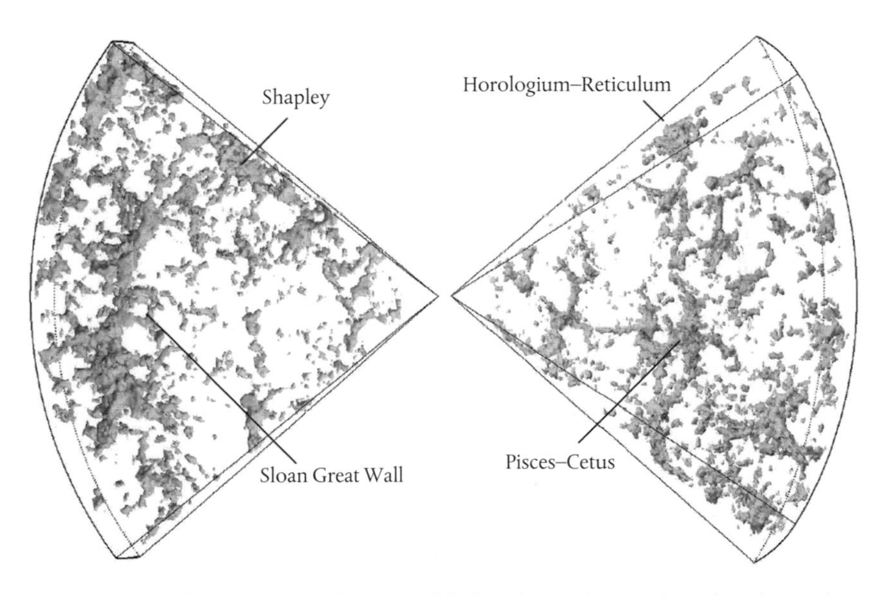

FIGURE 33 Sections of the map of galaxy redshift in the northern and southern hemispheres (relative to the plane of the Milky Way) measured in the 2dF Galaxy Redshift Survey. The map shows huge structures consisting of galaxy superclusters, among the largest known structures in the visible universe. The pattern of filaments, walls, and voids reflects the anisotropy in the distribution of matter in the early universe.

away with z = 0.072 (21 600 kilometres per second). This supercluster complex also includes the Virgo supercluster, in which we find the Local Group, the galactic cluster that includes the Milky Way and Andromeda galaxies.

The Sloan Great Wall is a filament of galaxy clusters which, until very recently, was the largest object in the visible universe. It measures 1.38 billion light-years in length and lies about a billion light-years distant, with a mean z of 0.078 (23 400 kilometres per second).

Between these filaments, walls, and superclusters lie great voids. These patterns are a legacy of the pattern of anisotropies in the cosmic background radiation illustrated in Figure 20. It's not possible to map these patterns to establish a correspondence, not just because there's insufficient angular resolution in the all-sky maps, but because the last scattering surface which produced these images is now about 40 billion light-years distant.

GALAXY TYPES, ROTATION CURVES, AND DARK MATTER

The kind of galaxy that will be formed in a dark matter halo will depend on the initial conditions of the halo and the accumulating cloud of baryonic matter. If the cloud sets out with a small amount of angular momentum—in other words, if it's spinning—then, in terms of what will later become visible to us, the result will be a flat thin disk or spiral galaxy.

There are several types of spiral galaxy, distinguished by the nature of their spiral arms. As the name suggests, barred spirals have pronounced, bar-like arms. In normal spirals the bars are less pronounced. Each type of spiral galaxy is further categorized based on the density of stars (and hence the brightness of the light) in the central bulge and the tightness with which the arms are wound around the bulge (Figure 34).

In addition to spirals there are elliptical galaxies, classified on a scale from 0 to 7 according to the dimensions of the ellipse, with E0 near-spherical and E7 (not shown in Figure 34) elongated along the major axis. When American astronomer Edwin Hubble first proposed this classification, elliptical galaxies were thought to be fairly young, developing over time into spirals. We now know that elliptical galaxies are rather old, and are most likely formed through the merger of neighbouring spiral galaxies.*

* I mentioned earlier that the Andromeda galaxy is moving towards our Milky Way galaxy at about 300 kilometres per second. This means that these galaxies will collide and merge in about 400 million years' time, likely producing an elliptical galaxy in the process.

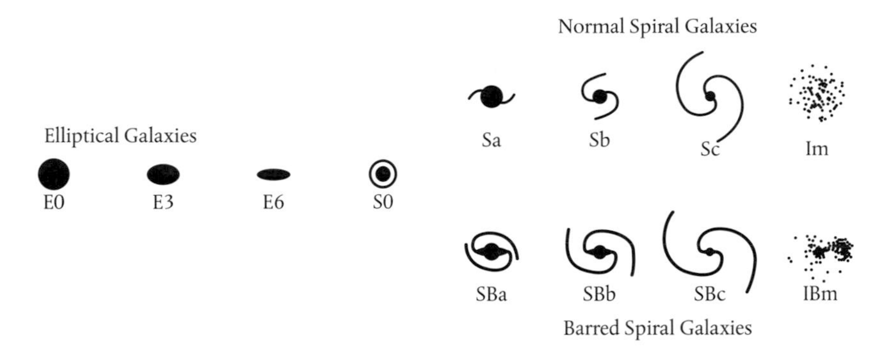

FIGURE 34 Hubble classified galaxies according to their shapes. Spiral galaxies are split into barred and normal spirals, with further subcategories of each (labelled a, b, and c) based on the density of stars in the central bulge and the tightness of the spiral arms. Elliptical galaxies are classified based on the dimensions of the ellipse. Further categories include the intermediate lenticular or So galaxies and the irregular galaxies. Adapted from R. G. Abraham, astro-ph/9809131 v1, 10 September 1998.

Hubble identified an intermediate classification, between ellipticals and spirals, which are called lenticular or So. Finally, irregular galaxies (designated Im and IBm in Figure 34) fit into none of these categories.

When a galaxy rotates, with its stars orbiting around its centre, Newtonian mechanics predicts a perfectly logical and reasonable relationship between the mass of the galaxy and the orbital speeds of stars at different distances from the centre. This relationship depends on the balance between the centrifugal force carrying the stars around the centre of the galaxy and the pull of gravity. Measuring from the centre, we would predict that the orbital speeds will rise to a maximum before falling away. Stars right at the edge of the galaxy feel the pull of gravity to a much smaller extent, and are expected to trundle around more slowly.

My use of words like 'predict' and 'expected' hint that this is not quite how it works out. The Swiss astronomer Fritz Zwicky was the first to sense that there was something fishy going on. In 1934, he published the results of observations of the Coma cluster, a large galactic cluster located in the constellation Coma Berenices, which contains over a thousand identified galaxies.

Although this is a cluster of galaxies, rather than a single galaxy, the principles are the same. He used the observations of the orbital motions of galaxies near the edge, combined with Newtonian mechanics, to estimate the total mass of the cluster. He then compared this to another estimate based on the number of observable galaxies and the total brightness of the cluster. He was shocked to find that these estimates differed by a factor of 400. The discrepancy was reduced by subsequent analysis, but it remains significant.

The mass, and hence the gravity, of the visible galaxies in the cluster is far too small to explain the speeds of the galaxies orbiting around the edge. As much as 90% of the mass required to explain the size of the gravitational effects appeared to be 'missing', or invisible. Zwicky inferred that there must be some invisible form of matter which makes up the difference. He called it 'missing mass'.

Missing mass was acknowledged as a problem but lay relatively dormant for another 40 years or so. Similar discrepancies began to appear in galaxies a little closer to home. Studies of the motions of stars in our nearest neighbour, the Andromeda galaxy, were first reported in 1939 by American astronomer Horace Babcock. Although the discrepancies were already visible in Babcock's work, he sought to explain them by more conventional physics. Further work on the 'rotation curve' of Andromeda followed in the 1960s, edging towards the inevitable conclusion.

The definitive work on the rotation curves of spiral galaxies, including Andromeda, was published in the 1970s and early 1980s by American astronomers Vera Rubin and Kent Ford and their colleagues. They reported the results of meticulous measurements of the orbital motions of stars at several positions measured from the centres of the galaxies. What they discovered for individual galaxies was identical to what Zwicky had discovered for the Coma cluster: stars at the edge of a galaxy orbit much faster than predicted based on estimates of the mass of the galaxy derived from its visible stars. The rotation curve flattens out with increasing distance from the centre of the galaxy, rather than falling away as expected. This is illustrated in Figure 35 for the case of the Andromeda galaxy.

Although these results were initially greeted with some scepticism, they were eventually accepted as correct. In fact, Peebles and his Princeton colleague Jeremiah Ostriker had earlier concluded that the observed motions of galaxies (including our own Milky Way) could not be modelled theoretically unless it was assumed that each galaxy possesses a large halo of invisible matter, effectively doubling the mass. In 1973, they had written: '. . . the halo masses of our Galaxy and of other spiral galaxies *exterior* to the observed disks may be extremely large'.[11] The following year they suggested that the masses of galaxies might have been underestimated by a factor of ten or more.

We now know the missing matter as dark matter. Every galaxy is shrouded in a dark matter halo which contributes 5–10 times the mass of baryonic

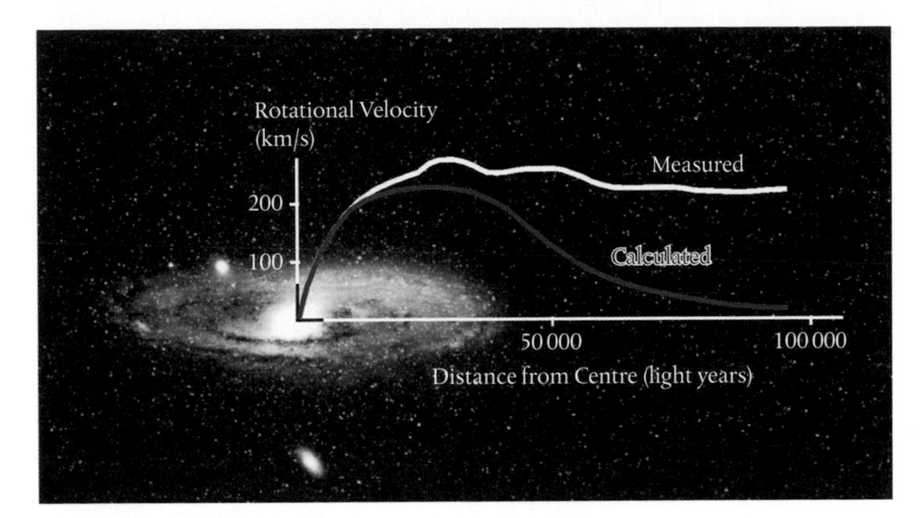

FIGURE 35 Simple Newtonian mechanics predicts that the stars in the Andromeda galaxy will orbit the centre with speeds that rise to a peak, before falling off with distance. However, the measured speeds flatten out with increasing distance. The measured behaviour is consistent with a dark matter halo that surrounds the galaxy.

matter. This is the evidence I promised you in Chapter 2 (and thanks for being so patient).

There is more evidence which we do not have space to consider here in detail. The properties and behaviour of the Bullet cluster in the constellation Carina, formed from two colliding galaxy clusters, can only be explained through the agency of their dark matter halos. A cloud of dark matter may act like a lens, distorting and displacing the images of galaxies behind it. This kind of 'gravitational lensing' has been used by the Planck mission to produce the first all-sky map of the distribution of dark matter in the universe.

Are there alternative explanations? Yes, there are. Modifications to general relativity have been proposed which can explain the 'missing' mass without having to invoke new, unknown forms of matter.[12] However, when all the evidence from observational cosmology and astronomy is weighed up, the case for dark matter remains pretty compelling. At least for now.

THE COSMIC EVENT HORIZON

All these statistics about superclusters, filaments, walls, voids, distances, and speeds beg a couple of obvious questions. Just how big *is* the universe? And what is the limit of what we can observe from our vantage point on Earth or from a satellite-borne instrument?

The answers to these questions are actually a bit complicated. We might be tempted to jump to the conclusion that, since the universe was created 13.8 billion years ago, then the limits of what we can observe must be fixed at the distance that light can travel in this time.

But, once again, this simple logic doesn't take account of the fact that the universe has been expanding for all this time. Suppose a burst of light is emitted from an early supernova, a few hundred million years after the big bang, or (say) 13.5 billion years ago. It takes this light 13.5 billion years to reach us, making the supernova 13.5 billion light-years distant, and we are tempted to think of this as a hard limit to our horizon. But as this light was making its way towards us, the spacetime it was moving in was expanding. The supernova that emitted the light is actually even more distant. If we take this expansion into account the apparent limit to our horizon is pushed out to about 46 billion light-years.

But there's a catch. If, as was once believed, the expansion is slowing, then we could in principle look forward to catching up with more galaxies that would eventually come into view. But, as we will see in the next chapter, the expansion is actually *accelerating*. This means that we are bound by a kind of cosmic event horizon. We will never be able to learn about events happening beyond this horizon, as objects emitting light are moving away from us faster than the light can cover the ground in between. Current estimates suggest that this cosmic event horizon is about 16 billion light-years distant.

'Space', declares *The Encyclopaedia Galactica* in Douglas Adams's *Hitchhiker's Guide to the Galaxy*, 'is big. Really big. You just won't believe how vastly hugely mindbogglingly big it is. I mean you may think it's a long way down the road to the chemist, but that's just peanuts to space.'[13]

The universe is indeed big, and it is easy to be bedazzled by the extraordinary dimensions of even the local structures that form and shape our cosmic neighbourhood. But we are reaching a stage in our story where it is necessary for us to draw our gaze away from distant horizons. We need to direct it instead towards a small region about 27 000 light-years from the centre of a perfectly ordinary SBc galaxy. Within this region a number of Population II stars have come to the ends of their lives, dying in spectacular supernova explosions that have scattered their baryonic matter, now heavily polluted by heavy elements.

As these clouds of gas and dust draw together yet again, an environment starts to take shape that we will one day call *home*.

5

SYNTHESIS

The Origin of Chemical Complexity

The universe now enters a long period of relative calm. The flurry of activity associated with the formation of the first stars ended just 600 million years or so after the big bang. This has given way to a much more sedate process involving smaller Population II stars, with lifetimes now spanning *billions* of years.

The large-scale structure of the universe is taking shape, with the distribution of superclusters, filaments, walls, and voids evolving towards the patterns that we see today.* The expansion of spacetime is decelerating.

A billion years pass. Then another. And another.

Now is a good time to zoom in on our cosmic neighbourhood. Galaxies emerge from their dark matter halos, growing in size and forming into what we will come to recognize from our vantage point on Earth as clusters of galaxies, then superclusters. Their patterns are constantly evolving.

Let's focus on a small collection of galaxy superclusters in a region of spacetime spanning up to half a billion light-years across. At the centre of this region lies what we will come to recognize as the Virgo supercluster, surrounded by

* Of course, when I say 'today', I mean the patterns that we observe today. Light from a galaxy that is measured to be a billion light-years distant has obviously taken a billion years to reach us. This light therefore reveals what the galaxy looked like when it was emitted, a billion years ago. The more distant the object, the greater the 'look-back' time.

the Hydra, Centaurus, Perseus-Pisces, and Coma superclusters. Lying beyond the Sculptor void we find the Sculptor supercluster, part of a wall of galaxies which stretches across a billion light-years.

The Virgo supercluster contains more than a hundred galaxy clusters and spans about 110 million light-years. It was first identified as a supercluster in 1958 by French astronomer Gérard Henri de Vaucouleurs and confirmed as such by redshift surveys in the late 1970s and early 1980s. It has a mass of about 10^{15} $M_{\odot}$, with about two-thirds of the visible galaxies contained in a thin elliptical disk. The remainder forms a spherical halo.

Let's zoom in a little further. At the centre of the Virgo supercluster lies the bright Virgo cluster, an otherwise unremarkable mix of about 1500 spiral and elliptical galaxies. It contains about a third of all the galaxies in the supercluster disk.

Out on the periphery of the disk, about 65 million light-years from the Virgo cluster, lies the Local Group of galaxies, a term introduced in 1936 by Edwin Hubble. This has evolved to become a modest collection of some 50 or so galaxies with a mass of about 1.3×10^{12} $M_{\odot}$ and a diameter of about 10 million light-years. The largest galaxy in this Group is the Andromeda galaxy, which contains about a trillion stars.

About 2.5 million light-years from Andromeda we find our own Milky Way galaxy. This is the second largest galaxy in the Local Group but may actually contain more dark matter than (and may therefore be at least as massive as) Andromeda itself.

THE MILKY WAY

The Milky Way galaxy started to take shape about 10 billion years ago, although most of its stars are much younger. Like all other galaxies, it emerged as a result of the cooling and condensing of baryonic matter trapped within a large dark matter halo. Figure 36 shows a sequence of stills from a movie of a computer simulation of the formation of a galaxy like the Milky Way. This shows quite clearly that the galaxy doesn't just appear fully formed, but rather evolves continuously from its origins just a few hundred million years after the big bang.

Today the Milky Way is an SBc barred spiral galaxy, with a thin disk between 100 000 and 120 000 light-years in diameter and just a few thousand light-years thick. Most of the galaxy's 100 to 400 billion stars are to be found in this disk (the actual number of stars depending on the total number of dwarf

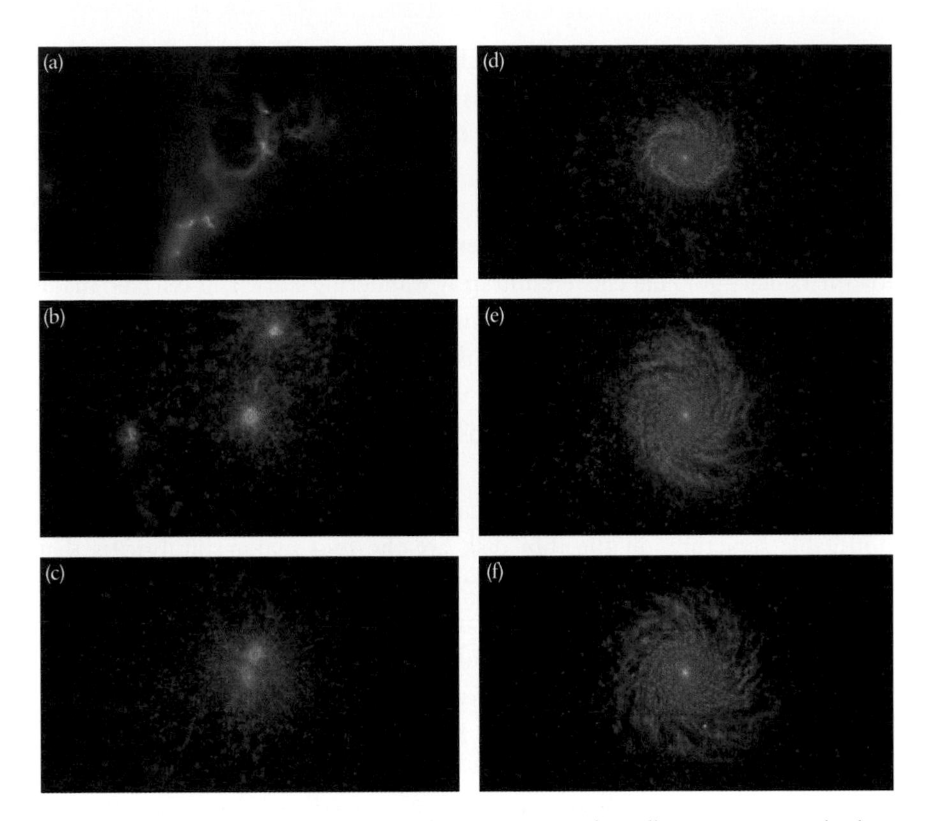

FIGURE 36 This computer simulation of the formation of a Milky Way type spiral galaxy shows the gradual accumulation of baryonic matter (mostly hydrogen and helium) within a large dark matter halo (a), about 500 million years after the big bang. Numerous proto-galaxies are formed which spin around, collide, and merge, (b). After about 4 billion years a single galaxy emerges with a well-defined centre, (c). A couple of billion years later the galaxy collides with another, (d), forming a larger galaxy which continues to accumulate baryonic gas and smaller satellite galaxies from its surroundings, (e). As we approach the present day, (f), the galaxy continues to evolve. These stills are taken from a movie of the simulation performed on the NASA Advanced Supercomputer by Fabio Governato and colleagues at the University of Washington.

stars, which are difficult to detect) with a mass of about 6×10^{11} $M_{\odot}$. Older stars are found on the outsides of the disk, with younger stars populating an inner layer just 300 light-years thick.

Some 3000 light-years above and below this thin disk there is another structure called the thick disk, which contains about 10% of the baryonic mass found in its thin companion. The thick disk may be the remnant of another, smaller, galaxy that got mixed up with the proto-Milky Way, or it may have formed through an act of gravitational theft, the Milky Way stripping the outer stars from a dwarf galaxy that strayed too close, leaving the core observable today as a so-called giant globular cluster (essentially a spherically shaped collection of stars) called Omega Centauri. Both these scenarios are consistent with the observed ages of stars in the thick disk, which are generally older Population II stars of low metallicity.

As we move even further out, we encounter a galactic halo of even older low-metallicity globular clusters and stray wandering stars. This is a halo made of visible baryonic matter, sitting within the larger dark matter halo. It seems likely that the outermost globular clusters were formed as part of separate dwarf galaxies that have since merged with the Milky Way. The inner clusters may have been formed from gas clouds in the early stages of the evolution of the galaxy.

Aside from the elongated elliptically shaped central bulge, the most noticeable structures in the thin disk are the two major spiral arms that wind anti-clockwise around the centre. In Figure 37, these are labelled the Scutum-Centaurus Arm and the Perseus Arm. Within this structure there are also some minor inner arms, labelled Norma and Sagittarius, and a minor outer arm, called (less imaginatively) the Outer Arm. Another minor arm called, variously, the Orion Arm, Orion-Cygnus Arm, or Orion Spur, sits (rather obviously) in the constellation Orion as seen from Earth. It's not clear if this is an independent arm of the galaxy or a spur connecting the Sagittarius and Perseus Arms. Either way, it appears rather innocuous, a quiet side-street in a bustling city-centre of a galaxy, itself one of hundreds of billions of very similar cities in the universe.

Imagine a circle with a radius of about 27 000 light-years measured from the centre of the galaxy (depicted as the dashed circle in Figure 37). Our attention now turns to a number of larger, and therefore rapidly ageing Population II stars sitting on or close to this circle. We wait another few billion years. The stars expend their hydrogen fuel, inflating to red giants before shedding their outer layers and burning helium. Then carbon. Then neon, oxygen, and silicon. The stars suffer core-collapse supernovae, spewing material into the interstellar medium, the space between the stars.

The debris from the death of these Population II stars is now greatly enriched with heavy elements. We are particularly interested in the composition of the resulting giant molecular cloud of gas and dust that now starts to cool and condense, ready to repeat the cycle all over again. For this is the cloud that will give birth to a third-generation Population I star which we call the Sun.

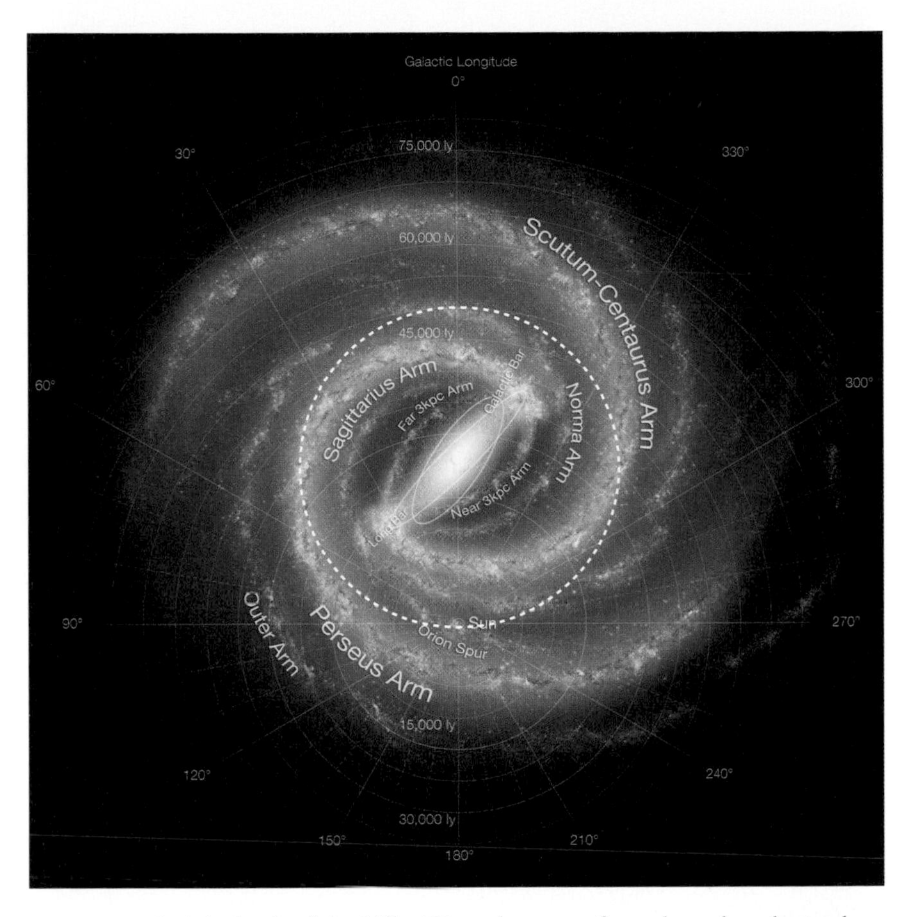

FIGURE 37 Artist's sketch of the Milky Way galaxy seen from above the galactic plane, based on data obtained at radio, infrared, and visible wavelengths. The Milky Way is believed to be a two-armed barred spiral with several secondary arms. The major arms are the Scutum-Centaurus Arm and Perseus Arm. The dashed circle has a radius of 27 000 light-years, and passes through the Orion Arm, which is where we will eventually find the Sun. Adapted from Ed Churchwell, Brian L. Babler, Marilyn R. Meade, Barbara A. Whitney, Robert Benjamin, Remy Indebetouw, *et al.*, *Publications of the Astronomical Society of the Pacific*, **121** (2009), p. 227.

CHEMICAL ELEMENTS

What we're going to get inside this giant molecular cloud quite obviously depends on what goes into it. The baryonic matter in the cloud has now been recycled through two generations of stars. To the background of hydrogen and helium atoms produced by primordial nucleosynthesis has been added an entire periodic table of elements formed through stellar nucleosynthesis and through the explosive release of energy in supernovae. This has produced an extensive palette from which complex molecular structures can now be created.

Hydrogen and helium still account for the bulk of the matter that is now assembling in the cloud. Their neighbours in the periodic table—lithium, boron, and beryllium—have much lower relative abundances (by 9–10 orders of magnitude) because they are not produced in significant quantities in either primordial or stellar nucleosynthesis. Once the triple-alpha process has bridged the carbon chasm, carbon, nitrogen, and oxygen are formed with relatively high abundance, as Figure 38 shows (note the logarithmic abundance scale).

Because ^{4}He is such an important building block in stellar nucleosynthesis, and because atomic nuclei with even numbers of protons and neutrons are especially stable, the element abundances from carbon through to calcium show a characteristic saw-tooth pattern. The ^{12}C contains six protons and six neutrons (or three ^{4}He nuclei). Add another ^{4}He nucleus to ^{12}C during helium burning and we get ^{16}O, which contains eight protons and eight neutrons. Fuse two ^{12}C nuclei together in carbon burning and shed a ^{4}He nucleus and we get ^{20}Ne, which contains 10 protons and 10 neutrons. Add another ^{4}He nucleus to ^{20}Ne during carbon burning and we get ^{24}Mg, which contains 12 protons and 12 neutrons.

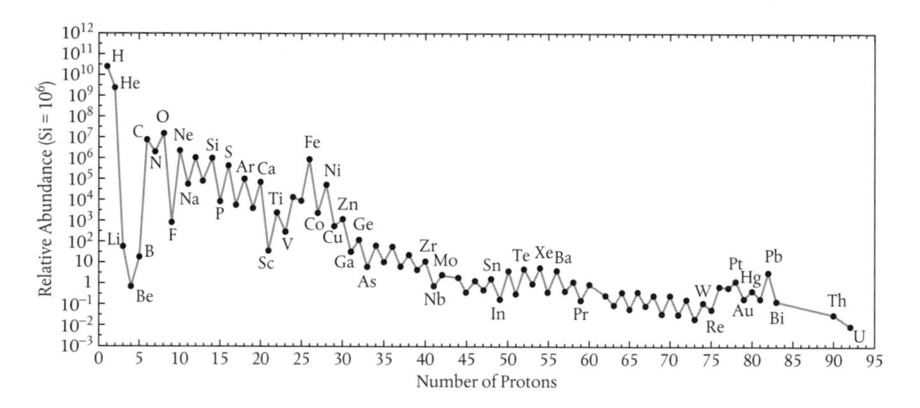

FIGURE 38 Abundances of chemical elements produced through primordial and stellar nucleosynthesis, in numbers of atoms per million relative to silicon. These abundances refer to our own Sun, and there is evidence to suggest that element abundances in newly formed giant molecular clouds may be a little different, but we can presume that by the time the cloud was ready to give birth to the Sun it possessed a pattern of abundances not much different from that shown here. Note the logarithmic abundance scale.

You get the picture. These heavier elements are all multiples of the ^{4}He nuclei that have been used to make them. Elements with odd numbers of protons and neutrons (such as ^{14}N, which contains seven protons and seven neutrons) are formed through different processes. This oscillating pattern continues up to ^{40}Ca, which contains 20 protons and 20 neutrons.

This building-block process continues in silicon burning, with successive additions of ^{4}He nuclei to ^{28}Si producing ^{32}S, ^{36}Ar, ^{40}Ca, ^{44}Ti, ^{48}Cr, ^{52}Fe, and ^{56}Ni. However, elements above ^{40}Ca are less stable and are susceptible to weak-force reactions, and the striking oscillation in the pattern of abundance is lost. At this point in the periodic table nuclear stability favours an *excess* of neutrons over protons. The most stable isotope of titanium is ^{48}Ti. Nickel isotopes ^{58}Ni and ^{62}Ni are both more stable than ^{56}Ni, which undergoes rapid weak-force decay, first to ^{56}Co, then to ^{56}Fe.

The end point of stellar nucleosynthesis is ^{56}Fe. Fusing another ^{4}He nucleus with ^{56}Fe costs more energy than can be released through the conversion of mass according to $E = mc^2$ and, as we saw in the last chapter, the iron accumulates in the core of a star until it undergoes a supernova explosion. This explosion rips the elements apart, stripping the material back to protons and neutrons and building them back up again, but this time through a mechanism of rapid neutron bombardment.

Neutrons are electrically neutral particles and so experience no electrical resistance as they hammer away at an exposed nucleus. As neutrons accumulate to produce heavier and heavier nuclei, some undergo weak-force decays, turning into protons. A sequence of new elements is produced and, as energy production in this process is no longer a requirement (there is now no star to hold up against the force of gravity), elements heavier than iron are formed. This process of nucleosynthesis through neutron bombardment is called the *r-process*, where 'r' stands for 'rapid'.*

Nevertheless, the relative abundances of heavier elements beyond iron falls away quite steeply, a feature that Figure 38 doesn't convey too well because of the use of a logarithmic relative abundance scale. Nuclei with even numbers of

---

* This implies that there must also be an *s-process*, where 's' stands for 'slow'. This is indeed correct. Free neutrons are generated in the interiors of stars throughout their evolution and can be picked up by elements formed in the core to produce heavier isotopes. This is a much slower process, however, with the neutrons being added one at a time with a pause between impacts for the isotopes to stabilize through weak-force decays. The heaviest nucleus that can be produced in the s-process is ^{209}Bi, a stable isotope of bismuth.

protons are still favoured, giving rise to oscillations, although the mechanism by which they are formed now depends on the combination of neutron capture and weak-force decay.

ASTROCHEMISTRY

Our creation story has been driven by the interplay between the four forces of nature. Gravity has played a huge role, both in the evolution of the universe to this point and in the formation of stars and galaxies. The strong nuclear force has played a significant role in nuclear fusion reactions, firing the nuclear furnaces that help to hold a star up against the force of gravity. The weak nuclear force has played an absolutely pivotal role, turning down quarks into up quarks and neutrons into protons and vice versa, thereby allowing heavier elements to accumulate in the interiors of stars and form in supernova explosions.

The electromagnetic force has also played an important role. It holds protons and electrons together in neutral atoms and was therefore responsible for recombination, which released the flood of cosmic background radiation. Stars and galaxies could not have formed were it not for the cooling processes aided by electromagnetism.

That said, there is a subtlety to electromagnetism that has not yet had a chance to reveal itself.

But even here, in the cold dark environment of a large cloud of gas and dust, conditions do not appear to offer much promise. This is the void of interstellar space. Densities of just a few hundred particles per cubic centimetre are lower than any vacuum we could ever hope to create in a laboratory on Earth. Temperatures are not much different from that of the cosmic background radiation. Nevertheless, it is here that the force of electromagnetism gives the creation story a rather fascinating plot twist.

Chemistry happens.

In fact, I have already mentioned the important role that molecular hydrogen plays in helping to cool the gas that may have helped to form the first stars. And, indeed, where there are lots of hydrogen atoms and electrons and the conditions are right, we will get hydrogen molecules. So let's now take a closer look at how this works.

Each hydrogen atom consists of a central proton and an electron which forms a kind of standing wave pattern around it (Figure 29). The force of electromagnetism between the positively charged proton and the negatively charged electron orbital keeps them bound together. When we bring two hydrogen atoms together, what happens is determined by the interaction between the electron orbitals. This is the domain of electromagnetism.

Figure 39 shows the lowest-energy 1s orbitals, and the associated energy levels, of two separated hydrogen atoms, one on the far left of the figure and the other on the far right. When we bring the two atoms together, the electron orbitals combine. As we start with two 1s orbitals, nature determines that we will get two new orbitals. We can think of one of these describing the result of a *constructive* overlap of the two 1s orbitals (similar to the constructive interference of overlapping waves) and the other a *destructive* overlap.

The constructive overlap of the electron orbitals produces a lower-energy molecular orbital, sometimes called a *bonding orbital*. The two electrons, one from each atom, are now spread over both atoms in a sausage-shaped orbital

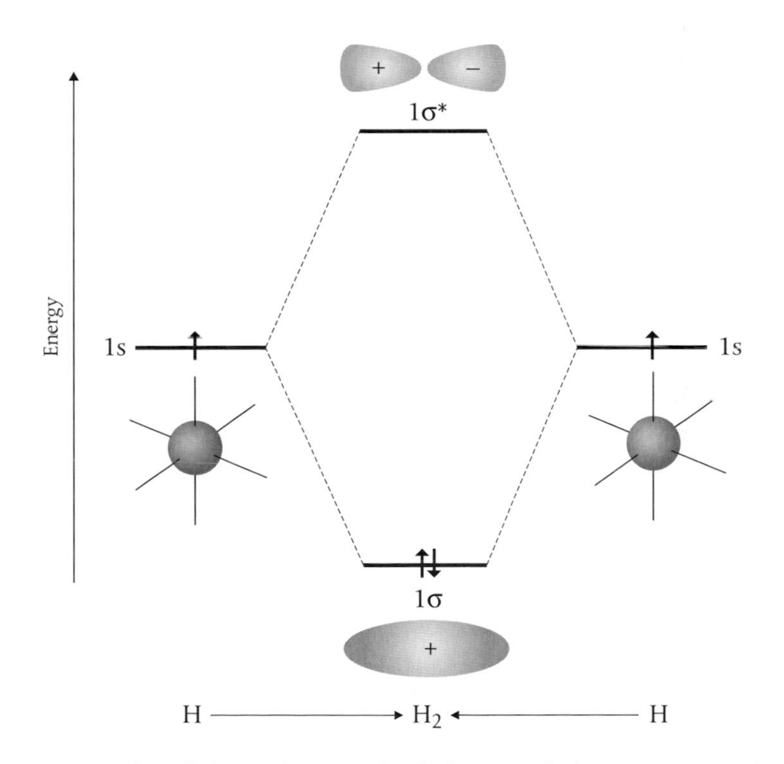

FIGURE 39 Overlap of the 1s electron orbitals from two hydrogen atoms produces a lower-energy 'constructive' bonding orbital (1σ) and a 'destructive' anti-bonding orbital (1σ*). The two electrons can now both occupy the bonding orbital, provided that their spins are paired. The result is a hydrogen molecule in which the two electrons form a single bond.

that shrouds the whole molecule. The extent to which the energy falls below the energies of the separate atoms is the *binding* or *bond energy* of the molecule, a measure of the strength of the chemical bond that's formed.

Quantum mechanics now demands that if two electrons are to share this one orbital, then they must take up opposite spin orientations, indicated in Figure 39 through the use of arrows (↑ indicating 'spin-up' and ↓ indicating 'spin-down'). We say that their spins are *paired*.

The bonding orbital is labelled as a sigma (σ) orbital. The orbital produced by the 'destructive' combination of the two 1s orbitals is a higher-energy *anti-bonding* orbital, labelled σ^* (pronounced 'sigma-star').

Once we have populated the molecular orbitals with the available electrons (paired as necessary), we can determine the *bond order* simply as the number of electrons in bonding orbitals less the number in anti-bonding orbitals divided by two. So, for molecular hydrogen, as illustrated in Figure 39, this formula gives two minus zero divided by two, or a bond order of one. A single bond forms between the two atoms.[1]

Figure 39 shows molecular orbitals forming from two 1s atomic orbitals, but what of the higher energy 2s, 2p, etc. orbitals? Only atomic orbitals of similar energies can participate in these molecular games, so a hierarchy is established. To the 1σ and $1\sigma^*$ molecular orbitals we can add higher energy 2σ and $2\sigma^*$ orbitals formed by combining the 2s orbitals. Even higher-energy $2p\sigma$ and $2p\sigma^*$ orbitals are formed by the overlap of the two dumb-bell-shaped 2p orbitals that lie along the H—H bond axis, with two lots of $2p\pi$ and $2p\pi^*$ orbitals formed from the 2p orbitals lying at right angles to this axis.* For simplicity I've not shown these orbitals and energy levels in Figure 39.

Molecular hydrogen is just the beginning, the very thin end of a very big wedge. We might be astonished at the complexity that nature can create by combining protons and neutrons to produce an array of 92 chemical elements, but this is nothing compared with the complexity that can be created by combining the electron orbitals of these different elements to form molecules.

* Actually, the hydrogen 2s and 2p orbitals mix it up a little bit, with the result that the order of the molecular energy levels is not quite as simple as I've suggested here.

INTERSTELLAR MOLECULES

To date, more than 140 different molecules have been detected in interstellar molecular clouds, ranging from simple diatomic molecules such as hydrogen (H_2), carbon monoxide (CO), molecular nitrogen (N_2), molecular oxygen (O_2), and sodium chloride (NaCl),[2] to polycyclic aromatic hydrocarbons such as anthracene ($C_{14}H_{10}$) and pyrene ($C_{16}H_{10}$) and the carbon allotropes buckminsterfullerene (the soccer-ball shaped C_{60}) and 70-fullerene (C_{70}). Illustrative examples are provided in Table 2.

Anyone who has made even a superficial study of organic chemistry—essentially the chemistry of carbon compounds—will be struck by the array of molecules in this table. What makes a vast array of organic molecules possible, including many associated with life, is the existence of a kind of toolbox of 'functional groups', groupings of atoms that determine the nature of the structures of the molecules and the chemical reactions they will undergo. These include the hydroxyl group (—OH), the carbonyl group (—CO), the amine group (—NH_2), the aldehyde group (—CHO), the hydroperoxy group (—O_2H), and the carboxyl group (—CO_2H), among others. Also apparent in this list are so-called aliphatic hydrocarbons, molecules consisting of carbon atoms strung together in chains (such as ethylene, $H_2C{=}CH_2$, and propylene, $H_2C{=}CHCH_3$, where '=' indicates a double bond*) and aromatic hydrocarbons in which the carbon atoms are formed in rings (such as benzene, anthracene, pyrene, and the fullerenes).

And, of course, the table includes glycine ($NH_2CH_2CO_2H$), the simplest of 20 amino acids containing amine and carboxyl groups known to be encoded in the universal genetic code and important in building proteins. We will meet glycine again in Chapter 8. I should point out that initial reports of its discovery in interstellar space in 2003 were *not* supported by further research efforts a year later, although glycine was reported to have been detected in the tail of comet Wild 2 in 2009 by NASA's Stardust mission, in quantities that suggest it might be very difficult to observe by the usual methods. So, despite its dubious status as a fully fledged interstellar molecule, I've not been able to resist the temptation to include it in this table.

* A double bond consists of one σ-molecular orbital and one π-molecular orbital, the latter resulting from the overlap of p-orbitals (or hybrids formed from mixing s and p orbitals—see later in this chapter) that sit at right angles to the plane of the σ-bond. A triple bond is formed from one σ-molecular orbital and two π-molecular orbitals.

TABLE 2: **Illustrative examples of interstellar molecules**

Number of atoms	Examples
2	molecular hydrogen (H_2), carbon monoxide (CO), molecular nitrogen (N_2), molecular oxygen (O_2), sodium chloride (NaCl), iron oxide (FeO), hydrogen chloride (HCl), hydrogen fluoride (HF), carborundum (SiC), titanium oxide (TiO)
3	carbon dioxide (CO_2), water (H_2O), hydrogen sulphide (H_2S), hydrogen cyanide (HCN), potassium cyanide (KCN), nitrous oxide (N_2O), sodium hydroxide (NaOH), ozone (O_3), sulphur dioxide (SO_2), titanium dioxide (TiO_2)
4	acetylene (C_2H_2), formaldehyde (H_2CO), hydrogen peroxide (H_2O_2), ammonia (NH_3), isocyanic acid (HNCO)
5	methane (CH_4), ketene (H_2C_2O), cyanoacetylene (HC_3N), formic acid (HCO_2H), ammonium ion (NH_4^+)
6	ethylene (C_2H_4), acetonitrile (CH_3CN), methanol (CH_3OH)
7	methylamine (CH_3NH_2), acrylonitrile (CH_2CHCN), ethylene oxide (*c*-C_2H_4O)*, vinylalcohol (CH_2CHOH), acetaldehyde (CH_3CHO)
8	acetic acid (CH_3CO_2H), glycoaldehyde (CH_2OHCHO), methylcyanoacetylene (CH_3C_3N)
9	dimethylether (CH_3OCH_3), ethylalcohol (C_2H_5OH), propylene (CH_3CHCH_2)
10	acetone ($(CH_3)_2CO$), glycine ($NH_2CH_2CO_2H$), ethylene glycol ($(CH_2OH)_2$)
>10	methylacetate ($CH_3CO_2CH_3$), benzene (C_6H_6), cyanopentaacetylene ($HC_{11}N$), anthracene ($C_{14}H_{10}$), pyrene ($C_{16}H_{10}$), buckminsterfullerene (C_{60}), 70-fullerene (C_{70})

*The prefix *c*- indicates a cyclic molecule.

HYBRID ORBITALS AND LONE PAIRS

The properties of functional groups are obviously derived from the nature of the atoms from which they are composed. But their reactivity is also determined in large part by their *shapes*. At first sight, the geometries of molecules—the lengths of chemical bonds and the angles between them—might not seem to be all that important. However, as we will see in Chapter 8, molecular geometry has a fundamentally important part to play in the chemistry of life.

It might not come as much of a surprise to learn that the shapes of molecules are fashioned by the spatial distribution of their molecular orbitals. This is a very subtle manifestation of the electromagnetic force (although it is, perhaps, not the most subtle, as we will see).

In the description I have just given of the molecular orbitals of hydrogen, I mentioned that only electron orbitals of similar energy can combine. We don't see 1s and 2s orbitals combining together, for example, as their energies are too dissimilar. But if the conditions are right, then combining different orbitals is much like combining different waveforms.

The American chemist Linus Pauling was the first to suggest that the shapes of some simple molecules are determined by the way that the electron orbitals of their constituent atoms mix together. For example, methane (CH_4) has a characteristic tetrahedral shape, with angles between each of the four identical C—H bonds (the H—C—H bond angles) of 109.5°. The central carbon atom contributes six electrons. Two of these sit tucked away in the lowest energy 1s atomic orbital (with their spins paired) and do not participate in bonding. Another two sit in the 2s orbital. The remaining two sit in each of two 2p orbitals, leaving a third 2p orbital vacant. We now need to overlap these orbitals with four 1s orbitals, one each from the four hydrogen atoms.

No matter how hard we try, it's impossible to get a tetrahedral shape from this combination. The 2s orbital is spherical—it favours no particular direction. The three 2p orbitals point along each of the three Cartesian (x, y, z) coordinates, implying bond angles of 90° (Figure 40(a)). Something's not right.

Pauling realized that the carbon 2s and 2p orbitals are of similar energies, and can therefore mix together. In fact, if we promote one of the electrons from the 2s orbital into the vacant 2p orbital and then mix all four of these orbitals together, the resulting *hybrid* orbitals will have one part s-character and three parts p-character. These are called sp^3-hybrid orbitals, pronounced 's-p-three'. And what shapes do sp^3-hybrid orbitals possess? You guessed it. Their major lobes point to each of the four apexes of a tetrahedron (Figure 40(b) and (c)).

All we need to do now is overlap a hydrogen 1s orbital with the lobe of each sp^3-hybrid orbital, creating four identical σ molecular orbitals corresponding to the four carbon-hydrogen bonds. The shape of the carbon sp^3-hybrids is carried through to the shape of the resulting molecular orbitals and so the molecule has a tetrahedral geometry.

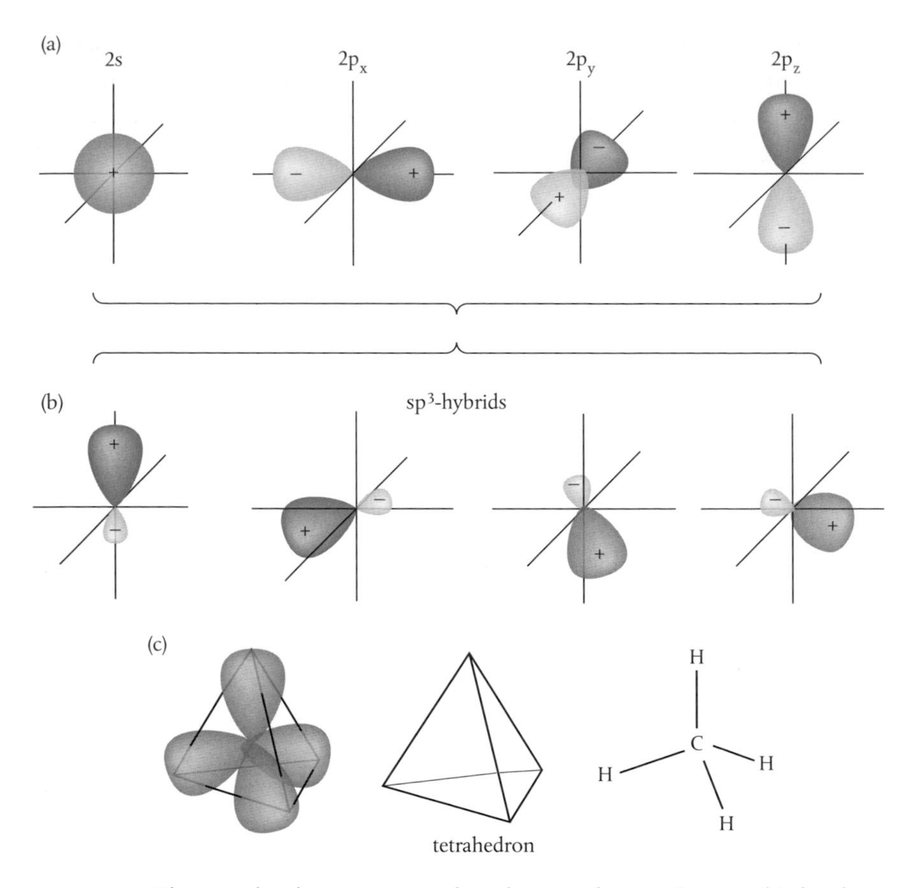

FIGURE 40 The central carbon atom in methane has two electrons in a 2s orbital and two electrons in each of two 2p orbitals, leaving the third 2p orbital vacant. But if one of the 2s electrons is promoted into the vacant 2p orbital, it is possible to mix all four of these together, (a). The result is four identical sp^3-hybrid orbitals, (b). The sp^3-hybrid orbitals produce a tetrahedral shape, (c).

One further subtlety. Molecular geometry is about chemical bonds and angles, but it is dependent on the geometry of the underlying molecular orbitals, and these may involve orbitals that are not directly involved in bonding. The ammonia molecule (NH_3) is described as having a pyramidal shape. If we remove the topmost hydrogen atom from the methane tetrahedron in Figure 40(c), then we would be left with a pyramidal geometry for the remaining CH_3 group. In fact, this is exactly the way to think about the structure of NH_3.

The nitrogen atom has seven electrons. As with carbon, two of these are 'buried' in the nitrogen 1s orbital, leaving five to participate in bonding, two in the 2s orbital and one in each of the 2p orbitals. As we now know, we can't get a pyramidal shape using the three 2p orbitals, so what happens is that four identical sp^3-hybrids are formed by mixing together the 2s and 2p orbitals. But this time there's an unequal distribution of electrons. The three hybrid orbitals containing a single electron each overlap with a hydrogen 1s orbital but the fourth hybrid orbital is already 'full' with two spin-paired electrons. This 'lone pair' of electrons simply points in a direction in space determined by the shape of the orbital it occupies. So, the overall shape of the ammonia molecule is tetrahedral, just like methane, but one of the bonds is replaced by a lone pair of electrons, leaving the remaining atoms to form a pyramidal shape.*

If we move along the periodic table from carbon and nitrogen then the next element we encounter is oxygen. We might be tempted to play the same kind of game again. Now we have eight electrons, two buried in the 1s orbital, two in the 2s orbital, and four to be distributed over the three 2p orbitals. By mixing the 2s and 2p orbitals to make four sp^3-hybrid orbitals, we would predict two lone pairs of electrons pointing up like a pair of rabbit ears, leaving the two remaining singly occupied hybrid orbitals to overlap with hydrogen 1s orbitals to form water (H_2O) with a molecular geometry that is rather unimaginatively called 'bent'.

Alas, this description is too simplistic. More sophisticated quantum theoretical calculations fail to predict the rabbit ears, and an alternative description appears more likely, based on three sp^2-hybrids and a pair of electrons occupying the unmixed 2p-orbital. This means that the two lone pairs of electrons are not equivalent: one is present in the unmixed 2p orbital and the other is contained in one of the sp^2-hybrids. This rabbit has lop-sided ears.

The bottom line is that subtle interactions between the electron orbitals of different atoms generate an astonishing variety of molecules with different shapes and patterns of reactivity. When wrapped around a central nucleus, the humble electron thus reveals itself to be an extraordinarily versatile elementary particle. It opens the door from the relative simplicity of atomic structures built from protons, neutrons, and electrons to the wonderful complexity of chemical structures built from molecular orbitals.

---

* There's a slight difference. In methane the four C—H bonds are identical, so the molecular orbitals space themselves out in a regular distribution, giving a uniform H—C—H bond angle of 109.5°. However, the lone pair of electrons in ammonia is not equivalent to the three N—H bonds, so the H—N—H bond angles are squeezed slightly, to 107°.

But molecules are delicate things and are easily decomposed (or 'photodissociated') by ultraviolet light from nearby stars. To a certain extent molecules in the interior of a cloud are shielded from this radiation by hydrogen molecules at its edges, which absorb light with particularly destructive wavelengths and prevent it from penetrating to the centre of the cloud.

The molecules are also shielded by grains of dust in the cloud and, as these dust grains have an important role to play in chemical synthesis, it's time we took a closer look at them.

STARDUST

We're familiar with dust. Household dust is an everyday nuisance, demanding the deployment of cleaning products at regular intervals. But household dust is the detritus of the living, consisting as it does of human skin cells, human and animal hair, and pollen, mixed with inorganic particles from our environment such as soil minerals, and textile and paper fibres. This dust provides an ecological niche within which dust mites can thrive, microscopic 'spiders' that feed on dead skin cells.

The dust grains in this rather amorphous mix have sizes ranging from a few tens of microns (millionths of a metre) up to a thousand microns or more. It goes without saying that this is *not* the kind of dust we find in interstellar space.

Much of this 'stardust' is generated by red-giant stars. These stars have reached a stage in their evolution where they have left the main sequence (Figure 27). Carbon, oxygen and silicon manufactured in the cores of these stars can get dredged to the surface by convection currents. At the surface these elements can cool and start to clump together to form small particles.

What kinds of particles we get will depend on the evolutionary history of the star. A red-giant star rich in carbon may produce small particles of graphite with dimensions of a hundredth of a micron, more like smoke than dust. Oxygen-rich stars may produce small silicate grains, chemical compounds of silicon and oxygen combined with heavier elements such as iron and magnesium, up to a micron in size.

For example, the mineral olivine (so-named because of its olive-green colour) is a mixture of magnesium silicate (Mg_2SiO_4) and iron silicate (Fe_2SiO_4). Magnesium-rich olivine has been found in meteorites, the tail of comet Wild 2 and the core of comet Tempel 1. It has also been observed in the dust swirling around young stars.

Other kinds of stardust particles so far identified include diamond, silicon carbide (SiC), titanium carbide (TiC), titanium dioxide (TiO_2), silicon nitride (Si_3N_4), corundum (Al_2O_3), spinel ($MgAl_2O_4$), hibonite (an oxide consisting of calcium, cerium, aluminium, titanium, and magnesium), and another silicate called pyroxene (consisting of a broad range of molecules containing calcium, magnesium, iron, and many other heavy elements).

These particles are all extended arrays of atoms or molecules that form a regular lattice structure, held together by yet another subtle manifestation of the electromagnetic force.

Driven from the outer layers of their parent red-giant stars by radiation pressure and ejected rather more forcefully in supernova explosions, the particles drift into the interstellar medium, providing surfaces on which the atoms of further elements—such as hydrogen, carbon, nitrogen, oxygen, and sulphur—can stick.* These elements react on the grain surfaces and, over time, the particles gain thin coverings of ice, frozen carbon dioxide, ammonia, and methane. Other molecules, such as carbon monoxide and hydrogen sulphide, may also be drawn to their surfaces.

All we need to do now is gently bathe the particles in ultraviolet light from nearby stars and we have ourselves an enormous number of tiny chemical *factories*.

Just as clouds of water vapour in the atmosphere block the light from the Sun, spoiling the beach holidays of countless sunbathers, so molecular clouds block the light from distant stars. The molecular cloud Barnard 68, in the constellation Ophiuchus, lies just 500 light-years from Earth. It was identified as a nebula by American astronomer Edward Emerson Barnard in 1919 and is one of a chain of such 'globules', suggesting that it was once part of a larger molecular cloud which became dispersed.

Barnard 68 has a mass estimated to be about 3 $M_{\odot}$, and measures about half a light-year in diameter. The temperature at the core of the cloud is about 8 kelvin, rising to 16 kelvin at the edge, and the density of particles at the centre lies in the range $0.9–4.3 \times 10^5$ particles per cubic centimetre.[3] Figure 41 shows a composite image of the cloud produced in 1999 by the European Southern Observatory,

* Striking images of billowing dust clouds formed at the centre of supernova 1987A, just 168 000 light-years from Earth on the edge of the Tarantula Nebula in the Large Magellanic Cloud, were produced using the Alma radio telescope array in Chile and revealed at a meeting of the American Astronomical Society in January 2014.

FIGURE 41 Barnard 68 is a molecular cloud in the constellation Ophiuchus with a mass of about 3 $M_{\odot}$ measuring about half a light-year in diameter. This is a composite of visible and near-infrared images obtained in 1999 and shows that the light from more distant stars in the Milky Way is completely obscured by dust particles in the cloud.

a collaboration of 15 European member states which operates ground-based observatories in Chile. As this picture shows, the light from background stars is rather dramatically obscured by dust particles in the cloud.

That the dimensions of the cloud are so well defined suggests that it is on the verge of collapsing to produce one or more stars, in about 100 000 years or so.

'We are stardust,' sang Joni Mitchell at the Big Sur Folk Festival in California in September 1969, 'We are golden. And we've got to get ourselves back to the garden.'[4] Mitchell wrote this song as an anthem for the Woodstock festival, which had been held a month earlier but which she did not attend. She based her lyrics on what her boyfriend Graham Nash had told her. The song later became a major hit for the folk-rock group Crosby, Stills, Nash, and Young.

Despite what you might make of late-60s counterculture, these are nevertheless prophetic words. There is much more to humans than just atoms and molecules, of course, but this is the stuff from which we are made and it is a simple fact that every atom of our mortal 'hardware' larger than hydrogen was forged in the interior of a star or in a supernova explosion before being hurled into interstellar space to form a giant molecular cloud.

GIANT MOLECULAR CLOUD

So, let's turn the clock back a bit and focus our attention on a specific giant molecular cloud. This is about 65 light-years in size with a mass likely to be of the order of thousands of $M_{\odot}$. It circles the centre of the Milky Way galaxy with an orbital radius of about 27 000 light-years, completing one round-trip every 220–250 million years, passing through the spiral arms as it does so (Figure 37). It contains a wide variety of chemical elements with relative abundances similar (but perhaps not yet identical) to those shown in Figure 38. It contains a broad range of molecules, the list of 140+ molecules represented in Table 2 most likely barely scratching the surface, with an average density of just a few hundred molecules per cubic centimetre. Temperatures in the cloud are just a little higher than the temperature of the cosmic background radiation.

The cloud also contains dust particles, ice-coated grains of graphite, and silicates among others, which make up less than one per cent of the total mass of the cloud but which provide an important shield against destructive ultraviolet radiation and which act as tiny chemical factories.

So, what do we do now?

We wait. This is a long pregnancy. The first stars may have been born within a few hundred million years of the big bang. Population II stars arrived soon afterwards, but these were smaller and more stable, spending billions of years on the main sequence. The death of Population II stars in this neighbourhood of the Milky Way has helped to create this giant molecular cloud, and we must now wait patiently for the cloud to succumb once more to the force of gravity.

The universe is now approaching its 9 billionth year.

Expectation builds. But as the cloud starts to collapse, the universe witnesses another important event. As the years have passed by, the expansion of spacetime has been gradually winding down, losing its tug-of-war with all the matter in the universe. A threshold is crossed. The matter is now sufficiently dilute that the effects of dark energy, the unseen vacuum energy of spacetime itself, can at last be revealed.

This is the dark energy that may have been left over from cosmic inflation. It has never gone away. It has been gently accelerating the expansion of spacetime all this time, but its effects have gone unnoticed until now. Matter has finally eased its foot off the brake.

THE ACCELERATING UNIVERSE AND THE RETURN OF THE COSMOLOGICAL CONSTANT

As I explained in Chapter 1, Einstein introduced the cosmological constant into his gravitational field equations for the universe because he thought the result should be static. Although others—such as Lemaître—argued in favour of an expanding universe, it was the astronomical observations of Hubble and Humason that eventually led Einstein to abandon his infamous 'fudge factor' in the early 1930s.

But there was still a problem.

It was understood that a flat spacetime could only be achieved if there was a fine balance between the expansion of spacetime and the amount of mass-energy in it. This balance is reflected in a density parameter, called omega (Ω). This is the ratio of the density of mass-energy to the critical value required for a flat universe. A closed universe has Ω greater than 1, an open universe has Ω less than 1 and a flat universe has Ω exactly equal to 1.

While cosmic inflation solved the flatness and horizon problems in principle, there was still the matter of accounting for all the mass-energy in the universe. Put simply, there wasn't enough. Even dark matter didn't help. In the early 1990s, the best estimate for Ω based on the observed (and implied) mass of the universe was of the order of 0.3. To keep track, I'll give this contribution the symbol Ω_M, where the subscript M stands for 'mass', both baryonic matter and dark matter. If this really is all there is (in other words, if $\Omega = \Omega_M$) the universe should be 'open'. With insufficient mass to halt the expansion, the universe would be destined to expand forever, and spacetime should be negatively curved.

Some astrophysicists began to mutter darkly that Einstein's fudge factor might after all have a place in the equations of the universe. They were beginning to appreciate that a mass density of 0.3 and a cosmological constant contributing a dark energy-density (which I'll call Ω_Λ) of 0.7 might be the only way to explain how Ω (now given by $\Omega_M + \Omega_\Lambda$) can be equal to 1.

The value of Ω determines the future evolution of the universe, as illustrated in Figure 42 for a number of different Friedmann-type models. In an expanding universe with no mass ($\Omega_M = 0$), there is nothing to change the expansion from its initial big bang rate, and so the average distance between points in space (where galaxies would be if they existed) increases linearly with time from the big bang. If the matter (including dark matter) is really all there is, then $\Omega_M = 0.3$, and we see small deviations from straight-line behaviour as the matter applies a gentle brake on the rate of expansion over the next 15 billion years. If we assume we've missed something and there really is much more matter in the universe than we've so far been able to identify, then we can be bold and set $\Omega_M = 1$. The brakes are applied a little harder.

In contrast, in a universe with $\Omega_M = 0.3$ and $\Omega_\Lambda = 0.7$, we obtain an upward curve in later years. Dark energy overcomes the decelerating effects of matter and starts to accelerate the expansion once more.

Of course, we can't see into the future. But notice how these different scenarios also predict different *histories* for the expansion of the universe. Universes with $\Omega_M = 0.3$ and $\Omega_M = 1$ predict that the universe should be younger than its 'Hubble

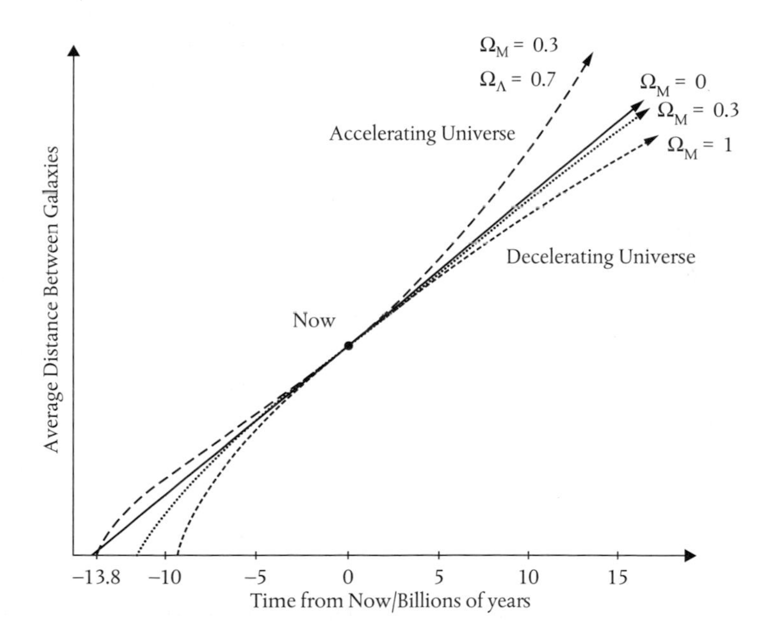

FIGURE 42 The future evolution of the universe depends on the density parameter Ω, which may be comprised of contributions from the density of matter (Ω_M) and the density of dark energy (Ω_Λ).

age', obtained by simple linear extrapolation backwards from the present. Coincidentally, $\Omega_M = 0.3$ and $\Omega_\Lambda = 0.7$ is entirely consistent with the age of the universe as currently estimated, but notice how the average distances between galaxies (and hence the overall size of the universe) in its early history is *greater* in this scenario than a simple linear extrapolation backwards in time would predict.

How does this help? We can't see into the future but, just as surely, we can't look back into the past. Ah, but we can. Light from distant objects shows us how these objects were behaving at the time the light was emitted. For example, light from the Sun shows us how the Sun looked about eight minutes ago, which is the amount of time it takes for the light to travel 150 million kilometres to the Earth. Light from the Andromeda galaxy shows us how it was 2.5 million years ago. This is the 'look-back' effect.[5] If we could look back to galaxies that emitted their light 5, 10, or even 13 billion years ago, we could map the universe's expansion history.

But the most distant galaxies are also the faintest, making it very difficult to measure their distances (so giving us a measure of average distance) and their redshifts (giving us a measure of their ages).

Unless, of course, stars within these galaxies go supernova.

Astronomers classify supernovae based on the different chemical elements observed in their spectra. Supernovae involving stars that have exhausted their hydrogen are called Type I. They show no spectral lines attributable to hydrogen atoms. The Type I supernovae are further sub-categorized according to the presence of spectral lines from other atoms or ions. Type Ia supernovae exhibit a strong line due to ionized silicon.

This is the signature not of a core-collapse supernova, but of a supernova resulting from thermal runaway within a binary star system, in which two stars are gravitationally bound to each other. One of the stars progresses through its red-giant phase but leaves a remnant that is insufficiently massive to trigger carbon burning. As we learned in the previous chapter, the result is a white dwarf with a core of carbon and oxygen.

But this dwarf has a neighbour. As the second star matures to its red-giant phase, the white dwarf may drag matter from its neighbour's bloated outer layers, so building itself up once more. When the mass of the dwarf exceeds 1.4 $M_\odot$, carbon burning is triggered in its core.

This is all too much for the hapless dwarf. The energy from carbon burning is released too quickly and the reborn star is torn apart. The resulting shockwave can travel at speeds up to 20 000 kilometres per second.

This is a scenario with highly predictable consequences. The dwarf has a predictable chemical composition and cannot go supernova until it accumulates a sufficient (and precisely determined) mass that gets it across the threshold for carbon burning. Consequently, the light emitted in the explosion has a characteristic luminosity and 'light-curve', meaning that its intensity decays over time (typically a few days) in a way that is also entirely predictable. Type Ia supernovae therefore serve as secondary standard candles.

Finding a Type Ia supernova and determining its peak brightness provides a measure of the distance of its host galaxy (called the luminosity distance). Measurements of spectral lines can be used to determine its cosmological redshift, and hence the so-called cosmic scale factor—a measure of the relative size of the universe—at the time the light was emitted.[6]

To understand the implications of such cosmological measurements, we need to study the different curves shown in Figure 42. In a universe with a rate of expansion that is constant throughout its history, Hubble's law applies. This is the straight line in Figure 42. We know that for a supernova with a relatively modest redshift, we can use the simple set of algebraic formulae described in the last chapter to determine a 'redshift distance'. We would expect there to be close correspondence between this and the luminosity distance, determined from the peak brightness of the supernova.

But when we look at supernovae with much larger redshifts, we look back to much earlier moments in the universe's expansion history, and different scenarios suggest different, non-linear relationships between redshift and luminosity distance. The situation gets much more complicated.

So, here's what we do. We use the redshift to determine the cosmic scale factor at the time the light was emitted. For example, a redshift $z = 1$ suggests a scale factor of 0.5; the light was emitted when the universe was half its current size, or the average distance between galaxies was half what it is today. We can now use the luminosity distance to tell us how long the light from the supernova has been travelling, and therefore how long it has taken for this expansion to happen.

In a decelerating universe (the dotted lines in Figure 42), the expansion rate was *faster* in the past and so it takes a little less time for the universe to expand from half its current size than simple linear extrapolation would suggest. This means that the travel time of the light from the supernova with a redshift $z = 1$ is *shorter* than we would anticipate from Hubble's law. In other words, the luminosity distance is shorter, and the supernova appears a little *brighter*.

In an accelerating universe (dashed line), the expansion rate was *slower* in the past and it takes more time for the universe to expand from half its current size. The travel time of the light is *longer* than we would anticipate from Hubble's law. The luminosity distance is longer and the supernova appears *fainter*.

So it comes down to this. For a given redshift or cosmic scale factor, in a decelerating universe a Type Ia supernova will appear brighter; in an accelerating universe it will appear fainter than a simple linear extrapolation would allow.

In 1998, two independent groups of astronomers reported the results of measurements of the redshifts of distant galaxies lit up by Type Ia supernovae. These were the Supernova Cosmology Project (SCP), based at the Lawrence Berkeley National Laboratory near San Francisco, California, headed by American astrophysicist Saul Perlmutter, and the High-z (high-redshift) Supernova Search Team formed by US-born Australian–American Brian Schmidt and American Nicholas Suntzeff at the Cerro Tololo Inter-American Observatory in Chile.

The results reported by both groups suggested that, contrary to expectations that prevailed at the time,* we live in an accelerating universe. The High-z team claimed: 'The [luminosity] distances of the high-redshift [Type Ia supernovae] are, on average, 10–15% *farther* than expected in a low-mass density universe . . . without a cosmological constant.'[7] In other words, looking back in time shows galaxies with average distances that are *larger* than we would expect from a simple linear extrapolation or from universes with $\Omega_M = 0.3$ and $\Omega_M = 1$. This is the dashed line in Figure 42.

The SCP team's results were entirely consistent. 'Our teams, especially in the US, were known for sort of squabbling a bit,' Schmidt explained at a press conference some years later. 'The accelerating universe was the first thing that our teams ever agreed on.'[8]

Adam Reiss, a member of the High-z team, subsequently unearthed a very distant Type Ia supernova labelled 1997ff that had first been discovered in the 'Hubble Deep Field'† in 1997 and which had been serendipitously photographed by the Hubble Space Telescope during the commissioning of a sensitive infrared camera, allowing its light-curve to be measured. It has a redshift of 1.7,

* The High-z team was formed specifically to measure the *deceleration* of the universe using Type Ia supernovae. On this basis, I guess they failed . . .

† This is a small region of space in the constellation of Ursa Major that was photographed extensively over a 10-day period in 1995 by the Hubble Space Telescope. Aside from a couple of Milky Way stars in the foreground, most of the 3000 objects visible in this region are distant galaxies.

implying that its star exploded when the universe was just 37% of its current size and consistent with a distance of 11 billion light-years. But whereas older and older supernovae had appeared fainter and fainter than simple extrapolation would suggest, 1997ff bucked this trend and appeared a little *brighter*. It was the first glimpse of a supernova that had been triggered in a period when the expansion of the universe had been decelerating.

The result suggested that about five billion years ago, the expansion 'flipped' (look back again at the dashed curve in Figure 42). As expected, gravity had slowed the rate of expansion of the post-big-bang universe, until it reached a point at which it had begun to accelerate again.

In 2002, astronauts from the Space Shuttle *Columbia* installed the Advanced Camera for Surveys (ACS) on the Hubble Space Telescope. Reiss, now heading the Higher-z Supernova Search Team, used the ACS to observe a further 23 distant Type Ia supernovae. The results were unambiguous. The accelerating expansion has since been confirmed by other astronomical observations and through modelling the temperature variations in the cosmic background radiation, which requires a cosmological constant contributing a value $\Omega_\Lambda = 0.68$ (Chapter 3).

6

SOL

The Origin of the Sun and its Planets

It would be fair to say that astrophysics—the branch of astronomy concerned with the physics of the universe—is a relatively mature science. Many of the physical principles needed to interpret astronomical observations and measurements are reasonably well understood, underpinned by the tried, tested, and trusted structures of quantum theory and general relativity.

But this maturity does not make the science any less awesome. I personally find it utterly incredible that centuries of speculative philosophizing (and theologizing) about the origin of the universe, stars, galaxies, and planets like the Earth has in the 20th and 21st centuries given way to rational descriptions based on relatively simple physical principles that can be compared directly with what we see and what we can measure. You don't need to be able to follow the sometimes subtle scientific and complex mathematical arguments in order to appreciate the underlying concepts. I believe anyone can grasp the essence of the science that describes the origin of the place we call home.

The science may be mature but, make no mistake, there's still lots of debate and much that we don't fully understand. Seeking answers to the astrophysical questions that remain—for example about aspects of the origin and evolution of the Milky Way and the origin of the Sun and solar system—is typically the purpose of Earth-bound telescopic studies and current and future planned satellite missions.* The subject of this chapter is a very active area of research.

We know that the Sun and solar system were formed by the gravitational collapse of a giant molecular cloud of gas and dust grains. But how, precisely, did this happen? What caused it? How did it work? You might have noticed that in previous chapters I haven't been very specific about this process. I've been content to declare that the clouds are drawn together by gravity and clump and fragment to produce dense cores which may then collapse to form protostars. That might be okay if I say it quickly and move on, but now we want to understand the origin of our own Sun, and this vague, arm-waving description isn't going to be good enough.

So, what do we know?

NEITH, THE 'MOTHER CLOUD'

I scratched around in the literature and searched on the internet, but I couldn't find any evidence that the giant molecular cloud that gave birth to the Sun about 4.6 billion years ago has ever been accorded a special name.† For sure, the cloud no longer exists, which is why we can only speculate about how it might have behaved, but it seems rather remiss not to give it a name. Henceforth I propose to name it the *Neith Cloud*, for the Egyptian goddess and mother of Ra, the Sun-god.

We can find clues to the likely behaviour of the Neith Cloud by observing similar clouds in other parts of the Milky Way galaxy. I mentioned Barnard 68 in the previous chapter. The Taurus Molecular Cloud, which lies in the constellation Taurus just 430 light-years from Earth, is larger and a little closer to home.

The Taurus Molecular Cloud measures 90 by 70 light-years and has a total mass of about 15 000 $M_{\odot}$. It has an average density of about 300 molecules per cubic centimetre and an average temperature not much different from that of the cosmic background radiation.

Because of its proximity, it offers possibilities for the investigation of both the nature of its contents, and their distribution, through detailed spectroscopic

* For example, NASA's James Webb Space Telescope will be the successor to the Hubble and Spitzer Space Telescopes. It has been in planning since 1996 and is targeting a 2018 launch. See http://jwst.nasa.gov/

† Note that from this point on, I'll tend to refer to the timeline of unfolding events measured *backwards* from the present rather than forwards from the big bang.

surveys. Although molecular hydrogen is an important constituent, under the conditions prevailing in the cloud, H_2 offers no spectral signature that can be used to observe it. There isn't enough energy around to excite Hα emission, for example (the cloud doesn't glow with a reddish hue). Studies have instead focused on emission from carbon monoxide, the second most abundant molecule in the universe after H_2, which is a little less shy.

Carbon monoxide offers emission lines in the millimetre (infrared) region corresponding to the excitation and de-excitation of rotational motions of the molecule. Studies of these emission lines are used to infer the density of carbon monoxide and, by inference, the density of molecular hydrogen in different parts of the cloud.

So, to conventional pictures based on visible light we can add pictures derived from the emission of infrared light from carbon monoxide. Further pictures of the masking (called 'extinction') of starlight from behind the cloud caused by dust grains that get in the way help to map the contents of the cloud in some detail.

The cloud appears rather cold, dark, and quiescent, and at first sight appears an unlikely breeding ground for new stars. But within the cloud about 13 'cores' have been identified and examined. These are essentially regions where the cloud has thickened, with densities more like 10 000 molecules per cubic centimetre, dimensions measured in tenths of a light-year, and masses more like one solar mass.

The cores are strung out like beads along filaments of gas. These same filaments also contain most of the newly formed stars in the cloud. It is thought that they are formed along lines of force generated by the cloud's magnetic field, just as the iron filings line up around a bar magnet.

Each filament of gas appears to behave like some kind of stellar production line. The gas and dust first condense to form a core, which may then collapse under its own gravity to form a protostar, then a fully fledged star.

Those looking for something a little more spectacular might prefer the Eagle Nebula, a region of very active star formation about 6500 light-years away, in the constellation Serpens, near Sagittarius. The Eagle Nebula includes the famous 'Pillars of Creation', vast columns of interstellar gas and dust photographed by the Hubble Space Telescope in April 1995 (Figure 43).

The columns in this picture are several light-years in length, composed largely of molecular hydrogen and dust grains. The 'fingers' protruding from the columns are dense enough to satisfy the Jeans criteria and so collapse to

FIGURE 43 The 'Pillars of Creation', huge columns of molecular hydrogen and dust grains in the Eagle Nebula in the constellation Serpens. These columns are several light-years in length. This false-colour picture is a composite of 32 images taken using four different cameras mounted on the Hubble Space Telescope. The colours correspond to emission from different atomic and ionic constituents in the cloud: green for hydrogen atoms, red for singly charged sulphur (S^+) ions, and blue for oxygen (O^{2+}) ions.

form stars, but at the same time the cloud is being irradiated by intense ultraviolet light from large new-born stars. This light is illuminating the gas that is evaporating from the cloud, shaping the columns, and giving the picture its ethereal quality.

This seems rather more majestic, and perhaps more fitting as a model for the Neith Cloud. So, we imagine the cloud aggregating, gathering itself together, perhaps swept by stellar winds from neighbouring red giant stars, and stirred by the shockwaves from nearby supernovae. Small clouds merge to form larger clouds. The clouds billow.

The alignment of electron spins within the atoms, molecules, and charged ions that make up the cloud may generate a magnetic field. Filaments of gas form along the magnetic lines of force. Now pressed a little closer together, the atoms, molecules, and ions collide, exciting low-energy molecular vibrational and rotational motions (such as rotations of carbon monoxide). The excited molecules emit infrared and microwave radiation, which is lost to the cloud. This provides a mechanism for cooling the cloud even further.

Meanwhile, the dust grains present in the cloud shield the fragile molecules from ultraviolet light and prevent the gas from evaporating completely. The atoms and molecules huddle closer together.

We now wait patiently for the cloud to start to fragment to form denser cores that meet the Jeans criteria in terms of mass, temperature, and density. Debates are raging about whether this happened naturally, in a smooth evolution of the cloud, or suddenly, triggered by a compression shockwave from a nearby supernova explosion.

In other words, was the birth of the Sun natural or induced?

EVIDENCE FOR A SUPERNOVA

The Neith Cloud is long gone, so how do we obtain evidence for what triggered the collapse of its cores (and one core in particular)? The answer lies in the exotic nature of the *isotopes* produced during r-process nucleosynthesis.

As I mentioned in the previous chapter, the r-process involves hammering small nuclei with a relentless stream of neutrons, building up larger and larger nuclei, which frantically try to stabilize by turning some of the captured neutrons into protons through weak-force reactions. The most unstable of the isotopes so produced will decay very quickly. But some isotopes are more ambiguous. They are unstable and relatively short-lived, but decay a little more slowly. This stability is reflected in the half-life of the isotope, the time it takes for half of a given number of atoms to decay.

Here's an example. The isotope ^{26}Al, a radioactive isotope of aluminium with 13 protons and 13 neutrons, has a half-life of about 720 000 years. It decays to produce ^{26}Mg, with 12 protons and 14 neutrons, by emission of a W^+ particle (and thence a positron and an electron neutrino), so turning a proton into a neutron. It is formed in some abundance in r-process nucleosynthesis in a supernova.

Now, it makes sense that the elemental composition of the solar system that we observe today depends on the elemental composition of the Neith Cloud at the time that a core within the cloud collapsed to form the proto-Sun.

If this collapse had been triggered by a supernova, then the Neith Cloud might have become contaminated with materials ejected by the shockwave, including a fair proportion of ^{26}Al. But, of course, 4.6 billion years later, any ^{26}Al that might have been present at this time will have all but disappeared through radioactive decay.

Here's the thing. The magnesium 'daughter' isotope ^{26}Mg is stable. It doesn't undergo any further nuclear transformation. So, any ^{26}Al present in the Neith Cloud at the time of its collapse will leave a calling-card in the form of an *enrichment* of the ^{26}Mg isotope within the material in which it had become incorporated.

In 1974, a team of geologists from the California Institute of Technology led by Gerry Wasserburg* investigated the ratio of ^{26}Mg to the dominant isotope ^{24}Mg in mineral† grains from the Allende meteorite, which crashed to Earth on 8 February 1969 in the Mexican state of Chihuahua.‡ This is a so-called *carbonaceous chondrite*, essentially rock formed by the accretion of dust grains within the Neith Cloud as the solar system was forming. The meteorite has an abundance of features called calcium-aluminium-rich inclusions (abbreviated as CAIs), consisting of minerals such as spinel (which contains $MgAl_2O_4$), melilite (a mixture of $Ca_2MgSi_2O_7$ and $Ca_2Al_2SiO_7$), pyroxenes, and anorthite ($CaAl_2Si_2O_8$, a calcium-feldspar), among others. An example X-ray photograph of a CAI in another chondrite is shown in Figure 44.

Isotopic analysis can be used to date such meteorites, and thereby confirm that these are rocks that formed at the same time that the solar system was forming. For example, dating based on the analysis of radioactive lead isotopes suggests an age for such meteorites of 4,567.2 ±0.6 million years.

Astute readers will have noticed that there is no magnesium in anorthite, so how can it contain any ^{26}Mg? This is precisely the point. If the Neith Cloud had contained small amounts of ^{26}Al at the time of the formation of the Sun

* Wasserburg was one of the 'four horsemen', members of NASA's Lunar Sample Analysis Planning Team (LSAPT or 'less apt') responsible for advising NASA on the collection, recovery, and analysis of lunar rock samples from the Apollo missions. The other horsemen were Jim Arnold, a chemist at the University of California, San Diego, Paul Gast, a geologist at Columbia University in New York, and Bob Walker, a planetary scientist at Washington University.

† Quick definition: minerals are naturally occurring substances with defined chemical compositions. Rocks are typically composed of one or more minerals, and may have no minerals at all (coal is an example of a non-mineral sedimentary rock).

‡ Just to be clear, a *meteorite* is the debris that is left after impact, it's called a *meteoroid* before impact and, if it heats up and glows as it passes through the atmosphere, it's a *meteor*.

FIGURE 44 X-ray picture of a calcium-aluminium-rich inclusion in the Efremovka meteorite which, like Allende, is a carbonaceous chondrite. The picture shows different phases in the rock corresponding to different mineral structures, such as melilite (mel), pyroxene (px), and anorthite (an). The false colours reflect the elemental composition: calcium is green, aluminium is blue and magnesium is red. Adapted from: Ernst Zinner, *Science*, **300** (2003), pp. 265–7.

and solar system, these would logically have become incorporated in the dust grains, and hence in the CAIs of small rocks that were swirling around.

Over time, the ^{26}Al would have decayed to ^{26}Mg, trapping the latter in the mineral structure and so producing a higher-than-expected ^{26}Mg/^{24}Mg ratio. These small rocks have persisted, relatively unchanged through the subsequent evolution of the solar system, orbiting the Sun and occasionally getting caught in Earth's gravitational field. They fall to Earth as meteorites like Allende.

The geologists reasoned that this ratio would be highest for mineral grains with high ratios of aluminium to magnesium. In fact, the slope of a straight-line plot of the ratio of ^{26}Mg/^{24}Mg vs the ratio of aluminium to magnesium (^{27}Al/^{24}Mg) is related to the ratio ^{26}Mg/^{27}Al. Since any excess of ^{26}Mg must come from the decay of ^{26}Al, the slope can be used to infer the ratio ^{26}Al/^{27}Al (Figure 45), a measure of the excess of ^{26}Al over the stable ^{27}Al isotope.

Put it this way. If there is no excess ^{26}Al, then we would expect this plot to produce a horizontal line with a slope of zero and a value corresponding to the

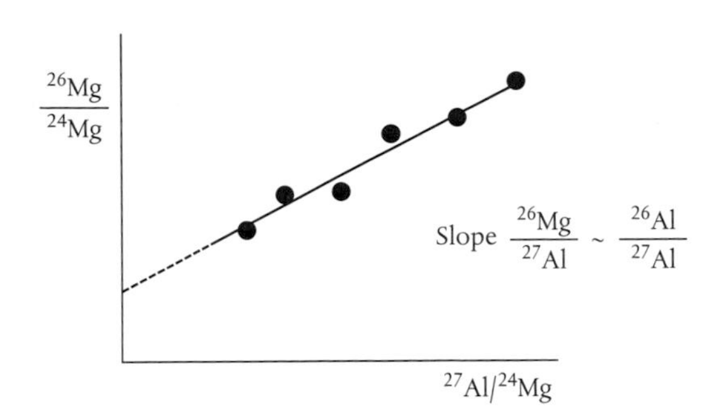

FIGURE 45 A plot of $^{26}Mg/^{24}Mg$ vs $^{27}Al/^{24}Mg$ for different minerals within a calcium-aluminium-rich inclusion can be used to infer an excess of the short-lived isotope ^{26}Al in the early stages of the formation of the Sun and solar system. The slope of this plot is related to the ratio $^{26}Mg/^{27}Al$. Any excess ^{26}Mg must be derived from the decay of ^{26}Al, so the magnitude of the slope also indicates an excess of ^{26}Al vs the stable isotope ^{27}Al.

ratio of naturally occurring $^{26}Al/^{27}Al$. A positive slope indicates an excess of the ^{26}Al isotope and provides a clue to the events unfolding during the formation of the solar system.

The results did indeed produce a positive slope, suggesting an overabundance of ^{26}Al, much more than could be reasonably expected from sources that do not involve a supernova. Subsequent measurements have served to demonstrate the importance of isotopes from supernova-derived r-process nucleosynthesis (such as ^{41}Ca and ^{60}Fe) in the dust grains of the early solar system.

The overabundance of ^{60}Fe is particularly telling, and has been called a 'smoking gun' in favour of a supernova trigger.[1] The ^{60}Fe isotope decays to ^{60}Ni with a half-life of 2.6 million years. Studies of the ratio $^{60}Ni/^{58}Ni$ vs $^{56}Fe/^{58}Ni$ in mineral samples from the Semarkona and Chervony Kut meteorites again show a positive slope, indicating an excess of ^{60}Ni and, by inference, an excess of ^{60}Fe.[2]

A recent (2011) study suggests that the ^{26}Al isotope would have needed to have become thoroughly mixed with the gas and dust present in the Neith Cloud within about 20 000 years. Of all the different explanations for how this might have been achieved, a supernova is the most promising. Computer simulations indicate that it would take just 18 000 years to enrich a 10 $M_{\odot}$ cloud core with ^{26}Al transported by the shockwave of a supernova explosion just 16 light-years distant.[3]

Like the Neith Cloud 4.6 billion years ago, the 'Pillars of Creation' in the Eagle Nebula are pregnant with potential stars. The clouds are also vulnerable to the shockwaves from nearby supernova explosions. And, indeed, observations by the Spitzer Space Telescope suggest that a supernova capable of destroying the cloud (and also triggering star formation) has already occurred. A rapidly expanding bubble of hot gas and dust is on its way towards the columns pictured in Figure 43. We are likely to be able to watch the drama unfold when the shockwave hits the cloud in about a thousand years or so.

Of course, as the Eagle Nebula is 6500 light-years away we would be observing events that have already happened. The 'Pillars of Creation' may already be long gone.

THE NEBULAR HYPOTHESIS

The notion that the solar system was somehow formed from a cloud of gas and dust—a nebula—has a long, rather tortuous but nevertheless illustrious history. The Swedish natural philosopher Emanuel Swedenborg is credited as one of the first to articulate this idea, in a book called *The Principia* which he published in 1734.

Swedenborg was a contemporary of Newton and heavily influenced by the ideas of French philosopher René Descartes, whom we will meet again in Chapter 12. There is much in the language of Swedenborg's description that we would struggle to recognize today, but he reasoned that in the early stages of the evolution of the solar system, material particles (which he called 'finites') trapped in a vortex around the Sun formed into a 'zodiacal circle', which 'surrounded the Sun like a belt or broad circle'.[4] The belt stretched and eventually broke into pieces, forming the planets.

Swedenborg went on to make some significant contributions to 18th-century physiology before experiencing a divine revelation in a dream in 1744. This encouraged him to turn his attentions to aspects of Christian religious mysticism, and it is his work in this area for which he is more widely known today.

The German philosopher Immanuel Kant was familiar with Swedenborg's ideas, and in the *Universal Natural History and Theory of the Heavens*, published in 1755, he elaborated the nebular hypothesis. He reasoned that as the nebula slowly rotated, it would have flattened into a disk. The Sun then formed in the centre and the planets formed in the plane of the disk, all rotating in the same direction.

The French mathematician and astronomer Pierre-Simon, marquis de Laplace independently developed a similar model which he described in his book *Exposition du Systeme du Monde*, published in 1796. He went a little further than Kant. According to Laplace, the nebula began as a near-spherical slowly rotating cloud—Figure 46(a). As the cloud collapsed, its speed of rotation increased, just as an ice skater spins faster as they draw their arms in closer to their body. The increased rotation caused the nebula to flatten along the equator—46(b) and (c). As the material in the flattened disk slowly gathered together (or *accreted*), it formed annular rings moving in orbit around the Sun, much like the rings around Saturn—46(d). Each ring then condensed to form a planet—46(e).

The nebular hypothesis dominated thinking about the formation and early evolution of the solar system for much of the 19th century. But there were problems with the model that couldn't be resolved at the time. The most important of these concerned the distribution of angular momentum.

We calculate the angular momentum of a rotating object by multiplying together the object's moment of inertia—a measure of its resistance to

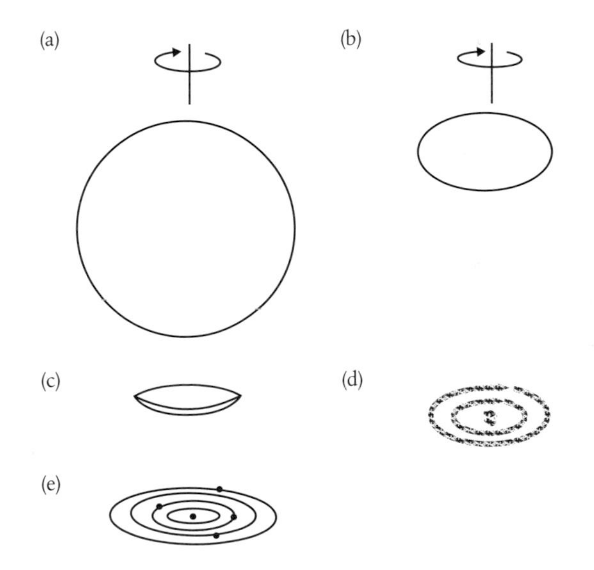

FIGURE 46 In the version of the nebular hypothesis due to Kant and Laplace, the formation of the solar system is initiated by the collapse of a near-spherical rotating cloud of gas and dust (a). As the cloud collapses it spins faster and starts to flatten (b), eventually forming a lenticular shape, like a lens (c). As the core collapses further to form the Sun, material in the flattened disk gathers together to form annular rings (d). These rings eventually condense to form the planets (e). Adapted from M. M. Woolfson, *Quarterly Journal of the Royal Astronomical Society*, **34** (1993), p. 2.

rotational acceleration—and its angular speed around the centre of rotation. The moment of inertia is proportional to the object's mass. The greater the mass, the greater the moment of inertia and thus the greater the angular momentum for a given speed of rotation.

The angular momentum of the Sun is associated with rotation around its spin axis. The planets have two sources of angular momentum, one associated with their orbital motion around the Sun and the second associated with their individual spins.* In what follows we will consider only the spin angular momentum of the Sun and the orbital angular momentum of the planets.

Now, much of the mass of the solar system is contained in the Sun which, as we know, is about 1.99×10^{30} kilograms, or $M_\odot$. The total mass of the solar system is 1.0014 $M_\odot$, indicating that the Sun accounts for more than 99.9% of the total mass. On this basis, we would surely anticipate that at least 99.9% of the solar system's total angular momentum should be found to reside in the Sun's spin.

And yet, when we calculate the actual angular momentum of the Sun and planets, we find precisely the opposite. The angular momentum associated with the Sun's spin can be approximated based on its mass, its equatorial radius (696,342 kilometres) and its sidereal rotation period (25.05 days). This gives a spin angular momentum of about 0.11×10^{43} kilogram-metres-squared per second.[5] The orbital angular momentum of the eight planets† of the solar system totals about 3.14×10^{43} kilogram-metres-squared per second. In other words, the Sun's spin accounts for only 3% of the solar system's total angular momentum. In fact, much of this angular momentum resides in the orbital motions of Jupiter (about 58%) and Saturn (25%), with small contributions from Uranus (5%) and Neptune (8%).

The simple fact is that the Sun is spinning way too slowly. For the Sun to exhibit a share of angular momentum consistent with its share of the mass, it would have to spin around on its axis much faster, rotating once every 77 seconds or so.[5] I'm guessing that a star spinning so rapidly wouldn't be very conducive to the formation of a stable planetary system.

There were other problems. A contracting core within the Neith Cloud may develop a magnetic field which could be expected to exert a force on the

* The Earth orbits the Sun with an average orbital period of 365.26 days, and spins on its axis with a period (called a sidereal rotation period) of 23 hours 56 minutes. The latter is measured by reference to background stars, so it differs from the 'solar day', which requires an extra 4 minutes of rotation to make up for the fact that the Earth is also moving along in its orbit around the Sun as it spins.

† Pluto was demoted by the International Astronomical Union to the status of a dwarf planet in 2006, although there is some talk that it might yet be reinstated.

charged particles in the surrounding cloud and inhibit further contraction. There was also the issue with residual hydrogen and helium gas in the spinning disk. Although the planets could be considered to be formed through the accretion of dust grains and ice particles, the content of this disk is still expected to consist largely of hydrogen and helium. These substances are too light and volatile to be captured in great quantities in planetary interiors and atmospheres, especially when these planets are formed close to the Sun. And yet there is only empty space between the planets. Where had all the interplanetary gas gone?

These problems seemed insurmountable, and the nebular hypothesis fell out of favour. Astrophysicists sought other explanations in which the Sun and planets somehow formed separately. In one theory the already formed Sun was thought to have captured the gas and dust necessary to form the planets from a neighbouring protostar that had wandered into its gravitational field.

But these alternative theories came with problems of their own. In the meantime a body of evidence was building to suggest that the Sun and planets had indeed formed at around the same time, 4.6 billion years ago. Studies of chondritic meteorites revealed the presence of round grains called *chondrules*, composed primarily of silicate minerals such as olivine and pyroxene, anything from a few microns to over a centimetre in diameter. The chondrules are formed by the rapid heating and melting of silicate grains at temperatures around 1000 kelvin. The resulting liquid silicate drops then became accreted with other dust grains and incorporated into small rocks. This is hard to explain if the (cold) dust grains are assumed to have accreted around an already formed Sun, but is exactly what might be expected to happen in a hot dense nebula.

There was also the undeniable regularity of the solar system. The planets orbit the Sun all in the same plane. They all rotate around the Sun in the same direction, which is the direction in which the Sun itself is rotating. This can't be coincidence, and suggests an intimate set of mechanical relationships between the Sun and planets that are difficult to establish in other theories.

Astrophysicists began to realize that the advantages of the nebular hypothesis outweighed its disadvantages. In the late 1960s, Russian astronomer Victor Safronov published what was to become a highly influential monograph, titled *Evolution of the Protoplanetary Cloud and Formation of the Earth and Planets*. This was translated into English in 1972. Astrophysicists began to find work-arounds for some of the more stubborn problems with the nebular theory.

The result is the Solar Nebula Disk Model, which incorporates modern theories of the physics of the formation and evolution of stars and planets, informed by a number of important astronomical observations.

YOUNG STELLAR OBJECTS

Irrespective of whether core collapse within the Neith Cloud was triggered by a nearby supernova or not, the simple fact is that gravity now has the upper hand. The Jeans criteria of mass, density and temperature are all met, and the core starts to implode. The core is essentially in free fall at this point. The speed of the collapse depends on the density of the core. A core with a density of a million molecules per cubic centimetre will collapse in about a million years.

As the material in the core is forced closer together, infrared radiation that would have previously escaped into space becomes trapped. The core heats up. We can now legitimately refer to it as a protostar.

We are obviously interested in one particular protostar—which I will call the proto-Sun—which is now forming within the Neith Cloud. This starts out with some initial rotation, which picks up speed as the collapse progresses. Some of the angular momentum is shuffled into the flattening disk of gas and dust that surrounds it, but any further collapse will require that the proto-Sun first sheds some of its growing angular momentum, to prevent it spinning too fast. It also needs to rid itself of its magnetic field.

One way of dumping a lot of angular momentum involves splitting the collapsing proto-Sun into two smaller protostars which then rotate around their mutual centre of mass. These may persist to form a binary star system or may disengage as they evolve and go their separate ways. Our solar system has not formed around a binary star, so we can be pretty certain that this was not the trick played by the proto-Sun.

Radioastronomy, infrared and microwave spectral studies show that many young stellar objects (YSOs) have found another trick. As they rotate they eject jets of hot gas from both poles. These *bipolar outflows* act like a kind of safety valve. As material is drawn into the collapsing protostar from the surrounding cloud and flattening disk around the equator, the build-up of angular momentum is relieved by ejecting rotating material from the poles.

Such protostars are luminous but not yet visible. They are still shrouded by outer clouds of gas and dust. We observe them largely by the long-wavelength radiation that is re-emitted by the gas and dust in the clouds. They are called YSOs of Spectral Class 0 (Figure 47).

We know that the ejected gas must be rotating, as it must carry angular momentum away from the protostar. This gas can be observed. T Tauri is a very young variable star about 460 light years from Earth in the constellation Taurus, first identified in 1852. It is only a million years old and has a highly erratic and unpredictable luminosity. It serves as a prototype for young stars in the mass range 0.2–2.0 $M_{\odot}$ that have not yet triggered nuclear fusion reactions in their cores and settled onto the main sequence.

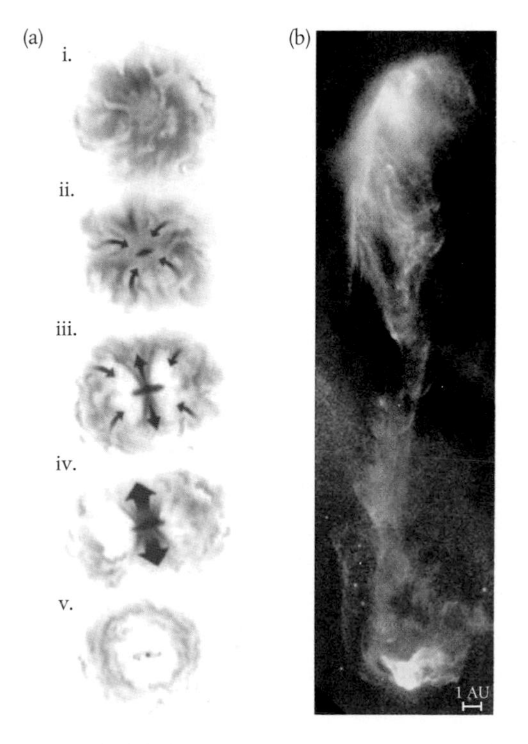

FIGURE 47 The diagram on the left shows the various stages in proto-stellar evolution, starting with a giant molecular cloud (i). The collapse of gas and dust within the cloud forms a rotating core (ii). As the core contracts it reduces its angular momentum by emitting jets of gas from its poles, called bipolar outflows (iii). The ejected gas forms an elongated nebula, called a Herbig–Haro Object, such as HH47 shown on the right in a photograph taken by the Hubble Space Telescope (the scale bar to the bottom right indicates 1 astronomical unit (AU), equal to the average distance from the Earth to the Sun, about 150 million kilometres). As the protostar matures to the T Tauri phase, convection currents produce strong winds which clear away volatile material from the protostar's accretion disk and outer cloud (iv). The protostar becomes visible, and follows the Hayashi track on its way to becoming a main sequence star (v). The diagram on the left is adapted from William H. Waller, *The Milky Way: An Insider's Guide*, Princeton University Press, 2013.

Astronomical observations of T Tauri in the 19th century revealed a nearby nebula. In the 1940s, American astronomer George Herbig and Mexican astronomer Guillermo Haro independently identified several such nebulae, which became known as Herbig–Haro Objects. An example is included in Figure 47. These are now understood to be the result of bipolar outflows from YSOs.

This would appear to be the mechanism by which the proto-Sun offloaded much of its angular momentum.

What of the proto-Sun's magnetic field? Astrophysicists have reached for a mechanism developed in plasma physics called *ambipolar diffusion*. The material being drawn into the collapsing proto-Sun consists mostly of neutral atoms and molecules but about one in every million entities is a charged ion, either a proton ($^{1}H^{+}$) or a free electron. As the magnetic field strength grows around the collapsing proto-Sun, these charged species become trapped in the magnetic field and resist the tug of gravity drawing them into the proto-Sun. Collisions between the ions and neutral entities can indirectly couple the latter to the magnetic field, which may then act as a brake on the collapse of all such material.

But if the density of ions in the cloud is low (as it is in this case), collisions between the ions and neutral entities are insufficient to couple the latter to the magnetic field in significant quantities. The neutral particles remain free to diffuse through the field, reducing the flow of magnetic force in the cloud. They drift on past the ions, and the collapse continues.

The proto-Sun is now about 10 000 years old. But it and other young protostars forming in the Neith Cloud are not all standing still. Supersonic turbulence within the cloud can disperse the outer clouds surrounding the protostars, thus inhibiting further accretion of material and delaying the collapse. It can also encourage collapse by driving more material in the direction of a hungry protostar. Turbulence can also bring protostars into proximity and they may then compete with each other for gas and dust from their accretion disks and outer clouds. Some protostars may be pushed out of the larger cloud completely and left to fend for themselves.

Eventually, the proto-Sun gathers sufficient material, and sheds enough angular momentum and magnetic field strength to settle into a more stable period in its evolution. Its outer cloud has now become thinner and more transparent, allowing the proto-Sun and its flattened accretion disk to become visible, primarily at shorter, near-infrared wavelengths. It has thus become a YSO of Spectral Class I (Figure 47). It is now about 100 000 years old.

ON THE HAYASHI–HENYEY TRAIL

The proto-Sun is shining, but this is not yet starlight. Its nuclear fires have yet to be ignited. Its luminosity is instead the result of a conversion of gravitational energy released by the accumulation and concentration of in-falling material into heat energy. The proto-Sun and its accretion disk heat up, and some of this energy is emitted in the form of infrared radiation.

The proto-Sun continues to evolve, accreting more material from its disk and from its outer cloud and contracting further. It convulses, as convection currents dredge hotter material from its centre to cooler regions near its surface. These convulsions produce strong winds which start to clear away the lighter gases in the accretion disk and any remaining material in the outer cloud.

As its core temperature rises to a few million kelvin, fusion of deuterium nuclei may occur. This will release some energy but the amount of deuterium in the core is small. The energy released is insufficient to hold the proto-Sun up against further gravitational contraction. This continues until the proto-Sun reaches its millionth birthday. It is no longer obscured by gas and dust swirling around it, and it is hot and active enough to emit visible light and X-rays. It is now a YSO of Spectral Class II, or a T Tauri star (Figure 47), with properties and behaviour characteristic of the class of young stars for which T Tauri is the archetype.

For a given mass, T Tauri stars are cooler and yet more luminous than their main sequence counterparts. On a Hertzsprung–Russell diagram, they lie above and to the right of the position on the main sequence that could be expected for a star of the same mass.* The proto-Sun's position on the surface temperature scale is determined by the interplay between its opacity and convection, as I discussed in Chapter 4. Cooler stars tend to be more opaque, trapping more radiation, producing strong temperature differentials between the core and the surface which drive turbulent—and destabilizing—convection currents. The proto-Sun remains stable by remaining hot.

How hot? In 1961, Japanese astrophysicist Chushiro Hayashi determined that there is a surface temperature boundary (called the Hayashi boundary), below which a young star will be unstable to its own convection currents. For

* Remember that on a Hertzsprung–Russell diagram surface temperatures are plotted with temperatures *decreasing* from left to right. See Figure 27.

the proto-Sun, this temperature is around 4500 kelvin. Lower temperatures (which lie to the right on the Hertzsprung–Russell diagram) are essentially 'forbidden'.

But the luminosity of a star depends on both its surface temperature *and* its radius,[6] and the proto-Sun is still contracting. As it contracts, the core temperature increases and it becomes less and less opaque. Temperature gradients become shallower and convection eases. The surface temperature changes very little. The end result is that the luminosity of the young star falls simply because it is shrinking.

As it evolves further, the proto-Sun fades but maintains an almost constant surface temperature. This means that on a Hertzsprung–Russell diagram it follows a near-vertical 'Hayashi track', heading downwards towards the main sequence line (Figure 48).

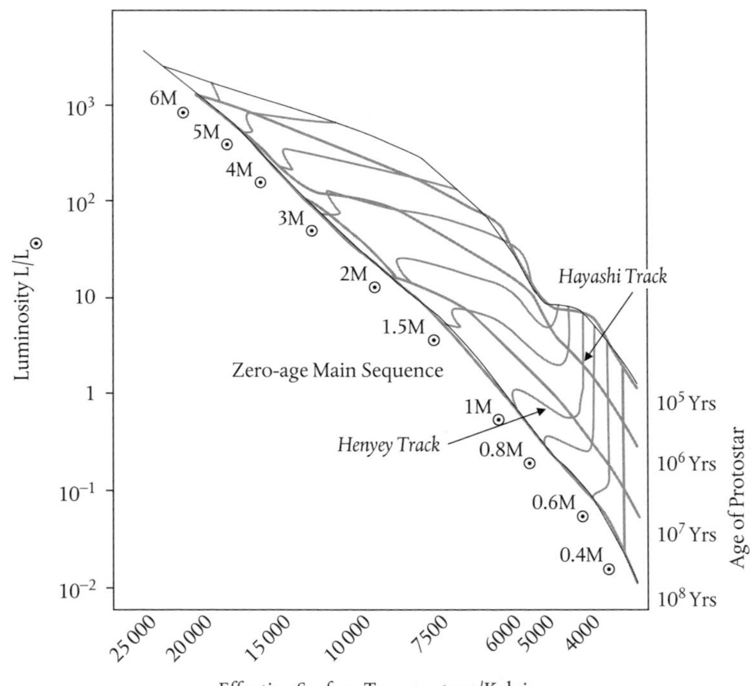

FIGURE 48 The evolution of young stellar objects is determined by the physical relationships established between mass, density, core temperature, opacity, temperature differentials, and convection. This figure shows Hayashi–Henyey trails for a range of stellar masses. The larger the mass, the faster the protostar evolves.

The proto-Sun is now ten million years old. Much of its accretion disk (which we should now refer to as a *protoplanetary disk*) is clear of gas and dust and much of the infrared radiation is now gone. It is still a T Tauri star, now classed as a YSO of Spectral Class III.

The proto-Sun is in the final stages of its evolution towards thermonuclear fusion, the zero-age main sequence (the point on the main sequence at which stars begin their lives) and stardom. But we must exercise some patience, as this final step takes many tens of millions of years.

The contraction now slows, although it does not stop. Convection has all but ceased. This means that we've come to the end of the Hayashi track, and the luminosity no longer falls. Instead, the subtle interplay of pressure, density, opacity and core temperature now raise the surface temperature. The luminosity increases slightly (roughly depending on the surface temperature raised to the power $^4/_5$), even as the radius continues to contract.

The proto-Sun moves leftwards and slightly upwards on the Hertzsprung–Russell diagram (Figure 48). This is the Henyey track, named for American astronomer Louis Henyey who published the results of calculations on the evolutionary tracks of pre-main sequence stars in 1955.

So, here we are. About 100 million years have elapsed since the Neith Cloud fragmented and cores started to form. As it contracted the young proto-Sun shed much of its gathering angular momentum through bipolar outflows, reducing the speed of rotation around its spin axis. It managed its magnetic field through ambipolar diffusion. It became a T Tauri star, following first the Hayashi and then the Henyey tracks towards the main sequence, fading and then brightening a little as its surface temperature increased.

The Sun is now ready to be born.

MEANWHILE, ON THE PROTOPLANETARY DISK

Before we take the final step, let's wind back and look at the events unfolding in the accretion disk, the flat pancake of gas and dust swirling around the proto-Sun's equator. It's called an accretion disk because material within it is being sucked into—accreted by—the proto-Sun sitting at its centre. But the term accretion is also applied to the process by which small dust grains and chips of ice in the disk collide and stick together, over time building up larger and larger grains.

The disk accounts for only about 0.2% of the mass of the collapsing cloud, but it is fundamentally important to us, and there's a lot going on in it. About 98% of the disk is made up of hydrogen and helium gas with heavy dust grains (containing metals and silicates) and ices (made of lighter molecules of water, ammonia, methane, etc.) accounting for the remaining 2%. The radius of the disk stretches some 15 billion kilometres from the proto-Sun at the centre.

As the proto-Sun heats up, it warms the disk from the inside out. The disk material is also packed more tightly closer to the proto-Sun and frequent collisions between the particles serve to increase their kinetic energy and raise the temperature. Some of this energy is released by emission of infrared radiation, but nevertheless a substantial temperature gradient builds up, with high temperatures in the inner parts of the disk and much lower temperatures in the outer, more distant parts.

At distances around 150 million kilometres from the centre, about the current distance from the Sun to the Earth, temperatures in this early stage of the disk's evolution would have been around 1000 kelvin. Further out, at distances of 400–700 million kilometres, the temperature would have dropped to something like 100–200 kelvin.

This temperature gradient will have a profound effect on the subsequent evolution of material in the disk. Metallic iron melts at a temperature a little above 1800 kelvin. Silicon dioxide (silica, SiO_2) melts at temperatures of the order of 1800–2000 kelvin. Forsterite (Mg_2SiO_4), the magnesium-rich end-member of the olivine family of minerals, melts above 2000 kelvin. Despite the rising temperatures in the inner part of the disk, the metal-rich grains and rocky silicate grains remain firmly solid.

This does not hold true for the particles of ice. We know that at normal atmospheric pressure water ice melts at around 273 kelvin and boils at 373 kelvin. Solid ammonia melts at about 195 kelvin and boils at about 240 kelvin. Solid methane melts at 91 kelvin and boils at 113 kelvin. Solid ice particles in the inner part of the disk are therefore vaporized, turned into gas and driven outwards by turbulent winds from the as yet unsettled proto-Sun.

In a sense, the material in the disk is *distilled*. The heavier, less volatile material is stubborn. It remains solid and stays put. Some of the lighter material migrates outwards until it encounters a region cool enough for it to condense, first to liquid and then back to solid ice once more. Astrophysicists have identified a 'frost line' (a term borrowed from soil science), about 400–700 million kilometres from the Sun. Beyond the frost line, temperatures are low enough for the volatile materials to re-form ice particles. If the disk starts out with a uniform mix of materials, then over time the concentration of heavier particles increases in the inner part of the disk as the lighter material evaporates and condenses in the outer part.

Two distinct regions of the disk are formed: an inner 'rocky' region inside the frost line and an outer 'icy' region. Of course, this is not a complete separation of these materials based on their volatility. If the disk starts with a uniform distribution of material, then we can expect to find heavier metal-containing grains and silicate grains in the cooler, outer part of the disk, too. In fact, as the water, ammonia, methane, and other volatile materials condense, they will likely do so around the heavy, solid grains and ice particles that persist here where the temperature has remained low.

PLANETESIMALS AND PROTOPLANETS

The proto-Sun becomes a T Tauri star about a million years after the initial core collapse and sets off on its journey along the Hayashi track. The dust grains and ice chips trundling around in the accretion disk collide and sometimes stick together. Sometimes they collide and fall apart.

But the grains are not moving in empty space. The disk is still filled with a fairly dense gas of hydrogen and helium. Grains that have now grown to more than a few millimetres, or centimetres, or even metres in diameter are pulled more strongly by the proto-Sun's gravity than the gas through which they are moving. They therefore experience a headwind. This slows them down and, now unable to resist the pull of gravity towards the centre, they spiral inwards.

Things are a little different beyond the frost line. Free water molecules that have been liberated from ice grains accumulate, and these interact with hydrogen and helium atoms, causing the pressure of the gas to fall. The pressure of a gas is related to the speeds with which its individual atoms or molecules are moving (in random directions). So, as the hydrogen and helium atoms slow down, they become more susceptible to the pull of the proto-Sun's gravity, and orbit faster. Small particles moving through the more rapidly orbiting gas no longer experience a headwind. Instead, they feel a tailwind, which boosts their speed and makes them less likely to be pulled towards the centre.

The upshot of all this is that dust grains and ice particles in the cooler, outer part of the disk are pulled towards the frost line and accumulate there.

The accretion of material continues and eventually objects with diameters of ten kilometres or so are formed. These are called *planetesimals*, literally small parts of what will go on to become planets. This happens in both the inner and outer parts of the disk but, because of the accumulation of material at the frost line, the process is accelerated here.

This is a process that's rather hard to envisage. It is our experience that throwing small stones together doesn't make them stick. Instead we have to imagine these small grains and ice particles floating alongside one another in the accretion disk, occasionally gently nudging or pressing together. Metal-containing grains may be drawn together by their magnetic fields. If silicate grains bumping along develop a small electrostatic charge then this may be enough to draw them together. In his monograph, Safronov modelled this process of accretion using coagulation theory, as applied to the study of colloids.

The formation of planetesimals signals a subtle change in the physics. Objects of this size are able to exert gravitational effects of their own.

The planetesimals within an orbital zone now compete with each other. This is a 'last rat standing' competition.* Neighbouring planetesimals collide and either merge or smash apart. The larger the planetesimal becomes, the greater its gravity and the greater its ability to pull smaller planetesimals towards it.

In the Kant–Laplace nebular hypothesis, material in the accretion disk was thought simply to accumulate in bands orbiting around the proto-Sun and these bands eventually came together to form planets. This sounds all rather orderly and directed, with a perfectly formed planetary system being the only logical outcome of the physics. In truth, the jostling now going on among planetesimals is far from orderly. It is chaotic, and there can be no guarantees about the outcome. The jostling continues as the larger planetesimals grow into *protoplanets* of the size of Earth's Moon, which has a diameter of about 3500 kilometres.

The disk is now officially a protoplanetary disk.

---

* Readers who have seen the 2012 Bond movie *Skyfall* will recall the last rat standing story, as told in a memorable scene by Bond's enemy Raoul Silva, played by Javier Bardem.

THE RISE OF THE GAS AND ICE GIANTS

Events are unfolding rapidly near the frost line. The accumulation and concentration of material has accelerated the formation of planetesimals and produced a protoplanet about the size of the Earth (with a diameter of about 12 000 kilometres). This now draws an envelope of hydrogen and helium gas around itself from the disk. In other words, it gains an atmosphere. To do this successfully, the gas must be cool enough to become trapped in the protoplanet's gravitational field. If the gas is too hot its atoms and molecules will have too much kinetic energy, and too much speed, and will escape.

Once again, the relationships between the opacity of the gas and temperature gradients between core and surface are critical. But as the protoplanet continues to accumulate its atmosphere, it becomes even more massive and the strength of its mutual gravitational attraction with the proto-Sun increases. The protoplanet may migrate from its original orbit, dragged inwards by the proto-Sun until it gets even closer to the frost line, where the tailwind helps to stabilize it, accelerating it to an orbital speed which allows it to resist any further pull.

The protoplanet continues to grow its atmosphere, clearing the gas from within its orbital path and creating an empty, near-circular band in the disk. The protoplanet is now large enough to ensure that any gas flowing into this vacuum is quickly swept up. It is well on its way to becoming the planet Jupiter. It is no coincidence that the largest planet in the solar system is formed close to the frost line, with an orbital radius today of about 780 million kilometres.

Jupiter is thought to have a rocky core (although its actual composition is unknown) surrounded by an atmosphere consisting of 75% hydrogen and 24% helium by mass. It has a total mass about 318 times that of the Earth but is much less dense.

The emergence of the first gas giant planet facilitates the formation of others. As it is forming, Jupiter creates an empty band in the disk around the proto-Sun. This acts as a barrier to material in the outer disk that is being pulled inwards by the proto-Sun's gravity, causing a build-up much like the build-up around the frost line. What stops this material from simply being drawn into the empty band and absorbed into Jupiter? Once again, it is the difference in gas pressures. The pressure of gas out beyond the band creates a tailwind, helping any planetesimals and protoplanets to resist the pull of gravity and giving them time to accrete.

But, as in life, there is a prize for being first. Protoplanets forming beyond Jupiter find that there is less gas available to accumulate in their atmospheres. Winds from the proto-Sun may have also depleted the amount of gas. Consequently Saturn is much smaller than Jupiter, with a mass 95 times that of the Earth. It is also thought to have a rocky core and has an atmosphere of mostly hydrogen, with a proportion of helium considerably smaller than that of Jupiter.

Although Uranus and Neptune now lie in orbits far beyond Saturn, astrophysicists do not believe they could have formed there. They are too big. It is thought that there would have been insufficient density of material available in the protoplanetary disk at their present distances for these planets to have formed quickly enough.

Instead, it seems likely that Uranus and Neptune formed much closer to Saturn and in a different order, with Neptune closer and Uranus further out. The density of planetesimals is thought to have been much greater here, and protoplanets of 15 and 17 Earth masses would have been relatively easy to assemble. But much of the gas in this neighbourhood has already been captured by Jupiter and Saturn, leaving little for these latecomers to acquire.

Consequently, these planets are not gas giants. They are ice giants.

Beyond Uranus lies an orbiting band of planetesimals and protoplanets too thinly spread to grow into fully fledged planets. This band of rocks and icy clumps will undergo some dramatic changes, as we will soon see, but we will eventually come to know it as the Kuiper belt, named for Dutch–American astronomer Gerard Kuiper. This was discovered in 1992 and today lies about 7500 kilometres from the Sun. Pluto, now classified as a dwarf planet, is the largest Kuiper belt object, with a highly eccentric orbit that brings it in as close as 4500 kilometres from the Sun.

BUILDING THE TERRESTRIAL PLANETS

The planets beyond the frost line have had something of a head start. After ten million years, as the proto-Sun comes to the end of the Hayashi track and switches to the Henyey track, the inner disk is a chaotic mess with something of the order of 100 rocky protoplanets with sizes ranging from Earth's Moon to the planet Mars, swimming in a thin soup of planetesimals.

To get any further, these protoplanets must collide and merge. Much of the gas is gone, drawn into the proto-Sun or blown away by intense winds, but the planetesimals may serve to perturb the orbits of the protoplanets sufficiently to set them on collision course. It is also possible that the gravitational effects of the newly formed planet Jupiter may have had a role in diverting the protoplanets into the path of further material that they could accrete.

One consequence of the early presence of Jupiter seems fairly clear. Relationships between the orbits of Jupiter and neighbouring planetesimals or protoplanets result in the establishment of *orbital resonances*. Think of this like the mechanics of pushing a child on a swing. The parent times his or her push to coincide with the period of the swing, so delivering extra momentum at precisely the right moment to increase the swing speed and amplitude. Similarly, coincidences in the periods of orbiting bodies* can magnify the effects of gravity and destabilize one or other of the orbits.

This is not an equal contest. Caught in an orbital resonance with Jupiter, a nearby planetesimal or protoplanet gains energy and is accelerated. Now these bodies acquire too much energy, and collisions between them are largely destructive, rather than constructive. Many of these bodies are dragged from their orbits and hurled inwards. Most are hurled out of the solar system completely.

Trying to form a terrestrial planet just inside the frost line proves to be an exercise in futility. As a result, the orbiting band of rocks and icy lumps remains unaccreted. We know it today as the *asteroid belt*. Today, about half of the mass in this belt is contained in just four large asteroids—Ceres (actually a dwarf planet with a diameter of 950 kilometres), Vesta (525 kilometres), Pallas (544 kilometres), and Hygiea (an oblong-shaped asteroid about 500 kilometres in length). In all, there are millions of asteroids in the belt, although they are spread quite thinly. We have good reason to believe that the asteroid belt was once much more densely populated.

Unaided by the accelerating effects of material accumulated around the frost line, the inner terrestrial planets take a little longer to form—100 million years rather than the 10 million years required to form the gas and ice giants. Over this time, these emerging terrestrial planets 'clear the neighbourhood' within their own orbits, mopping up any remaining planetesimals as they become the dominant masses (the last rats standing) within their orbital zones.

---

* For example, in a two-body system suppose one body completes four orbits in the time it takes the second body to complete just one. These bodies then exhibit a 4:1 orbital resonance.

We will eventually recognize these emergent planets as Mercury, Venus, Earth, Theia, and Mars, although they still have some evolving to do yet, as we will see.*

IGNITION

The universe is now 9.2 billion years old. The first stars have come, and gone. The Milky Way galaxy has formed within a dark matter halo, with baryonic gas and dust from the debris of the first stars which has condensed to form second-generation Population II stars. Many of these have also passed, seeding the interstellar medium within the Milky Way with heavy elements synthesized in their interiors or in spectacular supernova explosions. The clouds that drift and once more start to gather are also sprinkled with tiny dust grains and ice particles.

Now, one hundred million years after its initial collapse to a core within the Neith Cloud, this proud parent is ready to give birth to a third-generation Population I star.

The proto-Sun reaches the end of the Henyey track. Its core temperature rises above ten million kelvin, and the proton-proton chain of nuclear reactions is initiated. It is tempting to imagine that some kind of switch is thrown, and light finally bursts forth into the darkness as the proto-Sun becomes the Sun. But this is misleading. The proto-Sun was already bright, shining with light released from the hot gas near its surface as it collapsed inwards.

All that has really changed is that the Sun now has its own internal source of energy, producing radiation that can hold it up against further gravitational contraction. In fact, as the Sun now takes its place on the zero-age main sequence, it actually dims slightly (Figure 48).

As the Sun settles, stellar winds blow away the last of the remaining gas, leaving only empty space between the orbiting planets. Planet Earth does not yet have a moon, and the outer planets are not where we find them today. We have all the furniture, but now we need to get it into the right place.

* Theia? Really? Be patient—the answer is coming early in the next chapter.

7

TERRA FIRMA

The Origin of a Habitable Earth

How much of this planet-building story is established fact and how much is scientific hypothesis? We must always be aware that we're reconstructing a history concerning events that happened many billions of years ago. We apply established scientific theories to deduce what *might* have happened based on how the planets, Sun, stars, and giant molecular clouds appear to us today. It is inevitable that the story will change in some essential ways as our understanding of the universe evolves.

For example, although the Solar Nebular Disk Model is widely accepted by astrophysicists, there are some in the community who are happy to challenge it, offering their own alternative ideas. Despite the logic that would appear to connect a spinning flat disk of gas and dust with planets that share the same orbital plane and direction of rotation, a recent (2012) study suggests that it is nevertheless possible to arrive at the same results starting with a cold, three-dimensional cloud and simply applying the laws of thermodynamics.[1]

Until April 1992, all our reasoning about the formation of planetary systems was limited to the one example with which we are familiar—our own solar system. But this is now acknowledged to be the date on which was made the first, subsequently confirmed, observation of extrasolar planets or *exoplanets*—bodies orbiting stars other than the Sun.* Since then, many more have been found. At the time of writing, observations of 1889 exoplanets had been recorded, in 1188 exoplanetary systems, of which 477 contain multiple exoplanets. Details are available in an online Extrasolar Planets Encyclopaedia, maintained by astronomers at the Paris Observatory.[2]

This catalogue is necessarily biased towards giant exoplanets similar in size to Jupiter, and in many cases much larger, as these tend to be easier to find.† But the Kepler Space Observatory was launched by NASA in March 2009 specifically to search for Earth-like exoplanets. Until it malfunctioned in May 2013, Kepler had recorded about 3500 candidate exoplanets, most of them so far unconfirmed.‡

About one per cent of the entries in the exoplanet catalogue have masses between 0.5 and 2 times the mass of the Earth (given the special symbol $M_{\oplus}$). Some of the first Earth-size exoplanets were reported by Kepler mission astronomers in December 2011. A system of five exoplanets was observed to be orbiting a star designated Kepler-20, somewhat smaller and a little cooler than the Sun and lying about 950 light-years from Earth. Of these, three are large, with masses of about 9, 16 and 20 $M_{\oplus}$ (compare these to Jupiter, which has a mass of about 318 $M_{\oplus}$). However, Kepler-20e and -20f have masses in the range 0.39–1.67 $M_{\oplus}$ and 0.66–3.04 $M_{\oplus}$, and radii of 0.87 $R_{\oplus}$ and 1.03 $R_{\oplus}$, respectively, where $R_{\oplus}$ is the radius of the Earth. At the time of the announcement, Kepler-20e was the first exoplanet with a radius smaller than that of the Earth.

These exoplanets are Earth-size but not Earth-like. Kepler-20e orbits its star every 6 days and has a surface temperature estimated to be in excess of 1000 kelvin. The orbital period of Kepler-20f is 19.6 days and it has a temperature of a little over 700 kelvin. These exoplanets are too close to their star and therefore too hot to support liquid water on or just beneath their surfaces. Consequently, these exoplanets are not orbiting in their star's 'habitable zone' (sometimes referred to as the 'Goldilocks zone'), the ideal balance between stellar

---

* The International Astronomical Union (IAU) defines a 'planet' only in the context of our solar system and an older definition that might be applied to exoplanets is a little murky. I'm therefore going to continue to refer to them here as 'exoplanets' rather than 'planets'.

† Several methods are used to detect exoplanets, but the most popular is the *transit method*. This involves observing the dip in the luminosity of a star as an orbiting planet passes in front of it.

‡ In February 2014, the Kepler mission team adopted a new 'verification by multiplicity' approach which allowed 715 new exoplanets to be confirmed. In the same month, the European Space Agency elected to fund another satellite-borne telescope mission to search for exoplanets, called Plato (for PLAnetary Transits and Oscillations of stars), which will likely launch in 2024.

luminosity and orbital radius which ensures just the right amount of light and heat to maintain a life-affirming temperature.[3]

More promising candidates were revealed by astronomers in April 2013. Kepler-62 is also somewhat older, smaller and cooler than the Sun, lying 1200 light-years from Earth in the constellation Lyra. It was also found to have a system of five exoplanets containing two 'super-Earths', with masses that could not be determined accurately but which must be less than 36 $M_{\oplus}$ and 35 $M_{\oplus}$, and with radii 1.61 $R_{\oplus}$ and 1.41 $R_{\oplus}$, designated Kepler-62e and Kepler-62f, respectively. What's interesting about these exoplanets is that they have orbital periods of 122 and 267 days and surface temperatures of 270 and 208 kelvin. This puts them firmly in the habitable zone of their star (Figure 49).

How many such exoplanets might there be in the Milky Way galaxy? Kepler's field of view was limited to a patch of sky in the direction of the constellations Cygnus, Lyra and Draco, accounting for about 0.25% of the whole sky. A 2013 study of 42 000 Sun-like stars observed by Kepler suggests that 22% of these

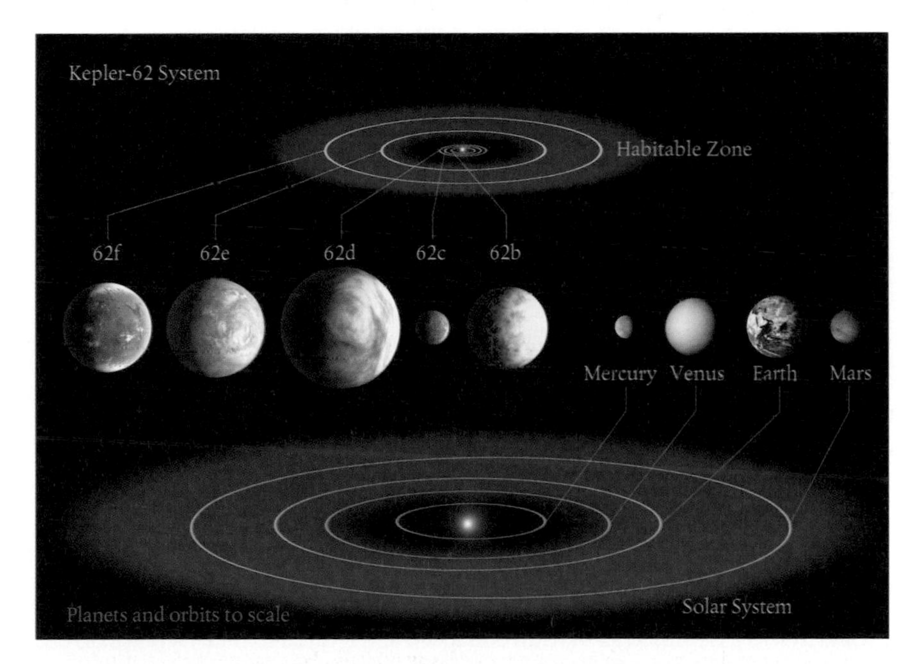

FIGURE 49 Artist's impression of the exoplanetary system of Kepler-62 and a comparison with the terrestrial planets of the inner solar system. The exoplanets Kepler-62e and -62f are 'super-Earths', with radii 1.61 and 1.41 times that of Earth, orbiting in the system's 'habitable zone' with orbital periods of 122 and 267 days, respectively. Both sets of planets are drawn to scale.

possess Earth-like exoplanets (Earth-size exoplanets orbiting in their stars' habitable zones). The astronomers deduced that there could be as many as 11 billion Earth-like exoplanets orbiting Sun-like stars in the Milky Way, the nearest of which could be just 12 light-years away.[4]

This brings us to the logical next question. If we can see exoplanets, can we see exoplanetary systems in the process of forming? After all, although observations of other systems do not necessarily show us how our own solar system originated, they do offer some significant clues.

The answer is yes. One of the first examples of a star system found to have a circumstellar disk of gas and dust is Beta Pictoris, in the constellation Pictor, about 63 light-years from Earth. This is a young star, thought to be only 8–20 million years old. It has already joined the main sequence, and possesses a flat disk of gas and dust that, from our perspective on Earth, is seen edge-on. The disk was photographed by the Hubble Space Telescope in 2003.

In November 2008, astronomers at the European Southern Observatory announced that they had found an exoplanet orbiting Beta Pictoris close to the plane of the disk. Beta Pictoris b has a mass of the order of 2000 $M_{\oplus}$. It is orbiting Beta Pictoris at a distance of 1.2–1.4 billion kilometres (that is, at about the distance of Saturn from the Sun), with an orbital period of 17–21 years. More recent studies suggest that the planet is younger than its star. We may indeed be watching a planetary system in the process of forming.

But perhaps the most striking image of a protoplanetary disk was published by astronomers at ALMA, the Atacama Large Millimeter/sub-millimeter Array, in Chile, an international facility jointly funded by European, North American, and East Asian agencies and the Republic of Chile. This is shown in Figure 50. It shows a young T Tauri star called HL Tauri (abbreviated as HL Tau), which is about a million years old and situated in the Taurus Molecular Cloud, about 450 light-years from Earth in the constellation Taurus.

There are many other examples of young stars observed to possess circumstellar disks that astronomers are happy to think of as protoplanetary disks, or *proplyds*.

Whether formed in a flat, spinning protoplanetary disk or a three-dimensional cloud of gas and dust, there's no questioning the simple fact that the Earth was assembled as the Sun collapsed and ignited to take its place on the main sequence, 4.6 billion years ago.

So, what happens next?

FIGURE 50 In this image from the ALMA observatory in Chile, the protoplanetary disk surrounding the young star HL Tauri can be clearly seen in the form of a series of concentric rings separated by voids. HL Tau lies about 450 light-years from Earth in the constellation Taurus, in the Taurus Molecular Cloud.

THE IRON CATASTROPHE

This is not so difficult to figure out. The physical properties and behaviour of a large ball of rock and metal are determined by the properties of the materials from which it is formed—mostly silicate minerals and metal oxides, some light, volatile chemicals—such as water—that may have become trapped in the rock structures, and trace amounts of other chemical elements formed through stellar nucleosynthesis.

The present-day Earth is composed largely of just six elements. These are iron (31.9%), oxygen (29.7%), silicon (16.1%), magnesium (15.4%), calcium (1.71%), and aluminium (1.59%). The remaining 3.6% is accounted for by all the other elements of the periodic table, up to uranium, present in quantities ranging from just a few hundredths right up to a few tens of thousands of parts per million.

What follows next is called *planetary differentiation*. The material accretes to form a larger and larger ball of rock and metal, and pressures and temperatures at the core start to rise. The accreting material also contains quantities of short-lived radioactive isotopes which, though small, release large quantities of energy as they decay, increasing the temperature still further. Things start to happen when the temperature rises above the melting points of the metal oxides and the silicates.

The critical parameters are density (mass per unit volume) and *miscibility*. Molten metal is denser than molten silicates. And, like oil and water, molten metal and silicates are immiscible—they don't mix together. The heavier metal is drawn by gravity down into the core as the lighter silicates float upwards towards the surface. This process starts slowly, with small globules of molten iron and other metals trickling down to the core, where it solidifies. But the trickle soon turns into a brook, then a stream, then a river. As it picks up more speed it turns into a torrent, sometimes referred to as the 'iron catastrophe'.

Why does the metal solidify in the core? In truth, it's not clear if the core is a solid or rather a plasma with a density characteristic of a solid. If solid it is, then it might be difficult to understand why it remains so, given that core temperatures in the modern-day Earth are likely to be of the order of 5700 kelvin. But we must remember that the *pressure* is much higher there, too.

The temperatures of phase transitions, such as melting (solid to liquid) and boiling (liquid to gas) are not fixed but depend on the surrounding environment, most notably the pressure. Water boils at 100°C (373 kelvin) in your kitchen at normal atmospheric pressure, but it boils at about 70°C (343 kelvin) at the top of Mount Everest, where the air pressure at 8800 metres, or 29 000 feet, is about 70% lower than the pressure in your kitchen. The higher the pressure, the higher the boiling point.

The same goes for melting points. These can rise by about 5–10 kelvin for every 1000 bars of pressure.* At a depth of just 120 kilometres below the Earth's surface, the pressure is about 40 000 bars, and the melting temperatures of rock can be up to 400 kelvin higher than at the surface. What happens is that

* One bar pressure is equal to normal atmospheric pressure at sea level.

at a depth of over 5000 kilometres the estimated pressure of 3.3–3.6 *million* bars is so high that even a temperature of 5700 kelvin is no longer high enough to maintain the metal in its molten state.

The interior of the Earth transforms from a randomly assembled, amorphous mix of silicates and metal oxides into a layered structure. An inner core forms composed largely (80%) of solid iron and other metals, such as nickel. In the present-day Earth it has a radius of about 1220 kilometres. This is surrounded by an outer core, a sea of mostly molten iron with a thickness of 2266 kilometres. Above this sits a mantle of solid (but moderately ductile) silicate rock about 2900 kilometres thick.

How do we know? Bear in mind that the evidence I'm about to summarize refers to the Earth as it is today, not the proto-Earth and, subsequently, the primitive Earth that is forming in the early solar system. As subsequent events unfold, we will see that there is no telling how much mass the Earth had accreted at this stage, but it is likely to have been somewhat less than the mass of the Earth today.

The boundaries of each of the layers produced by planetary differentiation in the Earth's interior are marked by sharp changes—or 'discontinuities'—in the *densities* of the materials that make up each layer. Clues to the presence of these discontinuities are available from seismological studies of shockwaves caused by earthquakes. These shockwaves pass through the interior of the Earth. There are two broad types—compression or pressure waves (P-waves) in which the material through which the wave is passing is displaced back and forth in the direction the wave is travelling, and shear waves (S-waves), in which the material is displaced at right-angles (up and down) to this direction. P-waves travel faster and tend to be detected first.

Both types of waves travel at speeds determined by the density of the material they pass through. As it turns out, S-waves cannot pass through liquid (or do so with greatly diminished amplitude, or 'height'), so do not penetrate the outer core to any great extent. S-waves produced by an earthquake on one side of the Earth will not be detected on the other side, unlike P-waves which do pass through the core. Careful studies of the timing of these shockwaves has allowed geologists to build up a picture of the Earth's interior structure.

The discontinuity that signals a change in density between the inner and outer core was discovered by Danish seismologist Inge Lehmann in 1936. Some texts refer to this as the Lehmann discontinuity, although this name tends to be reserved for another discontinuity that occurs just below the Earth's continental crust, and which was discovered by Lehmann in 1961. In Figure 51 I have labelled the inner–outer core discontinuity simply as the 'Inner Core Boundary'.

The most obvious discontinuity occurs at the boundary of the outer core and mantle, where the density jumps dramatically, from about 6 up to 10 grams per cubic centimetre. Not surprisingly, this was discovered a little earlier, in 1912, by German-American seismologist Beno Gutenberg, although the precise nature of the outer core was not established until 1926. This is an extremely sharp boundary, called the Gutenberg discontinuity.

We will take a closer look at the mantle and crust shortly, but before we leave the core let's consider the speed with which it is forming as the proto-Earth assembles.

The process of planetary differentiation is happening even as the proto-Earth is accumulating mass, and is triggered as soon as it crosses the threshold in terms of mass, density, and gravity. How quickly does the core form? To answer

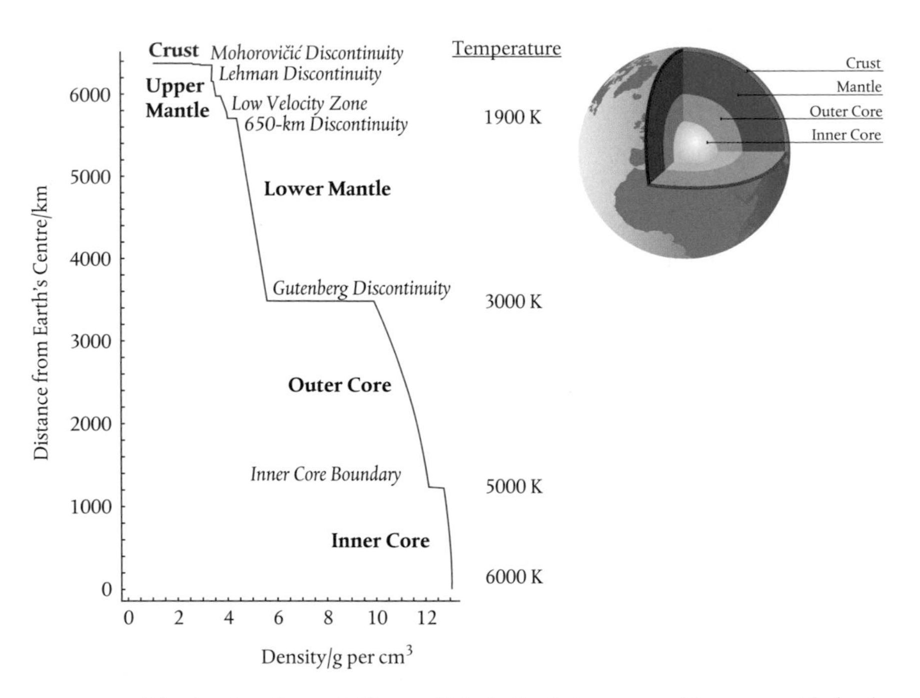

FIGURE 51 The density of material from which the Earth is composed increases with depth. The boundaries between each layer are marked by abrupt changes in density which can be detected in the patterns and timings of shockwaves from earthquakes recorded by seismometers. The inset shows the distribution of inner and outer core, mantle, and crust.

this question we need to reach once again for a short-lived radioactive isotope that can serve as a kind of internal clock.

The radioactive isotope of the element hafnium, ^{182}Hf, has a half-life of about 9 million years, decaying to an isotope of tungsten, ^{182}W. Any initial ^{182}Hf present in the newly forming Earth will therefore have decayed completely in about 100 million years.

Hafnium has a greater chemical affinity for the silicates in the mantle, whereas tungsten is more at home alongside iron and nickel in the core. Trace mixtures of hafnium and tungsten present in the bulk silicate materials as the Earth is forming will separate as a result of planetary differentiation. The tungsten will sink to the core, and the hafnium will stay in the mantle.

If this process takes longer than 100 million years, then the radioactive ^{182}Hf will have disappeared before the core and mantle properly separate. We would then expect that the ratio of ^{182}W to a reference isotope (such as ^{183}W) will be the same in both core and mantle and, for that matter, in chondrite meteorites that will spend the next 4.6 billion years floating about the solar system and will never experience the temperatures and pressures associated with planetary differentiation.

However, if the core forms more quickly, the tungsten migrates more quickly to the core, leaving the hafnium in the mantle and so raising the ratio of hafnium to tungsten in the mantle. As the ^{182}Hf in the mantle decays, it enriches it in ^{182}W, so raising the $^{182}W/^{183}W$ ratio.

The measured enrichment of ^{182}W relative to ^{183}W in bulk silicate thrust to the Earth's surface indicates that the core forms within about 34 million years. More recent isotope studies suggest a range for core formation of 27.5–38 million years and for silicate differentiation (and hence mantle formation) of 38–75 million years.

MANTLE AND CRUST

The mantle is composed of complex chemicals in complex mixtures, and phase transitions are no longer about crossing some simply defined temperature or pressure boundary. Instead, the mantle melts across a broad range of temperatures and pressures. For this reason the composition of the mantle has proved rather difficult to pin down precisely.

We do know that the mantle is split by another discontinuity at a depth of about 650 kilometres below Earth's surface (in Figure 51 this is referred to simply as the '650-km discontinuity'). Above this discontinuity lies the upper mantle; below it lies the lower mantle.

The principal phases of the lower mantle are thought to consist of minerals known as magnesium and iron silicate perovskites ($MgSiO_3$ and $FeSiO_3$), ferropericlase (also known as magnesiowüstite, a mixture of magnesium and iron oxides—MgO, FeO), calcium silicate perovskite ($CaSiO_3$) and corundum or alumina (aluminium oxide, Al_2O_3), and lots more besides. These minerals combine to match the characteristic density profile of the lower mantle, which runs from about 4.5 to 5.5 grams per cubic centimetre.

In the upper mantle just above the 650-km discontinuity lies a region of reduced seismic wave velocity known as the Low Velocity Zone (or LVZ), first identified by Gutenberg in 1959. It spans depths from about 410 down to 650 kilometres. Higher still, at a depth of 220 kilometres, lies the discontinuity named for Inge Lehmann. This does not appear to be a global discontinuity but rather sits specifically beneath the continental crust. These discontinuities are thought to arise from physical changes, rather than chemical differences between different layers in the upper mantle.

The composition of the upper mantle is a little easier to identify, as there are places on Earth where this has been thrust up above the crust and is exposed and accessible. It is composed largely of a coarse-grained igneous rock called peridotite, formed mostly from the minerals olivine (which, remember, is a magnesium-iron silicate) and pyroxenes (a range of silicates containing calcium, magnesium, and iron). Olivine accounts for about 55% of the mineral content, pyroxenes 35%. The remaining 10% consists of calcium and aluminium oxides.

Once again, we must abandon an intuition born of experience of a world on the Earth's surface in which atmospheric pressure remains reasonably constant. Geologists believe the upper mantle at this depth is partially molten, the result of a phenomenon known as *pressure-release (or decompression) melting*. The notion that a material will melt as its temperature falls is rather counter-intuitive. But, as with the molten iron in the outer core, the rock in the LVZ experiences a lower temperature but partially melts because the pressure is lower.

This partial melting makes the upper mantle rather plastic and, as we will see shortly, capable of supporting convection currents which move sections of the crust around.

Finally, we ascend to the crust itself. The partial melting of the upper mantle causes a further separation of the chemicals from which it is composed, with minerals of lower density rising to settle on the top. The separation of crust and upper mantle is marked by another seismic wave discontinuity, which is named for Croatian seismologist Andrija Mohorovičić and often referred to as the 'Moho'. It lies about 5–10 kilometres beneath the ocean floor and 20–90 kilometres below the Earth's continents.

With only 5–10 kilometres between the ocean floor and the Moho, we conclude that the oceanic crust is quite thin. It is formed largely from basalt. This is an igneous rock, which means it is formed by the extrusion and cooling of molten or semi-molten magma or lava. It can have a range of mineral compositions but typically consists of anorthite (a form of feldspar), pyroxenes, and iron ore. It may also sometimes contain olivine and quartz (SiO_2).

The crust beneath the oceans today was formed relatively recently, within the last few hundred million years, largely from upwelling of magma at the mid-ocean ridges (which we will take a closer look at shortly). The basalt produced by the primitive Earth would have formed in a different environment, and is likely to have had a different chemistry.

The continental crust is much older. It is formed in much the same way, but involves a series of melting–cooling cycles. Cooling magma produces basalt. As the basalt is reheated, it melts and produces a further chemical separation, cooling to form granite, a mixture of quartz, mica (sheet silicates), and potassium feldspar ($KAl_2Si_2O_8$). Reheating granite produces no further chemical separation and so granite is the end result of any number of melting–cooling cycles.

In truth, the formation of the continental crust is not quite as straightforward as this simple description implies, and geologists continue to debate the details.

It's difficult to grasp the nature and significance of this process of chemical separation by looking only at the mineral compositions of the upper mantle and crust. The mixtures are too complex. But we can simplify things a little by reducing the chemical composition of these layers to a series of simple 'building blocks'. We find that their composition is dominated by just five oxides—silica (SiO_2), magnesium oxide (MgO), iron oxide (FeO), aluminium oxide (Al_2O_3), and calcium oxide (CaO).

Of course, some of these chemicals exist separately in the peridotite of the upper mantle, the basalt of the ocean crust, and the granite of the continental crust. To this we add the ingredients of the more complex silicates. For example, the mineral diopside is a magnesium-calcium silicate pyroxene with the formula $MgCaSiO_4$, which we can think of as SiO_2 plus MgO plus CaO.

Now the consequences of chemical separation become a little easier to visualize. Figure 52 shows the chemical compositions of peridotite, basalt, and granite. The peridotite melts, separates and cools to produce basalt which is richer in SiO_2 and Al_2O_3 and considerably poorer in MgO. The average density

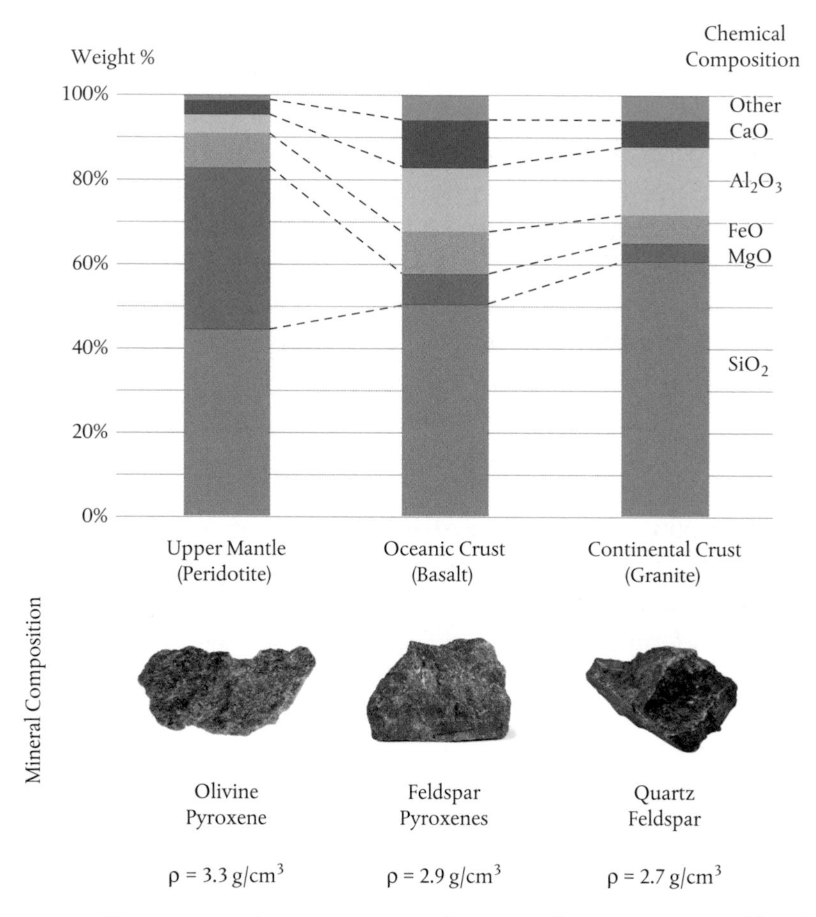

FIGURE 52 The upper mantle, oceanic crust, and continental crust are composed largely of three rock types, peridotite, basalt, and granite. These rocks have different mineral compositions, with the principal mineral types shown here. But these are complex mixtures of complex chemicals, and to get a sense of what happens it helps to look at the changes in chemical composition. The bar charts show the proportion (in weight per cent) of five oxides for each rock type. As these proportions change, the density (ρ) of the rock declines.

declines from 3.3 to 2.9 grams per cubic centimetre and the less dense basalt floats to the top.

Further melting, separation, and cooling then forms the granite of the continental crust, which is even more enriched in SiO_2 and Al_2O_3, and poorer in both FeO and MgO. The average density falls to 2.7 grams per cubic centimetre, and the granite rises to the top.

ATMOSPHERE

There is a further layer to consider. The planetesimals that accrete to form the proto-Earth are composed of rock that contains volatile chemicals, such as water, ammonia, and methane, trapped within it. As the process of planetary differentiation progresses and the rock is heated, these substances are released. The chemical decomposition of minerals in molten magma can also release volatile gases, and volcanic eruptions release water vapour, carbon dioxide, sulphur dioxide, nitrogen, and lots of other gases.

This process of 'degassing' creates the primitive Earth's first atmosphere. We can only really guess at its composition, but it is likely to have been pretty unpleasant.

Evidence that Earth did have an atmosphere this early in its history comes once again from a short-lived isotope. An isotope of iodine, ^{129}I, decays to produce ^{129}Xe with a half-life of 16 million years. Xenon is a so-called noble gas and has eight different isotopes that are stable or very long-lived. It doesn't readily form chemical relationships, and much prefers to find its way into the atmosphere than hang around trapped in rock structures. Iodine, on the other hand, is quite at home in the mantle.

If the atmosphere forms slowly, say over 100 million years or more, then pretty much all the ^{129}I will have decayed, allowing the ^{129}Xe so produced to equilibrate between mantle and atmosphere. We would expect to find very little xenon in the mantle, but because there has been enough time for equilibration we would nevertheless expect there to be no difference in the ratios of ^{129}Xe to other stable xenon isotopes. If, on the other hand, the atmosphere forms more quickly, much of the xenon enters the atmosphere *before* all the ^{129}I has decayed. As the ^{129}I now decays in the mantle, it enriches the mantle in ^{129}Xe compared to the atmosphere. Measured xenon isotope ratios reported in 1983 suggest that the atmosphere and mantle separated within about 30 million years, roughly consistent with the timing of the separation of core and mantle.

Let's take stock. About 90 million years have now passed since the Earth first started to condense in the Sun's protoplanetary disk. A noxious atmosphere sits above the primitive Earth's surface, which is relentlessly tortured, twisted, and ripped apart by volcanic activity. An ocean of hot magma glows red and spews a dense vapour of silicates. The Sun, still very early in its journey along the main sequence, produces a feeble light that cannot penetrate the smog. The planet continues to be hammered by large meteorites as the disk slowly clears of debris.

Welcome to the *Hadean*, the Earth's first geologic eon, aptly named for Hades, the Greek god of the underworld. This primitive Earth is an unpromising, inhospitable, and uninhabitable place.

And things are about to get much, much worse.

MOON

There is one difference that is *very* noticeable. This primitive Earth has no moon.

There are generally three ways in which a planet might acquire one or more companion moons. They can form in concert with the planet itself, co-condensing from the cloud of gas and dust sharing the same orbital path as it swirls around the proto-Sun. A second possibility is that bodies straying too close to a planet's gravitational field may be captured and become moons. Or, under the right circumstances, a moon can be formed from the oblique impact of two planets.

This last mechanism is proposed to explain how Earth acquired its moon. It is called the *giant impact hypothesis*. So, let's see how this plays.

This primitive Earth has a Mars-sized neighbour with an uncomfortably close orbit. In 2000, the neighbour acquired the name Theia, for the mythical first generation Greek Titaness who gave birth to Selene, the Moon goddess. Nobody can be sure of the circumstances which altered their orbital paths, but about 90 million years after the formation of the solar system, Earth and Theia were set on a collision course.*

Of course, the proto-Earth experienced many collisions with increasingly large bodies as it accreted, and it survived. But this is now a collision of two

---

* In a proposal published in 2004, Theia is formed in a 'trojan' orbit, along with the Earth, moving just ahead or just behind the Earth. As Theia accretes more mass, its orbit is disturbed by collisions with planetesimals, and this sets it on a collision course.

fully formed planets. Planetary differentiation has likely already produced core, mantle, crust, and atmosphere in both. If this were a head-on collision, it is quite possible there would be no survivors, just a vast quantity of scattered debris from which a new planet or planets might eventually form.

Fortunately, this is a glancing collision. Computer simulations published in 2004 explored different collision scenarios and focused on the ones that produced a moon of the right size, with a small iron core, consistent with what we know of the Moon today.[5] These simulations suggest that Theia strikes the primitive Earth at an angle of 45°,* with a relative speed up to 4 kilometres per second (Figure 53). The collision is oblique, but the effects are nevertheless pretty devastating. They serve to remind us that on this planetary scale, 'accretion' can be a very violent and destructive process.

The result is meltdown.

Theia is torn apart. Its iron core is ripped out and dragged down through the Earth's interior to merge with the Earth's core. Much of Theia's mantle is absorbed into the Earth's mantle, but the side that is furthest away from the impact is heated less by the collision and is hurled into space. Theia is thought to have had a mass of 0.11– 0.14 $M_{\oplus}$, and mantle with a mass of 0.018–0.026 $M_{\oplus}$ is now ejected, of which about 70% comes from Theia's far side.

Between 10 and 30% of this mass is vapour. Whatever atmosphere had formed around the primitive Earth is now gone.

About half of the ejected material is lost. The rest accretes to form the Moon, with a mass of 0.012 $M_{\oplus}$.

The Moon consists largely of mantle. The small amount of iron present in the material ejected by the collision now migrates to the centre as a result of planetary differentiation. It forms a small inner core of solid iron, thought to have a radius of just 240 kilometres, surrounded by an outer core of liquid iron 60 kilometres thick.

The chemical composition of the Moon's crust is somewhat different from the compositions of Earth's oceanic and continental crust, as shown in Figure 52, as might be expected if 70% of its content comes from Theia. Of course, there are no oceans on the Moon, but there is nevertheless a difference in composition between the *lunar maria* (or 'seas') and *lunar highlands*.

---

* A head-on collision would have a collision angle of 0°.

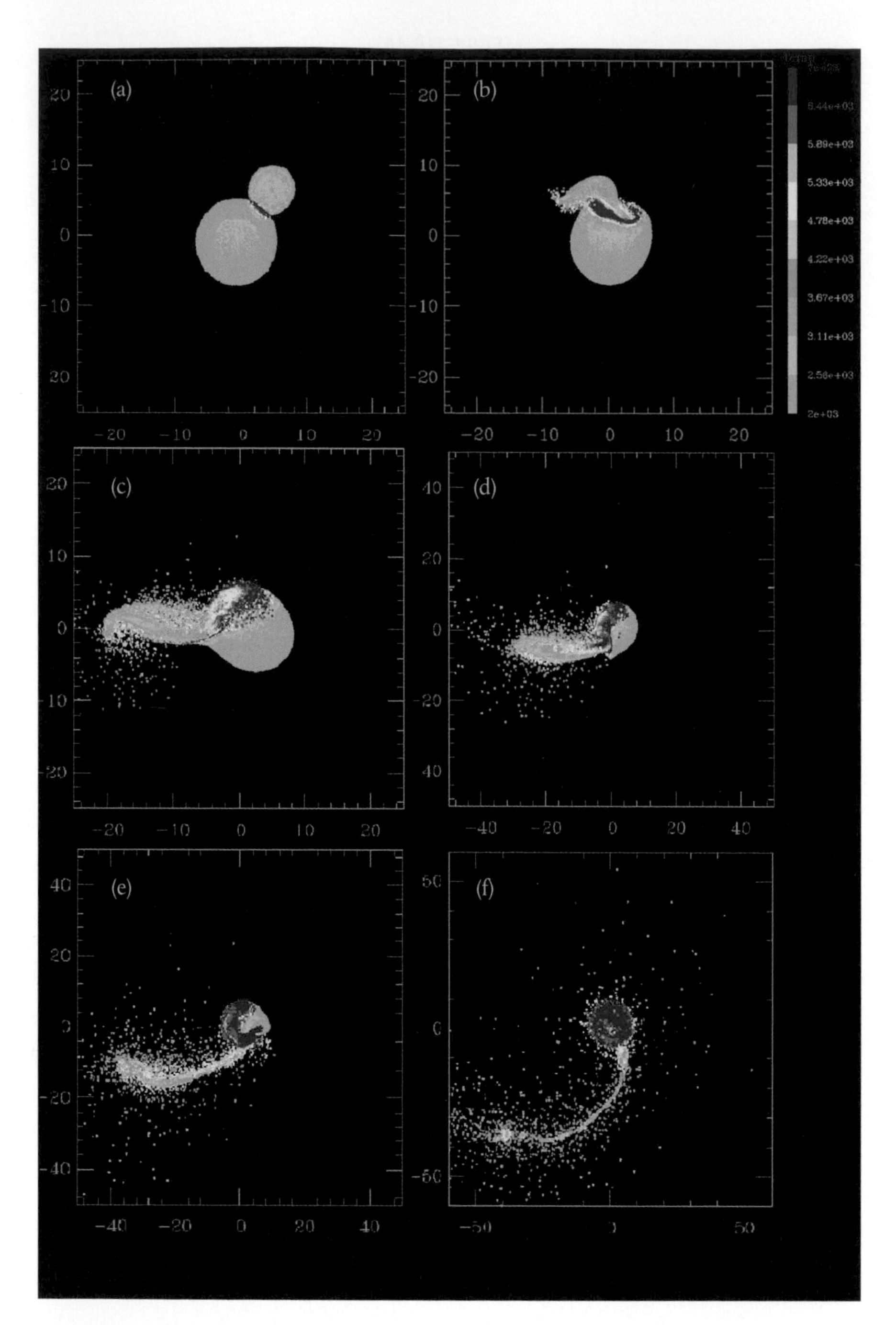

FIGURE 53 Computer simulations of a moon-forming impact reported in 2004 by planetary scientist Robin M. Canup at the Southwest Research Institute in Boulder, Colorado. These 'snapshots' show the unfolding collision and its effects at different times following the impact. (a) 6.5 minutes; (b) 19.4 minutes; (c) 51.8 minutes; (d) 84.1 minutes; (e) 129.4 minutes; and (f) 291.1 minutes. Note the change of distance scale (in thousands of kilometres) between frames (c) and (d) and (e) and (f). Temperatures are colour-coded from 2000 to 7000 kelvin. Adapted with permission from Robin M. Canup, *Icarus*, **168** (2004), p. 433.

The maria are the familiar dark patches visible on the Moon's near surface. They are composed essentially of solidified basalt from ancient magma flows, richer in FeO and titanium dioxide (TiO_2) compared to Earth's oceanic crust. They appear to be of the order of 3.1–3.9 billion years old, and so 600–1400 million years *younger* than the Moon itself.

As the name implies, the lunar highlands sit above the maria, just as Earth's continents sit above the ocean floor. The highlands are much richer in MgO, Al_2O_3, and CaO compared to Earth's continental crust, and appear to consist largely of anorthosite, rock composed mostly of the calcium-rich mineral anorthite, $CaAl_2Si_2O_8$. This rock is much older than the basaltic maria, more like 4.4 billion years old.

What we know of the Moon's geology comes from the analysis of about 390 kilograms of lunar rock samples returned to Earth by the Apollo mission astronauts. These missions were inevitably restricted to relatively accommodating terrain and a handful of accessible landing sites, so the samples may be far from representative. Nevertheless, the similarities—and the differences—with Earth's geology are striking.

During their second Extra-Vehicular Activity (EVA), Apollo 15 astronauts James Irwin and David Scott picked up a sample of anorthosite, initially thought to be a piece of the Moon's primordial crust. The sample was nicknamed the 'Genesis rock'. Subsequent analysis revealed that, although it is very old, it is actually some 400 million years younger than the Moon and formed after the Moon's crust first solidified.

Such samples have allowed planetary scientists to construct a plausible history. Shortly after forming, the Moon is covered by a huge ocean of hot magma. As this cools, the anorthite crystallizes and floats to the top, as the minerals olivine and the pyroxenes sink into the mantle, producing a light crust of anorthosite. The dark basalts of the lunar maria form later, following further planetary differentiation and upwelling of magma from within the mantle.

We need to explain why the lunar highlands are made largely of anorthosite, yet the Earth's continental crust is composed of granite. Simple physics supplies an answer. As the Moon's magma ocean cools the anorthosite crystallizes. But because the Moon is so much smaller, it has a much weaker gravitational field. Pressure differentials from the surface to the core are therefore much shallower than those in Earth's interior. Consequently, anorthosite is stable at much lower depths on the Moon than it is on Earth. A thick anorthosite crust forms.

Such a crust formed on Earth would be thinner, as the mineral anorthite becomes unstable at the kinds of pressures generated less than 40 kilometres below the surface. This thin crust is then at the mercy of impacts from meteorites and physico-chemical phenomena that the Moon's surface will never experience, which we call *weather*.

But there are puzzles, nevertheless. The pattern of darker maria and lighter highlands is utterly familiar to anyone following a few moments of moon-gazing. But from Earth we only ever get to see the Moon's near side. In contrast, the far side has virtually no maria, just highlands (Figure 54). How come?

Here's one possible explanation. In some more recent giant impact simulations carried out in 2011, the ejected material is presumed to form *two* moons, with the orbit of the much smaller second moon becoming unstable after a few tens of millions of years and slowly crashing into the Moon's far side. This potentially explains why the far side is so different, dominated by highlands with a crust that is about 50 kilometres thicker than the near, visible side.

The striking geological similarities between the Earth and Moon imply an intimate relationship that is not easily explained by the capture hypothesis. Aside from collisions with meteorites, rock recovered from the surface of the Moon has remained largely unchanged since the Moon formed, unlike rock on Earth. So, if the Earth and Moon had co-condensed then we would expect

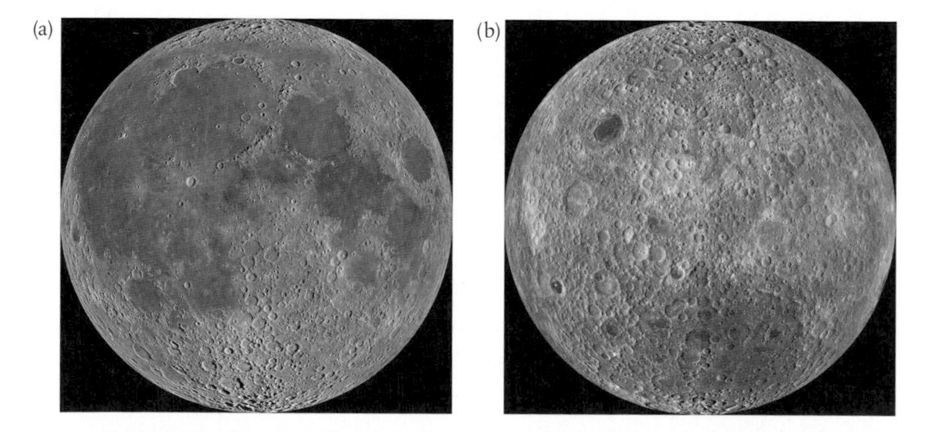

FIGURE 54 The familiar near side of the Moon (a) has lunar *maria*, or 'seas', which appear as dark patches of the surface, and lunar *highlands* which are lighter. In contrast the surface of the far side of the Moon (b) features only highlands and has a much thicker crust.

lunar rocks to be of a similar age to those of chondrite meteorites. But there is an age difference. Recall from the last chapter that chondrite meteorites are 4567.2±0.6 million years old. Analysis of lunar rock samples suggest that the Moon is 4430–4520 million years old, a difference of 47–137 million years. There's no doubting that the Moon came later.

But the Earth–Moon similarities are actually a little *too* striking. Isotopic analysis (e.g. of ^{16}O, ^{17}O, and ^{18}O) reveals *identical* isotope ratios in lunar rocks and rocks on Earth. At first glance, this would seem to be compelling evidence for the giant impact hypothesis. Yet isotope ratios in meteorites are a little different. And we anticipate that different planetary bodies would also show different isotope ratios. If about 70% of the Moon was formed from mantle on Theia's far side, then identical isotope ratios are rather curious. For sure, Theia could have had the same composition as the primitive Earth, with the same isotope ratios, but this is stretching coincidence further than most scientists are normally comfortable to admit.

Work-arounds have been found to account for these similarities, involving a period of turbulent mixing of the materials in the Earth–Moon system shortly after the impact. In one scenario, the Earth and Moon share a common atmosphere of metal plasma which forms a 'bridge' between the two, allowing material to be exchanged and equilibrated.

But there are other problems that the original giant impact hypothesis can't explain. A further elaboration of this idea, published in 2012, proposes that two similarly sized planets (both a little more than 0.5 $M_{\oplus}$) suffer an oblique collision at low speed.[6] This is a rather 'sticky' collision. The planets bounce off each other and collide again. Their iron cores merge, and their mantles mix and equilibrate. The now rapidly spinning, newly formed Earth produces spiral arms of liquid mantle, and the arms eventually coalesce to form the Moon.

If this were really the way it happened, then there was no primitive Earth, as such. Instead, Earth is born from the fusion of two other planetary bodies. For different reasons, in 2010 some planetary scientists proposed giving the primitive Earth another name. They suggested *Tellus*, for the ancient Roman Earth mother goddess.[7] So, let's steal that name and use it here. We can imagine that Tellus and Theia form with similar masses in painfully close orbits around the Sun. They suffer a glancing collision, from which are born Earth and its Moon, in a final act of planetary accretion.

This kind of collision scenario was rejected in the earlier simulations because it produces an Earth–Moon system with too much angular momentum. However, subsequent studies identified the possibility of an orbital resonance—called an 'evection'* resonance—between the Earth's orbit around the Sun and the Moon's orbit around the Earth, which can serve dramatically to reduce the rate at which the Earth spins on its axis.[8] The possibility of this resonance broadened the scope of possible impact scenarios that could produce something like the current Earth–Moon system.

Whatever the precise nature of the details, it is a fact that Earth acquired its Moon or was created along with its Moon about 90 million years after the solar system first began to form. If some kind of giant impact did indeed occur, then the planetary slate was wiped clean. The Earth isn't starting over from scratch, as core and mantle are already present in their new, expanded forms. But some further differentiation is likely, and the Earth now needs a new crust and atmosphere.

The giant impact has another legacy. The Earth's axis is tilted, and it is spinning faster than it does today. We can only speculate, but it seems that at this time in Earth's history, a day lasts only about ten hours. If the Earth has much the same orbit around the Sun as it does today, then a year has about 880 days. Months are much shorter too, as the Moon orbits the Earth much faster. It is also much closer. As the Earth's atmosphere re-forms through degassing and volcanic activity, a Moon twice the present size can be glimpsed in the primitive sky.

The Earth and Moon are now locked in an irresistible gravitational embrace. Angular momentum is drawn from the Earth and transfers to the Moon, pushing its orbit further out. Evidence for this exchange of energy is available from fossil corals. The days used to be shorter, with about 400 days in a year about 360 million years ago.

This is consistent with more recent evidence. During their first Moonwalk in 1969, Neil Armstrong and Buzz Aldrin left a small laser mirror on the surface. Bouncing pulses of laser light from Earth off this mirror allows the distance between them to be measured with great accuracy. In the 45 years that have elapsed, the Moon has receded from Earth at a rate of almost 4 centimetres a year.

---

* Latin for 'carrying away'.

WATER, OCEAN, AND CLIMATE

The atmosphere of the newly refurbished (or just-formed) Earth is produced within a few million years by degassing. The young Sun is still rather feeble, producing only 70% of the light and heat that it yields today. Once the surface settles down and the magma flows cool, we might expect the Earth to be a rather cold place.

Water vapour carried into the atmosphere along with other gases forms clouds. Water in these clouds condenses to form rain (or snow), just as it does today. The water collects to form oceans. On a cold Earth, the oceans will freeze to ice.

But there is evidence to suggest that the early Earth was actually much warmer; warm enough for water in the oceans to remain liquid. The oldest minerals found to date on the Earth's surface are small crystals of zircon, formed from zirconium silicate ($ZrSiO_4$), in the Jack Hills of Western Australia, about 800 kilometres north of Perth. These crystals have been determined to be 4404 million years old, give or take 8 million years, and had become trapped in much younger rock.[9]

These crystals are therefore between 26 and 116 million years younger than the Moon. What makes them especially interesting is that they were likely formed in molten granite. And granite, as we know, is a signature of continental crust. The existence of these ancient zircons suggests that the Earth had already formed a continental crust within 100 million years or so of its life after the impact that had created the Moon.

There's more. These are so-called 'detrital' zircons, mineral grains formed from rock through a process of weathering and deposited in sediment. Sediment? Oxygen isotope ratios in these zircons point to the existence of liquid water on the surface at the time of their formation. There is simply no known chemical process that can explain their formation within the crust or mantle. This is the oldest evidence for liquid water, and hence climate and weather on planet Earth.

This is a bit of a puzzle, referred to by planetary scientists as the 'faint young Sun paradox'. Once the magma oceans had cooled and volcanic activity had subsided, just how did the young Earth manage to be so warm? One possible solution is that the atmosphere provided a thick blanket of carbon dioxide, methane, and other 'greenhouse' gases that helped to trap the Sun's feeble rays and raise the planet's surface temperature.

Such gases are transparent to visible light, which passes through the atmosphere and warms the surface. But energy radiated from the surface as infrared (heat) radiation is absorbed by these gases and re-radiated in all directions, including back to the surface. The heat becomes trapped at the surface, raising the temperature. Much the same happens on Venus, which has experienced a runaway greenhouse effect and consequently has a surface temperature of 735 kelvin.

There are other potential solutions, but an early greenhouse Earth appears to be the most plausible.

MAGNETOSPHERE

The young Earth has an atmosphere and oceans of liquid water. Perhaps a moderate climate. It has weather. It is not yet the Earth that we will come to know, but it is well on its way to becoming habitable. There is one last component that we need to put in place to make it so.

The Sun is not as benign as it seems. In addition to life-giving heat and light, it also emits a rather unfriendly stream of charged particles—mostly electrons and protons—called the solar wind. This wind can strip an atmosphere from a young planet. Mars, which was never really large enough to hold on to a substantial atmosphere, lost what little it had due to the solar wind. Although there may have been liquid water on Mars at some stage in its early history (and there is some evidence to suggest this), the loss of its atmosphere meant the loss of water from its surface. For this reason, Mars is a barren, inhospitable place.

What makes Earth different is that it has a magnetic field which deflects much of the solar wind, protecting and preserving more fragile chemistry (and, ultimately, biology) that may develop on its surface or in its oceans.

This magnetic field is the result of electric currents established in the fluid outer core. Being composed of iron makes the outer core conducting. It is also convecting—heat is rising, carried by convection currents from the inner core which raise the temperature in the lower mantle. It is also fluid and rotating at a speed determined by the Earth's spin, such that Coriolis forces are established. These are 'pseudo' forces, not distinct in themselves but rather the effect of inertia within a fluid that is being dragged around a centre of rotation. They are more familiar in the context of Earth's atmosphere and oceans, directing flows of air and water that help to create different weather systems.

The interaction of these different forces in the interior of the Earth is complex, and computer models are only now being developed. But it is thought that Coriolis forces acting on the convecting fluid form rotating columns of molten iron within the outer core, called Taylor columns (named for British physicist Geoffrey Ingram Taylor). These columns are aligned with the Earth's axis of rotation.

As the molten iron moves down these columns, it carries an electric current. And an electric current generates a magnetic field. The outer core acts like a dynamo.

This is not a permanent magnetic field, of the kind we associate with a bar magnet, as it relies on the motions of charged particles that are established by rotation within the Earth's core. In fact, unlike the static field generated by a permanent magnet, the north and south poles of this dynamic field shift around over time, and the magnetic north pole is currently moving away from northern Canada towards Siberia. There have been times in Earth's history when the poles have reversed completely.

But, just as the 'lines of force' of the magnetic field of a permanent bar magnet extend into the space beyond the bar of iron, so the influence of Earth's magnetic field extends into the space around it.

The result is the inappropriately named *magnetosphere*, which would likely be spherical if it didn't interact with the solar wind. But it does and it isn't. In the direction pointing towards the Sun, the magnetosphere is shaped like a cushion or a flattened bubble, extending about 64 000 kilometres (ten Earth radii) out into space. Its boundary is called the magnetopause, and this is where the Earth's magnetic field and the field carried by the solar wind form an equilibrium. Further out beyond the magnetopause is the bow shock, a region of space in which the speeds of particles that make up the solar wind are slowed down.

On the side of the Earth pointing away from the Sun, the magnetosphere forms a long tail, stretching some 1.3 million kilometres behind it into space.

Within the magnetosphere is a complex sequence of layers of different energy, named the plasmasphere, the ionosphere, and the Van Allen radiation belts. High-speed charged particles from space (cosmic rays) can penetrate to the ionosphere and interact with the Earth's magnetic field to produce the spectacular aurora borealis (in the north) and aurora australis (in the south).

The magnetosphere forms a protective layer around the Earth, deflecting the solar wind and limiting its potential to cause damage. But it doesn't eliminate this potential entirely. Intense solar storms, involving ejections of large masses of hot gas from the Sun's surface, or corona, can enhance the solar wind sufficient to penetrate the magnetosphere. A solar storm in 1859 caused telegraph networks to short and produced aurorae displays that were visible in Hawaii. Another storm in 2003 damaged many satellites.

There is no evidence for the Earth's magnetic field older than about 3.5 billion years. But it seems reasonable to suppose that magnetism is an inherent physical property of an outer core of molten iron, once the core has stabilized.

THE NICE MODEL

It is a mistake to think that the solar system is an inherently stable structure, with fixed orbits that do not change over time, or that distant outer planets can have no influence on the inner, terrestrial planets. In fact, during the period in which the Earth–Moon system is forming, the outer planets are still in the 'wrong' order.

It's time to move the furniture.

We don't know the full details of what happened, but some recent computer simulations of the evolving early solar system suggest the following scenario.

Suppose that, about 90 million years after the birth of the Sun, the outer planets Jupiter, Saturn, Neptune, and Uranus occupy orbits with distances from the Sun of 818, 1230, 1730, and 2130 million kilometres, respectively. Further out, there lies a band of planetesimals called the Kuiper Belt, spanning a range of distances from about 2300 to 5100 million kilometres. This pattern is depicted in Figure 55(a).

About seven hundred million years pass. Earth stabilizes, its surface no longer rent by relentless volcanic activity. The magma ceases to flow and the surface cools. Atmosphere and oceans combine to produce a moderately friendly climate, though not one in which we could survive. Who knows? Perhaps a primitive form of early single-celled biological organisms develop.

We'll never know, because unfolding events are once again about to wipe the slate clean.

Jupiter and Saturn now enter a 2:1 orbital resonance, meaning that in the time it takes for Jupiter to orbit the Sun twice, Saturn orbits the Sun once. Their orbits become eccentric and the equilibrium of the outer planets is greatly disturbed by gravitational tidal forces. Neptune is driven out of its orbit by Saturn and is hurled into the Kuiper Belt, as shown in Figure 55(b).

Chaos.

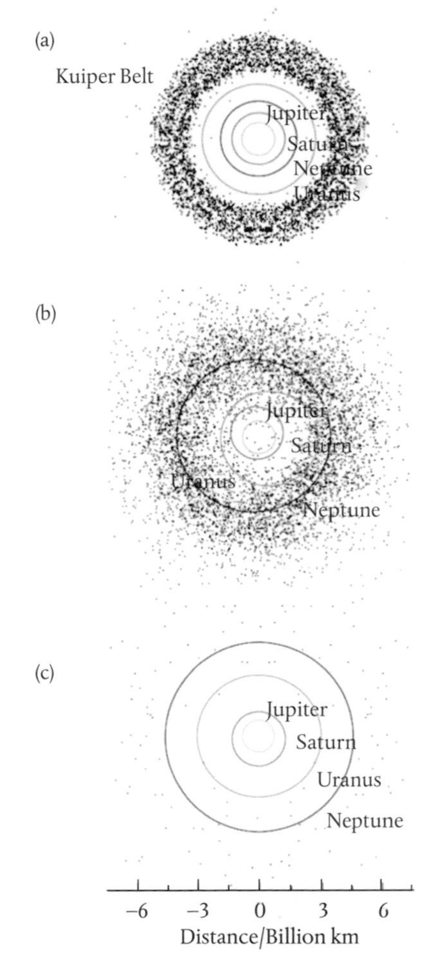

FIGURE 55 Computer simulations of planetary migration in the outer solar system demonstrate a connection with a period of intense cometary activity called the Late Heavy Bombardment. In (a) the outer planets line up in their initial sequence, Jupiter, Saturn, Neptune, and Uranus, surrounded by a thick band of planetesimals called the Kuiper belt. In (b), Jupiter and Saturn enter a 2:1 orbital resonance. Neptune is pushed out of its orbit by Saturn and is hurled into the Kuiper belt, dislodging planetesimals which are caught by the Sun's gravity and dragged in towards the inner planets (not shown in this figure). Jupiter is pushed closer to the asteroid belt, dislodging more planetesimals. The result is seen in (c), which shows the outer planets now in the 'right' order, and a greatly diluted Kuiper belt. Adapted from R. Gomes, H. F. Levison, K. Tsiganis, and A. Morbidelli, *Nature*, **435** (2005), p. 466.

Planetesimals are dislodged and scattered across the solar system. Many are hurled out to the distant reaches of the solar system, or are ejected completely. Jupiter's orbit is driven inwards slightly, knocking more planetesimals from the asteroid belt. These objects are now comets. Drawn inwards by the Sun's gravity they rain down on the inner planets in an episode known as the Late Heavy Bombardment.

We can see evidence for this bombardment on the cratered and pitted surface of the Moon which, the simulations suggest, was struck by something of the order of

14 billion billion kilograms of material from the Kuiper belt and asteroid belt over a period of about 150 million years. Earth's larger mass and gravitational pull would have ensured that it got rather more of this cometary material—the simulations suggest 1800 billion billion kilograms. Hard rain, indeed.*

These impacts are devastating and, once again, the clock is reset. The oldest rocks observed on Earth are measured to be about 3.98 billion years old, predating the Late Heavy Bombardment by no more than a few hundred million years. This is because the continental crust is constantly moving and renewing, and all the visible evidence of this bombardment has been wiped from the Earth's surface. And there is no evidence of life on Earth earlier than a few hundred million years *after* this episode. If life of any kind had managed to gain a foothold, this bombardment may have ensured that none survives.

The Kuiper belt is considerably depleted, and planetesimals pushed out to the further reaches of the solar system form what are known as the scattered disk and the Oort cloud. The latter is named for Dutch astronomer Jan Oort, and is thought to lie 0.75–7.5 billion kilometres from the Sun, at the very limit of the Sun's gravitational field. Although it has never been directly observed, some astronomers argue that the Oort cloud is the source of long-period comets with highly eccentric orbits.

This scenario is called the *Nice model*, first introduced in a series of papers published in 2005.† However, subsequent analysis of orbital patterns required to explain why the gas giants weren't pulled in closer to the Sun (which seems to have happened in many exoplanetary systems) appeared to be incompatible with the initial conditions assumed in the Nice model.

In more recent elaborations published in 2011, called *Nice II*, the details have changed in an attempt to remove some of the arbitrariness of the first model's initial conditions and the interdependence of the timing of planetary migration

* This bombardment may have delivered huge quantities of water to the Earth's surface, in the form of ice trapped in comets. We'll know more when data sent back from the Philae probe is analysed over the coming months and years. This is a probe placed on Comet 67P/Churyumov-Gerasimenko as part of the European Space Agency's Rosetta mission, launched in 2004. Rosetta rendezvoused with the comet in November 2014, a little over 500 million kilometres from Earth. The Philae probe landed successfully, but encountered a harder-than-expected surface and bounced a couple of times in the comet's weak gravity before coming to rest. It ended up in the shade of a boulder, its solar panels screened from sunlight, but was able to complete several scientific experiments and send the data back to Earth before its batteries ran down.

† Named for an international group of astronomers who collaborated at the Observatoire de la Côte d'Azur, in Nice, France, so not the biscuit, the 1960s progressive rock trio or the term indicative of general pleasantness.

with assumptions about the structure of the solar system's outer edge. The nature of the orbital resonance becomes a little more complicated, but a delayed instability of the kind that can trigger the Late Heavy Bombardment appears to be a generic outcome. In other words, no matter how we set it up, changes in the orbits of the outer planets and consequent bombardment of the inner planets appear to be a delayed, but inevitable, result.

At least the outer planets are now in the 'right' order, although it should be apparent from these simulations that there's nothing preordained about the planetary orbits and what we see in the solar system today is largely the result of physical mechanics and a goodly dose of chance.

The planets form a pattern determined by the mechanical interplay of orbital motion, spin, and gravity. In the 17th century, Johannes Kepler established the existence of a relationship between orbital period and radius. This became known as Kepler's third law: the ratio of the cube of the mean radius to the square of the period is approximately constant for all the planets in the solar system.[10]

Is there nevertheless a pattern in the relative orbital *positions* of the planets? There is an approximate numerical relationship between the orbital radii, called the Titius–Bode law, named for 18th-century German astronomers Johann Daniel Titius and Johann Elert Bode. Some contemporary astronomers have dismissed this 'law' as misleading numerology, but it seems to me that, despite being shifted about through the vagaries of chance orbital resonances, the planets must then settle into orbits that are reasonably stable, both within themselves and between neighbouring planets.

Anyway, here it is for what it's worth. According to the Titius–Bode law, we assign Earth an orbital radius of 10 'units', with each unit therefore corresponding to about 15 million kilometres. Mercury then has an orbital radius of $4 + n$ units, where $n = 0$, or roughly 60 million kilometres. For Venus $n = 3$ and the orbit is 7 units (105 million kilometres), for Earth $n = 6$, Mars $n = 12$ (giving a total orbit of 16 units or 240 million kilometres).

You get the basic idea. To estimate the orbit of the next planet, we double the value of n and add 4. For Ceres, the largest asteroid in the asteroid belt, $n = 24$ (420 million kilometres); for Jupiter $n = 48$ (780 million kilometres); for Saturn $n = 96$ (1500 million kilometres); for Uranus $n = 192$ (2940 million kilometres), and for Neptune $n = 384$ (5820 million kilometres). Roughly speaking, planets beyond Mars are about twice as distant from the Sun as their inner neighbour.

The relationship between the orbits predicted by this simple numerical law and the actual orbits of the planets is shown in Figure 56. The law does a reasonably good job for all the planets except Neptune, predicting orbits that are generally within 5% of the actual values. Given what we believe might have happened to Neptune, a discrepancy of 22% between prediction and actual orbit may not be altogether surprising.

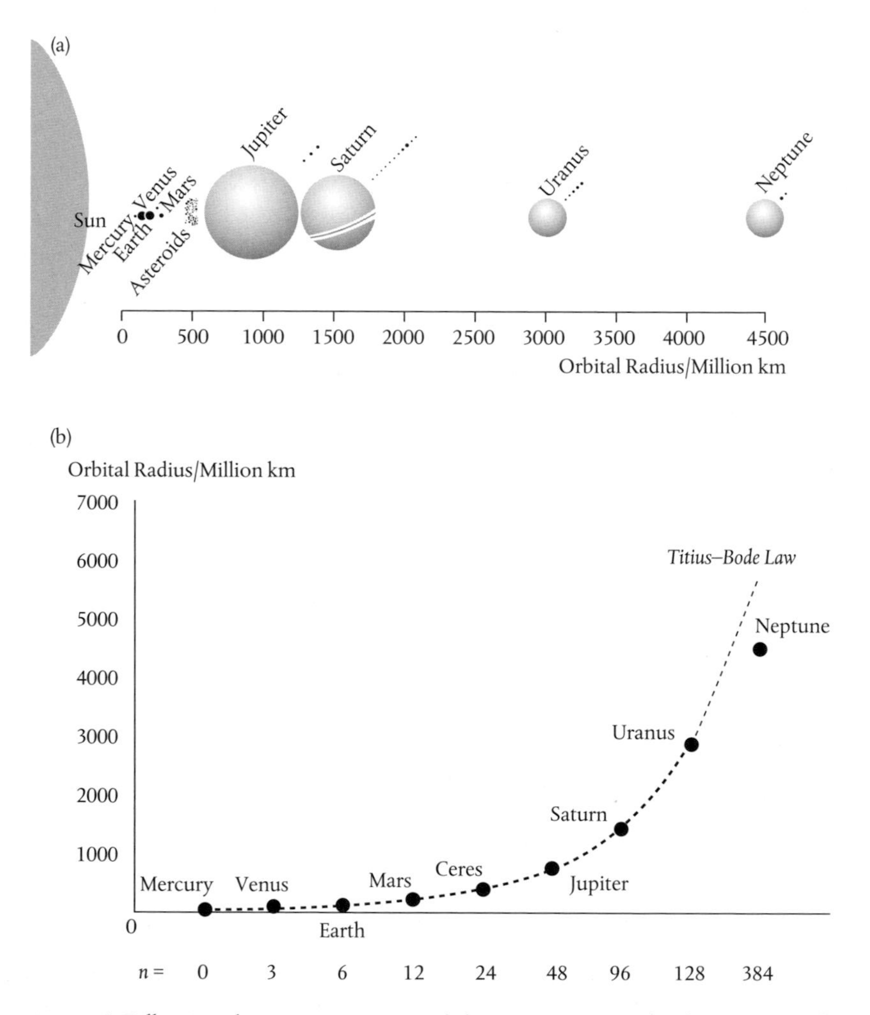

FIGURE 56 Following planetary migration and the Late Heavy Bombardment, 700 million years after the solar system was first formed, the planets take up the orbits that are familiar to us today. These are shown in (a), with the planets drawn to scale (though not consistent with the linear distance scale used to show the orbital radii). In (b), the planetary orbits (black circles) are compared with the predictions of the Titius–Bode law (dashed line).

The Late Heavy Bombardment is the solar system's last hurrah. It marks the end of the Hadean and the beginning of the Archean eon, although not all geologists agree with this demarcation. Aside from the Jack Hills zircons, there is no geological evidence from periods earlier than early Archean. We can nevertheless assume that the Hadean ended about 4 billion years ago.

PLATE TECTONICS

If we accept the evidence of the Jack Hills zircons at face value, then 4.4 billion years ago planetary differentiation of the giant impact-modified Earth was already well advanced. The planet had oceans of liquid water sitting above a largely basaltic ocean crust, and minor land masses formed by a continental crust composed largely of granite.

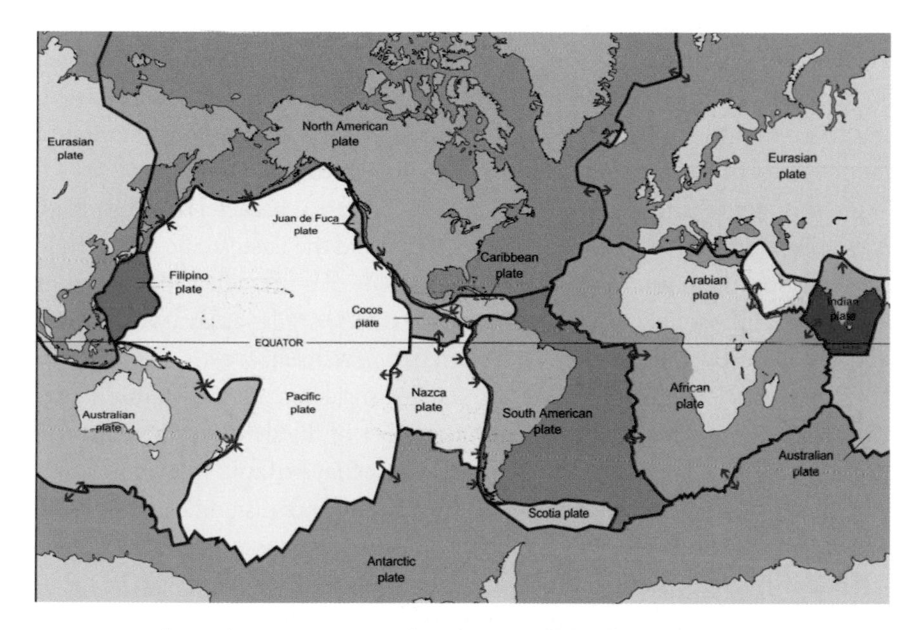

FIGURE 57 The Earth's uppermost mantle and crust (collectively called the lithosphere) breaks up to form a series of tectonic plates which move around the surface. This figure shows the plates as they appear today. 'Divergent' plate boundaries are illustrated by arrows pulling in opposite directions (←→). These appear in the middle of the oceans, the two most prominent being the Mid-Atlantic Ridge and the East Pacific Rise. Note that these divergent plate boundaries tend to run north to south, suggesting an intimate relationship with the Earth's rotation. Along these boundaries the seafloor spreads, renewing the crust with fresh material. At a 'convergent' boundary (→←) the crust of one plate is sucked down into the upper mantle, thrusting the crust of the other plate upwards to form a mountain range.

We know that pressure-release or decompression melting causes partial melting of the upper mantle. Convection currents are established within the mantle, driven in part by the energy released by radioactive isotopes of elements such as uranium and potassium. These combine with Earth's gravity, rotational motion and possibly gravitational influences exerted by the Sun and Moon in a complex mechanism which results in movements of the more rigid and inflexible upper mantle and the crust (ocean and continental) sitting above it. These uppermost layers of the planet's surface are referred to collectively as the *lithosphere*. As a consequence of convection churning beneath it, the lithosphere breaks up into a number of *tectonic plates*. The places where plates meet are called *plate boundaries*.

The plates don't just drift randomly around the planet's surface like ice floes. The forces acting on them set up a conveyor-like motion. At a 'divergent' plate boundary the seafloor spreads, renewing the crust with fresh material. At a 'convergent' boundary, the crust of one plate is sucked down into the upper mantle, thrusting the crust of the other plate upwards to form a mountain range. The amount of crust lost is roughly balanced by the new crust that is formed, so the Earth's surface doesn't noticeably shrink or grow larger. The plates are illustrated in their current configuration in Figure 57, their relative motions and hence divergent and convergent boundaries illustrated using arrows.

As this figure shows, divergent boundaries are characterized by deep, *mid-ocean ridges,* such as the Mid-Atlantic Ridge and the East Pacific Rise. These are undersea mountain ranges formed either side of a rift valley. Magma from beneath the lithosphere wells up through the rift, emerging as lava which cools and spreads, producing basalt which forms the new ocean crust.

The existence of plates makes the Earth's surface a very dynamic place, with plate movements responsible for a number of highly destructive natural phenomena—earthquakes and volcanoes. These may be tame when compared with the turmoil of the Hadean, but they are certainly not insignificant. Volcanic activity on a massive scale will have some big repercussions in our later story.

The Earth is still young, about 800 million years old if we reference the age of chondrite meteorites, or 90 million years younger still if we imagine that Earth was formed from the collision of Tellus and Theia. But, it seems, it has a warming atmosphere and oceans, a fully formed crust with drifting continents and a protective magnetosphere.

The Earth is now habitable.

And then something really rather remarkable happens.

8

THE COSMIC IMPERATIVE

The Origin of Life

My ambition in this book is to tell the scientific story of creation in a way that seems logically connected and consistent and hopefully not too difficult to follow. By logically connected and consistent I mean that each creation 'episode' in the history of the universe thus far is seen to follow naturally and inevitably from the application of reasonably well-understood physical and chemical principles applied to whatever it was that had been created in the preceding episode.

So, working backwards, the Earth was created in the thin disk of gas and dust that swirls around the collapsing Population I protostar that we will come to call the Sun. The Sun was formed in the collapse of the Neith Cloud that was in turn created by the slow gravitational drawing together of materials ejected from dying Population II stars. Population II stars . . . and so on.

Of course, we hit a discontinuity, both of time and of explanation, right at the 'beginning' of this story. But, once the universe was off to a good start, our scientific theories are generally adequate in providing plausible explanations (or speculations, or at least educated guesses) for how one thing then leads to another. Put another way: in the scientific story of creation, the universe and everything in it is built from the 'bottom up'.

Heads up. We are about to run into another substantial discontinuity of explanation. Simple logic dictates that the next episode in the creation story requires life to emerge spontaneously from the chemical ingredients available on the primitive, but now habitable, Earth, ingredients that happen to be far from their equilibrium state.

For sure, this is an assumption, but one that is not completely without foundation. With the exception of the inert noble gases and chemical elements that have a particular affinity for the rocky substances of the Earth itself—iron, magnesium, silicon, and aluminium—the relative abundances of the elements found in all biological systems map quite accurately to the relative abundances of these same elements in the Sun and in chondrite meteorites.[1] There are some notable exceptions,* but it does not require a great leap of imagination to conclude from this that life has simply made best use of the materials that were available on Earth.

It hints at a potentially profound truth. Life on Earth, like the universe itself, was also built from the bottom up.

So, let's go with the flow. Let's assume that complex biology is a necessary consequence of starting with the simple chemistry of the primitive Earth and the energy that is freely available to these chemical systems. Again, simple logic would seem to dictate the need for at least three steps in this particular creation episode. First, we need to make use of the available energy to construct the basic organic building blocks known to be important in biological systems—amino acids, nucleic acids, sugars, bases, and lipids (I'll explain what these are later in this chapter). Secondly, we need to assemble these building blocks into complex molecular systems capable of self-replication, metabolism and evolution. In a third step, we then need to pack the molecular systems into simple cellular structures.

It is here we hit the discontinuity. Any theory of the origin of life will need to accommodate and explain each of these steps, or find some convincing—and scientifically acceptable—argument as to why they are misconceived or are not necessary. These three steps, which get us from simple chemistry to incredibly complex, so-called systems chemistry of the kind found in living things,

* The relatively rare element molybdenum is required by just about all life forms—including ourselves. Aficionados of Douglas Adams' *The Hitchhiker's Guide to the Galaxy* will be pleased to note that this is element 42. (Thanks to Michael Russell for pointing that out.)

are already daunting. Theoretical ideas for these three steps abound, and we'll examine some of these in this chapter.

But getting complex chemical structures packed into a small bag of water protected by a thin cell membrane does not necessarily give us biology. It does not give us the last universal common ancestor (affectionately abbreviated as LUCA), the primitive, single-celled organisms from which all life forms on Earth are thought to be descended.

The missing ingredient involves *abiogenesis*, the spontaneous generation of life from non-living complex chemistry. This must have happened on Earth sometime between about 4.4 and 3.5 billion years ago, perhaps as the planet settled down after the Late Heavy Bombardment. If this did indeed happen, then it likely never happened again. Never, in the entire history of science, has life ever been observed to spring from anything other than life.

The gap in our understanding is huge. Although we experience it every day of our lives and we can easily recognize it when we see it, we don't really know what life *is*. There is no comprehensive, scientifically verified theory or 'standard model' for life. Every attempt to develop a definition for what life is has proved broadly futile, as exceptions to the 'rules' demanded by such definitions are often all too easy to find in the planet's rich diversity of living things.

The problem of abiogenesis has tried and taxed some of humanity's best scientific minds. In an influential monograph published in 1970 titled *Chance and Necessity*, the French Nobel laureate Jacques Monod argued that life emerged on Earth in a freak accident. He wrote:[2]

> The ancient covenant is in pieces; man knows at last that he is alone in the universe's unfeeling immensity, out of which he emerged only by chance. His destiny is nowhere spelled out, nor is his duty. The kingdom above or the darkness below; it is for him to choose.

In their book *Lifecloud*, published in 1978, English astrophysicist Fred Hoyle and astrobiologist Chandra Wickramasinghe argued for a form of *panspermia*, in which preformed bacteria and viruses are borne to Earth in comets. In 1981, English molecular biochemist Francis Crick, who shared the 1962 Nobel Prize for physiology or medicine with James D. Watson and Maurice Wilkins for their discovery of the structure of deoxyribonucleic acid (DNA), concluded that the only possible explanation for the emergence of life on Earth is *directed panspermia*. In other words, life on Earth was deliberately seeded by alien life forms, something that fans of Ridley Scott's 2012 *Alien*-prequel *Prometheus* will know all about.

Of course, panspermia—directed or not—simply begs another question. If life on Earth originated elsewhere in the universe, then how did *that* life emerge from the inanimate chemical ingredients that we assume to have been available in the universe's early history?

The problem is so entrenched that the only possible stand that can be taken is a bold one, I think, just as Monod, Hoyle, Wickramasinghe, and Crick have demonstrated.

So, here's how I propose to move forward. There are grounds to think that bacterial life was already flourishing on Earth just a few hundred million years after the Late Heavy Bombardment, based on a chemistry that, though complex, depends on chemical elements that would have been readily available on or beneath the surface of Earth's continents and oceans. I'm going to take this to suggest that life is therefore a perfectly natural result of the properties and behaviour of the chemicals that were lying around and the conditions that prevailed on the primitive Earth. In this scenario, life is *necessity*, not chance or a miracle, or panspermia. By 'necessity', what I really mean is that assembling the chemical ingredients on a warm, wet, rocky planet like Earth under the conditions that prevailed in its early history mean that life will emerge with a high probability, bordering on certainty.

Given the prevalence of Earth-like exoplanets, accepting this assumption necessarily implies that life must exist elsewhere in the universe. As Belgian biochemist and Nobel laureate Christian René, viscount de Duve once argued, life is a *cosmic imperative*:[3]

> Consideration of these . . . processes suggests that the origin of life may have been close to obligatory under the physical–chemical conditions that prevailed at the site of its birth.

But don't be under any illusions. Don't ask me to prove this cosmic imperative, or explain how and why it happened. I'm making a big assumption here and I'll happily admit that the scientific evidence for this is really all rather vague.

So, with this assumption made and out of the way, what do we think we know?

LIFE'S BUILDING BLOCKS

Despite the extraordinary diversity of life on Earth today, all life forms rely on broadly the same set of cellular structures and biochemical systems that are used to acquire energy, metabolize, grow, and replicate. Underpinning these fundamental biological processes is a set of bio-molecular mechanisms that are involved in (among other things) capturing, transferring, and storing energy and holding, transferring, and replicating what we have learned to think of as genetic information that specifies what kind of life it is.

In an attempt to keep things reasonably simple, I'm going to focus on two types of fundamental building blocks. These are chemicals called amino acids, the constituents of proteins, and nucleotides, involved in the complex molecular structures that carry genetic information and which code for the construction of proteins.

We'll begin with amino acids. These form a class of chemical compounds which consist of two functional groups, an amine group (—NH_2, hence 'amino') and a carboxylic acid group (—CO_2H, hence 'acid'). These compounds are distinguished from each other by possession of a third group which forms a side-chain, usually given the generic symbol 'R'. With some exceptions, we can write a general formula for amino acids as H_2N—CHR—CO_2H.

There are about 500 known amino acids, but there is a core of just 20 that are intimately involved in biological systems.* These can be further categorized according to the nature of their side-chains, and a breakdown is provided in Figure 58.[4]

I'm showing these chemical structures for reference—we won't need to remember them. But there is one further general feature of these structures that will prove to be rather important in what follows. Recall from Chapter 5 that the subtle interplay of the various atomic orbitals belonging to carbon, nitrogen, and oxygen atoms results in the formation of molecular orbitals (and lone pairs of electrons) which point in specific directions in space. The hydrogen 1s orbitals then simply latch onto the lobes of these orbitals to form C—H, N—H, and O—H bonds.

---

* Just to be clear—there are more amino acids involved in biochemical systems but there are just 20 that are manufactured based on genetic codes stored in DNA and ribonucleic acid (RNA).

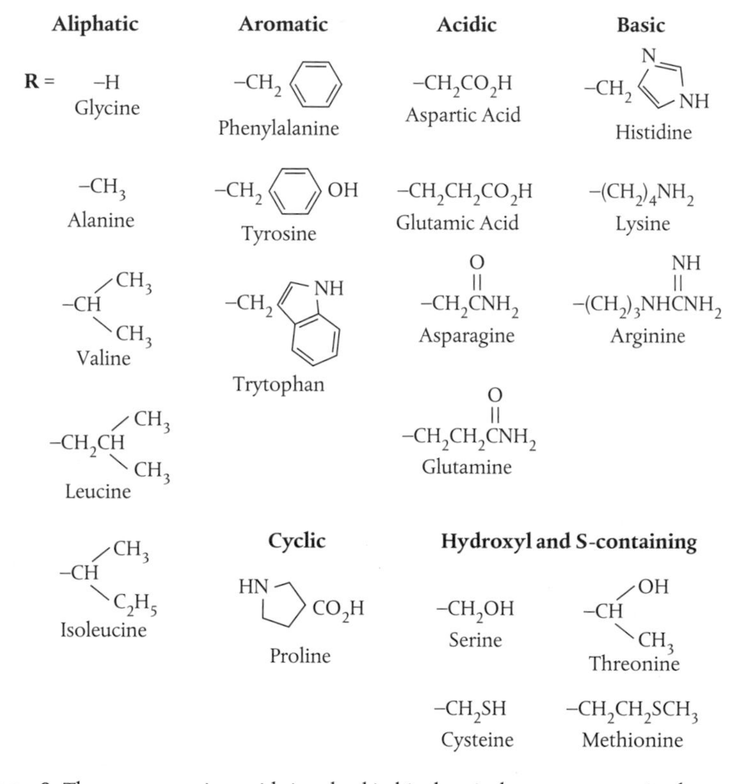

FIGURE 58 The core 20 amino acids involved in biochemical systems, organized according to the nature of their chemical structures. Most (19 of 20) conform to the general formula given at the top, the exception being the cyclic amino acid proline. See endnote 4 for an explanation of the cyclic structures which feature in proline, phenylalanine, tyrosine, tryptophan, and histidine.

So, let's consider the spatial properties of the amino acid alanine. Figure 59(a) shows that we can distribute the various functional groups and the side-chain (in this case R is —CH_3) around three of the four apexes of the tetrahedron formed by the central carbon atom, leaving a single hydrogen atom to occupy the fourth apex. This is quite straightforward.

But now we notice that there are two ways we can do this. For example, we could place the hydrogen atom at the far right apex and the amine group near-right, as shown. Or we could place the amine group far right and the

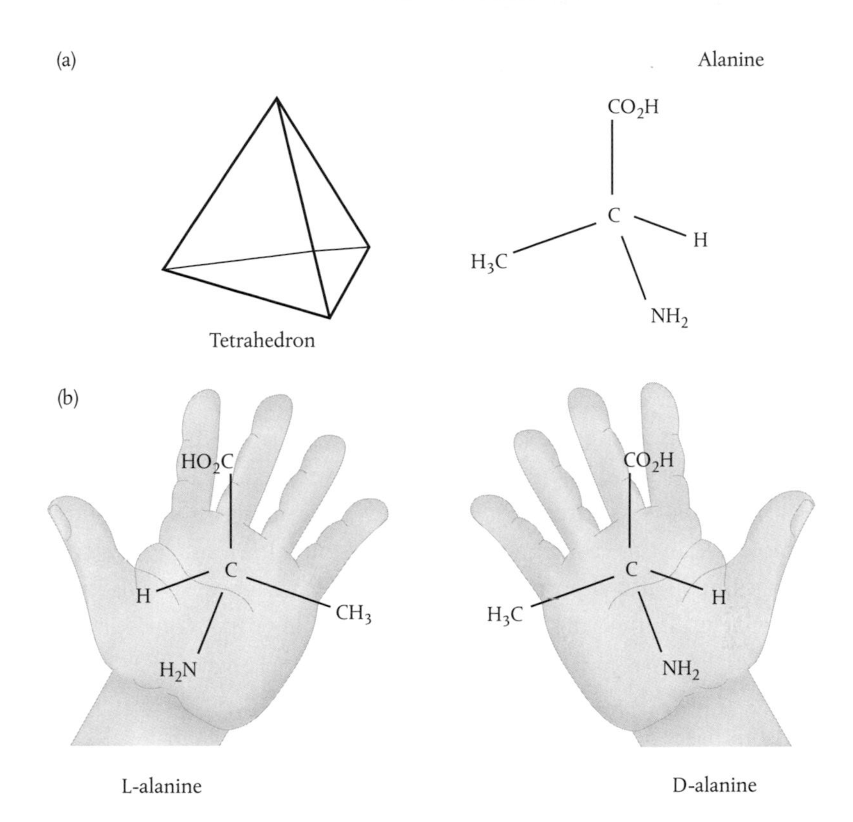

FIGURE 59 The amino acid alanine has four different atoms or functional groups bonded to the central carbon atom, each of which takes up a place at the apex of a tetrahedron (a). There are two different ways of distributing these groups, which represent mirror image forms (b). Such molecules are said to be chiral, and the different forms are called enantiomers.

hydrogen atom near-right. Does this make a difference? The answer is yes, it does. These two possibilities are mirror images of each other, as shown in Figure 59(b).

In your mind, pick up the version of alanine on the left-hand side of Figure 59(b), rotate it and superimpose it on the version on the right. The amine group on the left points out of the plane of the paper and, by the time we have rotated it, it is pointing into the plane of the paper. But the amine group on the right is also pointing out of the plane. We conclude that we can't superimpose these structures. Although these molecules contain precisely the same functional groups (they are both alanine), they are nevertheless *different molecules*.

The molecules are said to be *chiral*, and their different forms are called *enantiomers*.

These different forms of alanine can be distinguished by the effect they have on plane polarized light. A solution of pure enantiomer will rotate the plane of polarization of light passing through it. By convention, the enantiomer that rotates the polarization to the left is labelled L- (for *levorotatory*). The enantiomer that rotates the polarization to the right is labelled D- (for *dextrorotatory*).

Here's the curious thing. With the exception of glycine, which does not exhibit chirality (because it has two identical C—H bonds linked to the central carbon atom so can't possess mirror-image forms), all the amino acids involved in bio-molecular processes are exclusively of the L-form.

The second type of building blocks I want to introduce at this stage are collectively known as *nucleotides*. These are the units from which ribonucleic acid (RNA) and deoxyribonucleic acid (DNA) are constructed.

Nucleotides possess three key ingredients. The first is a specific kind of sugar molecule, in this case a pentagonal ring containing a single oxygen atom and either one hydroxyl group (—OH, called deoxyribose) or two hydroxyl groups (ribose). To this ring is attached a phosphate group, consisting of one ($—CH_2OPO_3^{2-}$), two ($—CH_2OPO_3^-PO_3^{2-}$), or three ($—CH_2OPO_3^-PO_3^-PO_3^{2-}$) phosphate units. The third ingredient attached to the sugar is a so-called nitrogenous base.

The nitrogenous bases come in two varieties, called purines and pyrimidines. The purines include adenine and guanine. The pyrimidines include cytosine, uracil, and thymine. These ingredients, and their assembly to form a single nucleotide, called adenosine monophosphate (AMP), are shown in Figure 60. Again, these structures are shown for reference, we do not need to remember them.

Although it's not obvious from the diagram in Figure 60(a), the ribose unit within the nucleotide also exhibits different chiral forms. The carbon atom lying just to the left of the oxygen atom in the ring is bonded to four different groups. There is the C—O bond in the ring itself. The second bond in the ring connects it with a —CH(OH) group. A third bond connects it with the phosphate group $—CH_2OPO_3^{2-}$. Finally, the fourth bond is to a hydrogen atom (C—H), which is not shown (but is implied) in the figure.

Now, we know from our experience with alanine in Figure 59(b) that when four different groups are bonded to a central carbon atom like this, it is possible to arrange the groups in two mirror-image forms that cannot be superimposed. The same goes for ribose (and deoxyribose). Intriguingly, while it is the L-enantiomers of amino acids that are involved exclusively in proteins, it is the

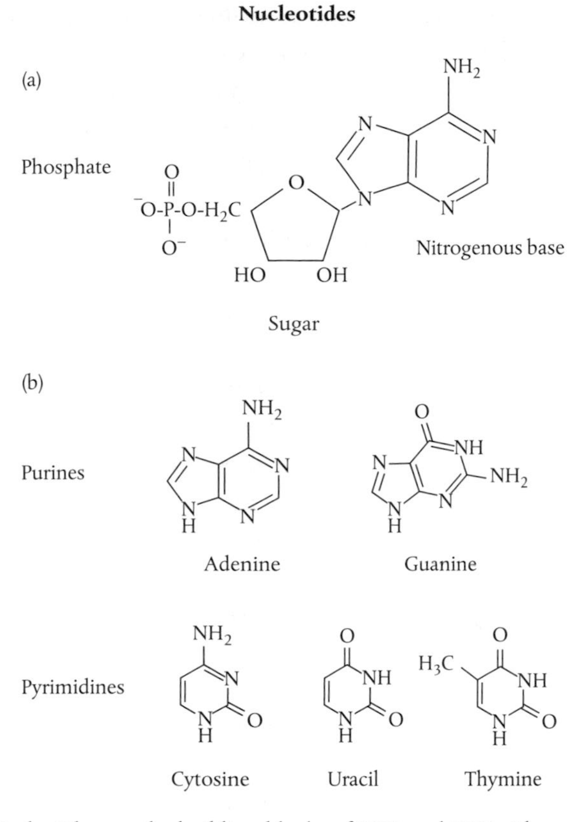

FIGURE 60 Nucleotides are the building blocks of RNA and DNA. They consist of a pentagonal ribose (or deoxyribose) sugar unit, a phosphate unit with one, two, or three phosphate groups, and a nitrogenous base (a). The bases are purines (adenine, guanine) or pyrimidines (cytosine, uracil, and thymine), shown in (b). The nucleotide shown in (a) possesses an adenine base unit and a single phosphate group and is called adenosine monophosphate, abbreviated as AMP.

D-enantiomers of ribose and deoxyribose that are involved in building nucleotides and thence RNA and DNA. We'll be coming back to this later.

Figures 58 and 60 begin to convey something of the complexity of the molecular structures that play important roles in biological systems (although, as the saying goes, you ain't seen nothin' yet). Yes, it's starting to get complicated, but notice how life is founded on a chemistry containing relatively few chemical elements. These are complex chemicals consisting of many chains and rings, but only a handful of elements are involved—predominantly carbon, nitrogen, oxygen, and hydrogen, which between them account for more than 98% of the mass of all living cells, with a little phosphorus and sulphur. We will see in later chapters how other elements, such as iron, get in on the act.

Now, it doesn't take a great leap of imagination to conclude that life could not have got started on Earth unless there was an abundance of these chemical building blocks from which more complex molecules could then form, from the bottom-up.

So, how were these building blocks first produced?

EXOGENOUS DELIVERY: BUILDING BLOCKS FROM SPACE

Perhaps we already have part of the answer to this question. Look back at the Table 2 in Chapter 5. This selection of interstellar molecules includes glycine, the simplest of the 20 amino acids of biological interest. Although the evidence for the existence of interstellar glycine is relatively weak, it seems quite reasonable to suppose that it's nevertheless there, perhaps the result of chemical reactions taking place on the surfaces of interstellar dust grains.

Could there be more? Again, it doesn't seem beyond the bounds of possibility that even more complex amino acids and perhaps the key ingredients for constructing nucleotides might also exist in interstellar space. These compounds haven't yet been detected in molecular clouds. But they *have* been found in meteorites.

The Murchison meteorite is a carbonaceous chondrite that fell to Earth at nearly 11 a.m. local time on 28 September 1969 near the town of Murchison, Victoria, in Australia. As it fell it broke up in the atmosphere, scattering debris over an area of about 13 square kilometres. More than 100 kilograms of the meteorite were recovered.

An analytical study of a 10-gramme sample published in 1970 revealed the presence of glycine, alanine, glutamic acid, valine, and proline. Subsequent studies found tyrosine, methionine, and phenylalanine, and suggestions of an excess of L-alanine over D-alanine. The existence of even a slight excess of the L-enantiomer is puzzling, as chemical synthesis in interstellar space might be expected to produce equal quantities of the L- and D- forms, leading some scientists to suggest that the samples had become contaminated with earthly sources of amino acids which are exclusively of the L-form. But the amino acids serine and threonine, which would have been indicators for such sample contamination, could be found only in trace quantities. This strongly suggests that the other amino acids are not of terrestrial origin.

How might these substances be formed? It is obviously difficult to be definitive, but the evidence from these analyses suggests that the amino acids might have been produced through a sequence of chemical reactions first identified by German chemist Adolf Strecker in 1850. Let's again take alanine as an example. In the Strecker synthesis we start with acetaldehyde (CH_3CHO) and react this with ammonium ions (NH_4^+) in the presence of a source of cyanide (HCN or KCN). The product is a substance called methylaminonitrile (H_2N—$CH(CH_3)$—CN). This reacts further with water, replacing the —CN group with —CO_2H to produce alanine. Note that all the starting chemicals for this synthesis are listed as interstellar molecules in Table 2.

The story doesn't end with amino acids. A further study of samples from the Murchison meteorite published in 2008 revealed the presence of the pyrimidine base uracil and the purine xanthine (a distant chemical relative of guanine, not involved in building RNA or DNA but still of some biochemical significance).

Of necessity, these studies were targeted: the scientists were specifically looking for compounds believed to be relevant to 'prebiotic' chemistry, the logical chemical precursor to biology. A further sophisticated but non-targeted analytical study published in 2010 revealed that the Murchison meteorite contains tens of thousands of different organic substances.

If a single meteorite can carry such a great variety of organic compounds, including many of the building blocks essential to the origin of life, then is it possible that these substances were delivered to Earth exogenously (meaning from 'outside', from interstellar space)? Is the reason that we don't see any evidence for the existence of life on Earth until after the Late Heavy Bombardment due, at least in part, to the fact that this bombardment was responsible for delivering the necessary building blocks in sufficient quantities?

Estimates of the quantity of prebiotic organic substances available on the primitive Earth are sensitively dependent on assumptions about the nature of Earth's early atmosphere. However, in one such estimate, a total of one hundred billion kilograms of organic material is thought to have been made available to the primitive Earth *each year*, of which the dominant source is exogenous delivery by comets.[5]

The amount of biomass on Earth today is estimated to be of the order of a million billion kilogrammes of carbon. If these estimates of the amount of available prebiotic organic material are order-of-magnitude correct, then a quantity equal to today's biomass would have accumulated on the primitive Earth within just ten thousand years.

The notion that the building blocks of life on Earth might have been delivered 'pre-packaged' from space is certainly appealing, but perhaps we shouldn't get too carried away. Delivery is, after all, only the beginning. The building blocks would need to survive the impact (and perhaps a more likely scenario is that these compounds are synthesized in the shock of an impact). They then need to be 'unpacked'. This is less difficult if the packaging is a meteorite consisting mostly of ice, more difficult if it is rock that must be eroded away in a warm sea.

And, what then? Any organic chemicals leaching from meteorites will likely become dispersed in Earth's vast oceans; oceans that are being churned by the gravitational influence of a Moon that is much closer than it is today.

Unfortunately, our second step in the journey to life requires that we string the building blocks together to form long chains, leading eventually to proteins and nucleic acids. This coming together is going to be very difficult if the building blocks are greatly diluted in a large volume of water.

THE PRIMORDIAL SOUP AND THE MILLER–UREY EXPERIMENTS

If we assume that the building blocks were, in part or in full, assembled on Earth then we need to make some further assumptions about the conditions that prevailed on Earth at the time. It's one thing to have all the chemical ingredients available; it's quite another to bring these ingredients together under conditions that will promote the necessary chemical reactions.

Charles Darwin gave some hints about how he thought this might have happened in a letter to his friend John Hooker in 1871:[6]

> It is often said that all the conditions for the first production of a living organism are present, which could ever have been present. – But if (and Oh! What a big if!) we could conceive in some warm little pond with all sorts of ammonia and phosphoric salts, – light, heat, electricity, etc., present, that a protein compound was chemically formed ready to undergo still more complex changes, at the present day such matter would be instantly devoured or absorbed, which would not have been the case before living creatures were formed.

Darwin's 'warm little pond' is a compelling metaphor, one that was translated and expanded in the early 1920s by Russian biochemist Alexander Oparin, and in 1929 by British-born Indian polymath J. B. S. Haldane. This became known as the 'primordial soup' hypothesis. Oparin envisaged a primitive Earth with a strongly reducing atmosphere consisting of methane, ammonia, hydrogen, and water vapour. Haldane proposed that the primitive oceans would have acted like vast chemical laboratories, turning the inorganic compounds present in the primitive atmosphere into a 'hot dilute soup' of organic materials from which biochemistry eventually sprang.

It was all very speculative, but in 1953 the results of some experiments by American chemist Stanley Miller, working in the Chicago laboratory of American Nobel laureate Harold Urey, lent the idea some considerable support.

Miller was inspired to study prebiotic chemistry, initially against the wishes of Urey, his PhD thesis advisor. He and Urey devised a relatively simple experiment in which methane, ammonia, hydrogen, and water vapour circulating around an apparatus is subjected to periodic electric discharges sparking between two electrodes.

Now Darwin had understood that it wasn't enough just to introduce these chemical ingredients to each other. To drive these inorganic substances to react together to form organic chemicals would require an input of energy in some form to get the reactions started. In his metaphor he proposed 'light, heat, electricity, etc.' In the primordial soup hypothesis, the faint young Sun may not have provided much incentive in terms of ultraviolet light and heat.* But an energy source may have been readily at hand in the form of bolts of lightning in a stormy primitive atmosphere.

In the Miller–Urey experiments, the electric discharges were intended to simulate the effects of lightning. The circulating gaseous mixture was cooled and collected in a flask containing liquid water (simulating the ocean), which was also steadily heated and slowly vaporized, the vapour then passing back between the electrodes for a further round of discharges. In this way the collection of simple inorganic compounds was constantly circulated between an electrically charged 'atmosphere' and an 'ocean'. The question was: what kinds of chemical substances might be produced after a week of continuous circulation through the apparatus? Miller wrote:[7]

> During the run the water in the flask became noticeably pink after the first day, and by the end of the week the solution was deep red and turbid.

---

* More contemporary estimates suggest that ultraviolet light from the Sun may have been a significant source of energy and therefore of prebiotic organic substances.

Miller treated the resulting solution with toxic chemicals to ensure that it could not be subsequently contaminated with living organisms. Analysis showed that after just one week, the experiment had produced glycine, alanine and (possibly) aspartic acid.

Miller wrote up the results in a paper, but Urey declined to put his name forward as co-author, as he wanted to ensure that Miller got full credit for the work (tricky, given Urey's vast reputation). After a short hiatus, Miller's paper was published in the 15 May 1953 issue of *Science* magazine. His fame was assured.

There followed further experiments and more painstaking analyses. Miller found nine different amino acids, plus some biologically significant hydroxy acid analogues of amino acids in which the amine group is exchanged for a hydroxyl group (e.g. lactic acid, $HO—CH(CH_3)—CO_2H$, which is the hydroxy acid relative of alanine). He was also able to deduce that the sequence of chemical reactions that had produced the amino acids was essentially the Strecker synthesis.

He secured his doctorate in 1954.

Miller revisited his experiment in the early 1970s. Analytical techniques had come a long way in the intervening years, and he now found 33 amino acids in the resulting solution, including more than half of the 20 biologically significant compounds.

Shortly after his death in 2007, several sample vials containing dried residues were discovered in his laboratory. These dated back to a number of experiments that Miller had conducted during the period 1952–54. Also among these were vials from experiments that he had performed in 1958 on mixtures of methane, ammonia, hydrogen sulphide, and carbon dioxide which, for some reason, had never been analysed. These earlier samples revealed 22 amino acids, many more than Miller had been able to identify at the time. The 1958 experiments produced 23 amino acids, including methionine.

The logic seems inescapable. The building blocks of life were spontaneously generated through natural physico-chemical processes in the Earth's primitive atmosphere and oceans.

But, there's a big 'but'. The problem with this kind of prebiotic chemistry is that what you get out necessarily depends both on what you put in and the conditions you apply, and nobody can be certain of the composition of Earth's primitive atmosphere or the conditions that prevailed four billion years ago.

Dissenting voices were raised. Who said that the Earth's primitive atmosphere must be reducing? Arguments were advanced that any methane and ammonia would have been quickly destroyed by ultraviolet light from the young Sun, and that the atmosphere was instead dominated by molecular nitrogen (N_2) and carbon dioxide (CO_2).* Miller–Urey type experiments with nitrogen and carbon dioxide produce rather poor yields of amino acids, yields that can be partially restored if we make some further assumptions about the content of the primordial oceans.

It's all a bit uncertain. And, irrespective of the conditions, the primordial soup hypothesis doesn't solve the dispersion problem. The Miller–Urey experiments produced a startling range of amino acids because the apparatus used serves to concentrate these compounds in a relatively small water reservoir as they circulate. But on the primitive Earth, any amino acids or nucleotides thus produced were likely to have been greatly diluted in the oceans.†

HYDROTHERMAL VENTS

Look back at the map of the Earth's tectonic plates shown in Figure 57, and particularly the divergent plate boundaries that lie in the middle of the Atlantic and Pacific Oceans. Dotted here and there along these mid-ocean ridges are a series of *hydrothermal vents*. Seawater drawn down through faults or porous rock is heated to temperatures ranging from 60°C to over 460°C by the hot magma before being shot back into the ocean through fissures in the crust. Despite being raised to such high temperatures, the water does not emerge as steam. The pressure at these depths can be 300 times normal atmospheric pressure, and so the water emerges as a liquid or a so-called supercritical fluid, with the properties of both gas and liquid.

When chemical elements dissolved in the rather acidic superheated water hit the more alkaline cold water of the deep ocean (with a temperature around 2°C) they quickly precipitate as minerals, forming cylindrical 'chimneys'. The rapid cooling isn't enough: chemistry is also required. Metal sulphides and oxides captured by vent fluids and brought to the surface would not precipitate until they were exposed to alkaline ocean water.

---

* Remember from the last chapter that a primitive atmosphere rich in carbon dioxide is thought to be necessary to resolve the faint young Sun paradox.

† Unless, of course, some other way can be found to concentrate them. In one suggestion, life's building blocks are concentrated in a type of volcanic rock called scoria, which contains tiny chambers (vesicles) formed by dissolved gases when the rock was in its molten state.

Because these chimney structures are composed of sulphide minerals they tend to be black in colour, and vent a blackened superheated water. They are called 'black smokers', and were first discovered in 1977 along the East Pacific Rise. In contrast, 'white smokers' emit lighter-coloured minerals rich in barium, calcium, and silicon.

Okay. This is all very interesting. But what does it have to do with the origin of life?

These hydrothermal vent systems are today home to a rich diversity of biological communities. These include so-called chemosynthetic bacteria, able to 'digest' hydrogen sulphide, combine it with oxygen dissolved in the ocean and synthesize organic substances in the absence of sunlight. These organisms form the lowest level of a food chain that supports larger organisms such as snails, clams, mussels, shrimp, tubeworms (up to eight feet long!), crabs, fish, and octopi. At the time of their discovery, nobody really expected to find life thriving so deep beneath the Pacific Ocean.

More than three and a half billion years ago this life would not have existed. But does the presence of these thriving communities today belie a simple truth? Could it be that hydrothermal vents offered just the kind of energy source, access to chemical ingredients, and physico-chemical conditions necessary to get life kick-started?

Opinions vary. Some scientists (such as Stanley Miller) have argued that black smokers are just too inhospitable and the conditions too improbable to identify these as places where life might have originated.* These vent systems are relatively young, with most only a few hundred years old, surely too short a time-span to have served as incubators for the origin of life. Complex organic substances are relatively fragile and likely to break up at the kinds of temperatures that prevail here. And, once again, we encounter the same problem of dispersion that plagues the exogenous delivery and primordial soup hypotheses.

Others, most notably Michael Russell (now at the Jet Propulsion Laboratory, part of the California Institute of Technology) and his colleagues, argued in 1989 that a different kind of vent system would be better suited, a system somewhat cooler and more alkaline than the black smokers. Although arguments based on geology and geochemistry could be devised in favour of the

* This doesn't necessarily contradict the observation of life in these places today. The larger organisms found close to black smokers require oxygen, and four billion years ago there would have been no oxygen in the oceans.

existence of such systems, none had been found. They were first discovered in December 2000.

These *alkaline hydrothermal vents* are not driven by volcanic activity beneath the seafloor. Instead, seawater percolates down through fissures in the ocean floor, to depths of 7–8 kilometres, where temperatures are around 200°C. Here the heated water comes into contact with uplifted upper mantle peridotite. These react together to produce another kind of rock called serpentinite.

To get a sense for what happens, let's pick some specific chemical ingredients. Olivine is a principal mineral component of peridotite (Figure 52). It consists of magnesium and iron silicate—Mg_2SiO_4 and Fe_2SiO_4. Magnesium silicate reacts with water to produce the mineral serpentine, $Mg_3Si_2O_5(OH)_4$, and magnesium hydroxide. Iron silicate reacts with water to produce magnetite (an iron oxide), silica, and molecular hydrogen (H_2). In the presence of carbon dioxide, produced by volcanic activity along the plate boundaries and then dissolved in the Earth's primitive oceans, these 'serpentinization' reactions will also produce methane and other organic compounds.

These reactions release a lot of heat energy which drives the now strongly alkaline fluid back to the surface to form a vent, but now with more modest temperatures of 40–90°C. Calcium carried in the alkaline fluid precipitates as calcium carbonate on contact with the cold ocean water and forms white chimney deposits. Magnesium in the vent fluid also precipitates, giving the chimneys a coating of magnesium hydroxide.

Whereas black and white smokers are typically found within a few kilometres of the 'spreading centres' of mid-ocean ridges, the first alkaline hydrothermal vent system—called 'Lost City'—was discovered near the undersea mountain Atlantis Massif, 15 kilometres from the Mid-Atlantic Ridge.* It consists of about 30 carbonate chimneys, each 30–60 metres tall (Figure 61). This system is at least 100 000 years old, possibly older. Some scientists have speculated that this kind of vent field might survive for hundreds of thousands of years.

So, how might alkaline hydrothermal vents make organic compounds? All origin-of-life scenarios require the 'fixing' of carbon from some simple

* Oceanographers tend to wax lyrical when naming vent systems: Loki's Castle, Neptune's Beard, Faulty Towers, Animal Farm, Atlantis II Deep, Bailey's Beads, and many more. A database of confirmed and unconfirmed vent fields is maintained by InterRidge, an international not-for-profit organization which supports mid-ocean ridge research. See http://vents-data.interridge.org.

FIGURE 61 The Lost City alkaline hydrothermal vent field features a collection of about 30 carbonate chimneys each between 30–60 metres tall. This picture shows a five-foot-wide ledge on the side of a chimney which is topped with dendritic carbonate growths.

inorganic source—such as carbon dioxide—and its incorporation into more elaborate chemical structures, and all life today depends on carbon fixation of some kind. We assume that the primitive Earth had plenty of carbon dioxide dissolved in its oceans, perhaps augmented by methane. Now, fixing carbon from carbon dioxide requires some hydrogen.

In most scenarios, the hydrogen has to be stripped from water or other simple inorganic substances, such as hydrogen sulphide. This isn't easy. But alkaline hydrothermal vents are among the very few places on Earth where *molecular* hydrogen is abundant, a side-product of serpentinization. Getting molecular hydrogen to react with carbon dioxide is not straightforward but it is a heck of a lot easier.

There's more. The chimneys themselves are porous. They contain a complex of interconnecting micropores with sizes of the order of a tenth of a millimetre, a kind of miniature Cretan Labyrinth more elaborate than Daedalus himself could ever have imagined. The pores are about 100 times the size of single living cells.

It seems we may now have all the ingredients. We have a steady flow of hydrogen and carbon dioxide bubbling up through a vent system. These combine to produce more complex organic chemicals which become concentrated in the chimney pores, thereby solving the dispersion problem.

I should point out that the hypothesis that life originated in the environment of alkaline hydrothermal vents has many detractors (and the proponents themselves argue about the details). I can only repeat that there is no such thing as a 'standard model' for the origin of life. But I personally find quite compelling the idea of generating a broad variety of organic chemicals in an environment that may serve to trap and concentrate them, through geothermal processes that are startlingly inevitable in Earth's early oceanic crust.

So, bear with me and let's see where this takes us.

THE EMERGENCE OF A PRIMITIVE METABOLISM

Nobody—to my knowledge—has yet attempted to model a late Hadean/early Archean-eon alkaline hydrothermal vent field in the laboratory in order to study the chemical substances that are produced in its mineral pores, although progress in this direction is firmly underway.*

Much of what we think might have happened in the first few chemical steps towards life in this environment is based on our understanding of its geochemistry, supplemented by what we do know of the biochemical mechanisms used by anaerobic microorganisms called acetogens and methanogens. These organisms fix carbon from carbon dioxide and use hydrogen as an energy source, producing acetate ($CH_3CO_2^-$)† and methane, respectively, as waste products. There is some further support from laboratory studies for some of the chemical reactions that may have been involved, demonstrating the general feasibility of the approach. But, make no mistake, there is also a lot of (very well educated and entirely plausible) arm-waving.

Our first challenge is to get molecular hydrogen and carbon dioxide to react together. They will happily do this, but at an interminably slow pace that is

* Nick Lane and his colleagues, at University College London, have built such an 'origin of life reactor' in the laboratory in the university's aptly named Darwin Building. Initial proof-of-concept experiments appear promising—see Hershy *et al.*, *Journal of Molecular Evolution*, **79** (2014), pp. 213–27.

† Acetate ions that pick up a proton (H^+) form acetic acid, CH_3CO_2H, the principal ingredient of vinegar.

not helpful in driving a fledgling biochemistry. One possible solution is to call upon the services of one or more *catalysts*, chemicals that participate in the reaction, providing an alternative pathway (think of it as a lower-energy shortcut), before disengaging relatively unchanged. The reaction also needs an injection of energy—a kick—to get it started.

The catalysts are thought to be iron-nickel-sulphur minerals that would have been components of vent systems in the Earth's young, and oxygen-free, oceans.* Examples include a form of greigite containing nickel impurities, with the formula Fe_5NiS_8. Perhaps it is no coincidence that very similar iron-nickel-sulphur structures (such as Fe_4NiS_5) lie at the heart of catalysts (which, in the world of biochemistry, we call *enzymes*) used by some present-day anaerobic bacteria to interconvert carbon dioxide and carbon monoxide.

Over the last 20 years or so, German chemist and international patent lawyer Günter Wächtershäuser has been building a body of evidence for the importance of iron-sulphur and iron-nickel-sulphur minerals in the origin of life. He was encouraged to publish his ideas by Austrian philosopher Karl Popper in 1988. His theory, called the 'iron-sulphur world', is consistent with the notion that life got started in a hydrothermal vent system† by first developing a cycle of chemical reactions that support a primitive metabolism.

The metabolic cycle used by most living cells today is called the citric acid cycle (or Krebs cycle, named for German émigré biochemist Hans Krebs, who first elucidated the mechanism in 1937). Aerobic organisms use the citric acid cycle to oxidize acetate (from fats, carbohydrates, and proteins, otherwise known as 'food') to produce various biochemically important substances, including chemicals that can be readily converted into amino acids and a nucleotide called adenosine triphosphate (ATP), which serves as a kind of universal energy 'currency', transporting energy around the cell. Carbon dioxide is a waste product, which is 'exhaled'.

Acetogens run this cycle in reverse, fixing carbon from carbon dioxide and producing acetate as a waste. Proponents of the theory that life originated in alkaline hydrothermal vents have therefore argued that the flow of hydrogen and carbon dioxide over an iron-nickel-sulphur catalyst trapped in the

* The alkaline hydrothermal vents observable today do not contain iron in their chimney structures.

† Although Wächtershäuser's target is a 'black smoker' environment.

chimney pores likely yields a variety of chemical substances that assemble and react together to form a crude reverse citric acid cycle.

What drives this cycle is thermodynamics—the hydrogen and methane in the alkaline vent fluid and the carbon dioxide and other oxidized chemicals in the acidic ocean are far from their equilibrium state. Simply bringing them together under these conditions provides enough energy to power their reactions. The relative speeds (called 'kinetics') of these reactions will also be important factors. The reactions most thermodynamically and kinetically favoured win out, and start to dominate the chemistry taking place inside the pores. Once the cycle becomes closed, it will run for as long as there is a steady supply of starting ingredients and catalysts.

This primitive metabolic cycle may be rather creaky and inefficient to start with, but it's not hard to imagine that a ready supply of different chemicals and a few hundred thousand years of trial and error will eventually produce a system of chemical transformations that is relatively stable and sustainable. For example, acetyl phosphate ($CH_3CO_2PO_3^{2-}$) and pyrophosphate ($HP_2O_7^{3-}$) may serve for a time as an energy currency, but as soon as ATP comes along the cycle goes up a gear and ATP takes over the role.

There's one problem. Running the citric acid cycle in a 'forward' direction produces ATP. Running it in the reverse direction *consumes* ATP. To keep the reverse cycle running, a regular and consistent supply of ATP is needed. Where does it come from?

ATP works as a unit of energy currency by transferring its 'activated' phosphate group, releasing the energy bound up in the phosphoanhydride (P—O—P) bond. Losing a phosphate group produces adenosine diphosphate (ADP). In living cells, ADP is 'phosphorylated' back to ATP, ready to do its duty once more.

There are a couple of biochemical mechanisms that living cells deploy to do this, but one of the most important involves something called *chemiosmosis*. Biochemical reactions produce an electrochemical gradient in which a region of low proton (H^+) concentration is separated by a region of high proton concentration by a cell membrane, rather like the separation of electrical charge in a battery. The membrane is semi-permeable, and as protons 'leak' from high to low concentration regions they drive the reaction between ADP and inorganic phosphate to form ATP.

Working with collaborators Allan Hall and Roy Daniel in 1993 (before the discovery of Lost City), Russell suggested that the environment of a submarine alkaline hydrothermal vent sitting in an ocean made acidic by dissolved carbon dioxide provides a natural electrochemical gradient.

When carbon dioxide is dissolved in water it reacts to produce carbonic acid (H_2CO_3), which readily gives up its hydrogen atoms as protons to form bicarbonate (HCO_3^-) and carbonate (CO_3^{2-}) ions. An ocean saturated with carbon dioxide is therefore a region of high proton concentration. In contrast, the fluid emerging from the vent is strongly alkaline—it is low in proton concentration. Russell and Hall argued in 1997 that the resulting gradient can drive the replenishment of pyrophosphate (and, later, ATP) in the chimney pores and keep the primitive reverse citric acid cycle spinning.

If this is right, then the pores in the vent chimneys become tiny chemical factories. Hydrogen and carbon dioxide, perhaps augmented by methane, flow up through the pores dissolved in the vent fluid, where their reaction is catalysed by iron nickel sulphides. A variety of simple organic substances accumulate and eventually a primitive metabolic cycle is established which is fed by the electrochemical gradient between ocean and vent fluid and which converts ADP back to ATP. As the cycle spins, the concentration of organic substances builds up. These substances combine in a series of chemical reactions to produce amino acids, then nucleotides formed from ribose sugar units, inorganic phosphate and purines and pyrimidines.

Anyone who has taken organic chemistry lab classes will likely have experienced on occasion a profound sense of disappointment. Instead of the 'white anhydrous powder' that should have been the product of a laboratory synthesis, all you're left with is a quantity of viscous, sticky goop.* If they're not watched closely, organic substances seem to have an innate tendency to do this. As a veritable cornucopia of organic compounds are produced in the chimney micropores, it therefore seems quite likely that the pore surfaces become quickly coated with 'goop'.

Perhaps within this goop there is a class of naturally occurring molecules called lipids. These consist of long chains of carbon and hydrogen terminated by active end groups, such as —CO_2H (fatty acids). In a sub-class called *phospholipids*, long chains are joined together by an intermediate phosphate group. Such substances serve as building blocks for cell membranes, and will become important towards the end of our origin-of-life story.

So, what happens next?

---

* This happens in the kitchen, too. Maybe it's just me.

THE IRRESISTIBLE RISE OF RNA

There's a line of argument, backed up by computer simulations, that suggests that temperature gradients established between the warm vent fluid and cold ocean water will cause an accumulation of any nucleotides produced in the chimney pores. Now, when it takes place in living cells what happens next typically requires an enzyme. But there are good grounds for believing that it will happen spontaneously on its own, albeit less efficiently.

The nucleotides *polymerize*. They string together like beads on a thread, the phosphate unit of one nucleotide latching on to a hydroxyl group on the ribose unit of its nearest neighbour.

These molecules are already complicated, so we need a simplified way to picture them. In Figure 62(a), I have turned AMP on its side so that the base unit is pointing straight up. The various sub-units that make up AMP are represented schematically on the right. We represent the ribose unit as a simple pentagon. The phosphate unit, represented as a circle, hangs off one apex of the pentagon.

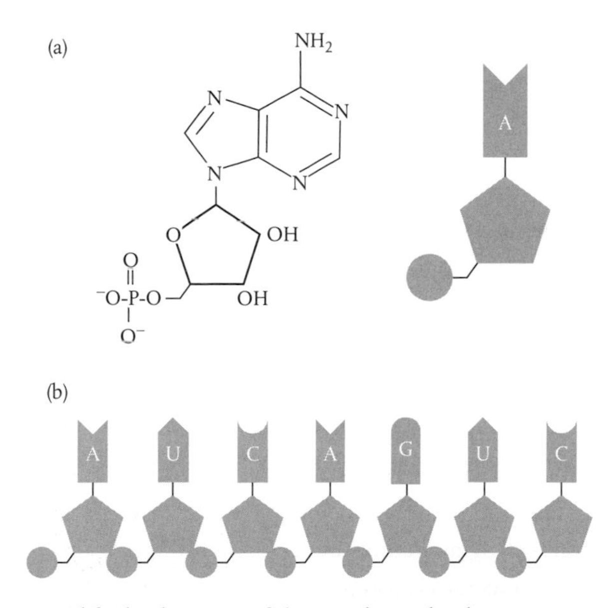

FIGURE 62 To simplify the depiction of the complex molecular structures involved in the chemistry of life we substitute simple geometric shapes to represent the different chemical sub-units. For example, in (a) the ribose unit of adenosine monophosphate is represented as a pentagon, and the phosphate unit as a circle. The adenine base is given a specific shape to distinguish it from guanine, uracil, and cytosine. When formed in sufficient quantities, the nucleotides polymerize to form RNA (b).

There are four different possible base units, which for reasons that will become clear shortly, we represent in terms of different shapes, each labelled A, G, U, or C, for adenine, guanine, uracil, and cytosine.

A short section of the resulting polymer (RNA) is depicted in Figure 62(b) using this simplified way of representing the different sub-units. The joined-up ribose-phosphate units form a kind of 'backbone' to the structure. And, although the short section I've shown here is depicted as linear, in reality the shapes of the molecular orbitals involved and the resulting geometry cause the polymer to bend into a spiral, like a stretched spring or screw thread, with the base units pointing inwards towards the axis. The spiral shape is called a *helix*.

We now encounter what I tend to think of as the most subtle manifestation of the electromagnetic force that it's possible to imagine. The structures of the purine and pyrimidine bases exhibit an extraordinary complementarity. The molecular orbitals associated with each of the N—H bonds in the —NH_2 group in adenine are not uniformly distributed. The central nitrogen atom tends to hog the electrons, leaving the hydrogen atoms without their 'fair share'. As a result, the resulting molecular orbital leaves the positively charged hydrogen atom nuclei (which, remember, are just protons) very slightly exposed.

Now let's recall our discussion of molecular orbitals involving oxygen atoms from Chapter 5. We concluded that the four electrons in the oxygen atom that are not involved in bonding would form 'lone-pair' orbitals. In water (H_2O), the lone pairs form lop-sided rabbit ears. Much the same kind of logic prevails when we consider the carbonyl (—C=O) groups in uracil. The lone pairs protrude from the oxygen atom, providing lobes of exposed negative charge.

You can guess what's coming. The slightly positively charged hydrogen atom in the amine group of adenine latches onto a negatively charged lone pair of electrons protruding from the oxygen atom in the carbonyl group of uracil. The result is called a *hydrogen bond*. This is much weaker than a conventional chemical bond, and is therefore often drawn in chemical structures as a dashed line.

When adenine encounters uracil, two hydrogen bonds are formed, as shown in Figure 63. These bind the base units together. I think you can see from this picture that such hydrogen bonding depends on the specific compositions and geometries of the molecules involved. For example, we can form one hydrogen bond between the amine group of adenine and the carbonyl group of cytosine, but we can't form another, which makes this combination of bases less stable.

In fact, the four bases form two *base pairs*. Adenine pairs with uracil and guanine with cytosine, also shown in Figure 63. If you look at these two pair

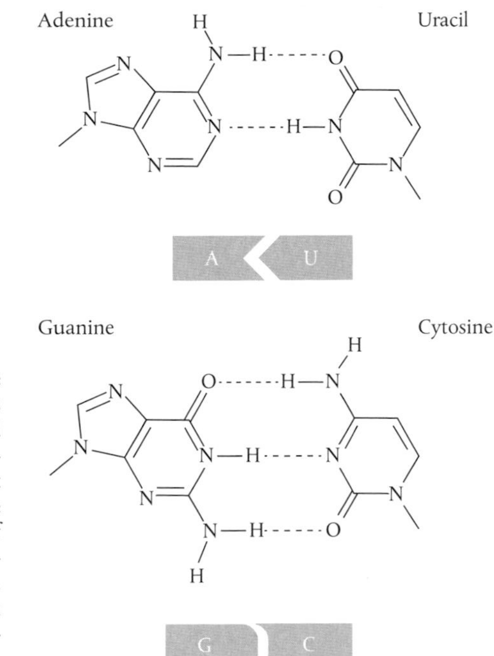

FIGURE 63 The electrons sitting in the molecular orbitals of the amine groups (—NH_2 and N—H) in both purines and pyrimidines are shared unequally, leaving the hydrogen atoms with a slight positive charge. These link up with lone pair electrons on the oxygen atoms of the carbonyl groups and the ring nitrogen atoms to form hydrogen bonds. Although these are much weaker than ordinary chemical bonds, they are sufficient to hold the structures together.

structures closely, you will see that they are very similar. The only difference is that the amine group in guanine and the carbonyl group in cytosine provide a third hydrogen bond.* Now we see the reason for the different shapes that I adopted to depict the base units in Figure 62. These shapes make clear the different pair combinations, A with U, G with C.

Now it starts to get *really* interesting.

Once assembled, a strand of RNA can catalyse its own replication. It pulls in free nucleotides, lining up A with U, U with A, G with C, and C with G. Drawn together in such close proximity, the nucleotides polymerize (Figure 64), forming another RNA strand which we can think of as a kind of photographic

* This reminds me of a moment from one of my all-time favourite television science dramas. *Life Story*, produced by the BBC in 1988, dramatizes the discovery of the structure of DNA. One memorable scene shows James D. Watson (played by Jeff Goldblum), playing around with molecular models of the base units. He chews gum mechanically as he does so (he's an American, and this is 1953). He pairs adenine with thymine (chemically identical to uracil but for a –CH_3 group). Then he pairs guanine with cytosine. He places one pair on top of the other, and realizes that the patterns are almost equivalent. He stops chewing. If you've ever wondered what it might feel like to make a scientific discovery, track down the drama on DVD and watch this scene.

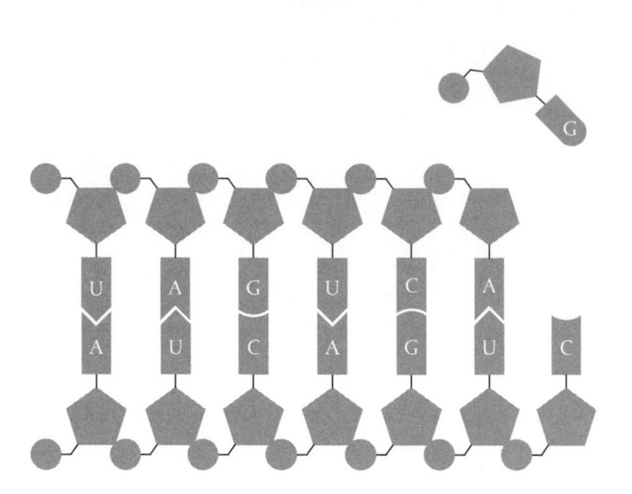

FIGURE 64 Hydrogen bonding between the complementary base pairs allows a strand of RNA to assemble a kind of photographic negative of itself in which A is replaced by U, C with G, etc. When these strands separate, the negative can go on to assemble another positive. In this way, a strand of RNA can catalyse its own replication.

negative of the original. The sequence of base units in the original strand (e.g. AUCAGUCCGAUC . . . , of which the first seven are depicted in Figure 64) produces a negative (UAGUCAGGCUAG . . .). The hydrogen bonds holding these two strands together are relatively weak and easily broken. Once the strands separate, the negative can assemble another strand which replicates the original.

The production of RNA has now become *autocatalytic*. One strand produces two. Two produce four. Four produce eight. Eight produce sixteen. Such doubling is extraordinarily powerful. You can get some sense of this power by considering the following. Place a coin on the top left square of a chessboard. Place two coins on the adjacent square. Four on the next, and so on until you have covered the whole board. How tall is the pile of coins on the 64th square? Well, it has roughly 2^{64} coins, or about 18 billion billion (so, the chances are you ran out of money fairly early on in this exercise). Let's say the coins are each two millimetres thick. The pile on the 64th square is therefore about 37 thousand billion kilometres tall.

What this means is that even if the production of RNA is initially slow and inefficient, once it is formed there's no going back. Rather intriguingly, if—by accident or some quirk of thermodynamics and kinetics—the RNA is formed from D-enantiomers of the nucleotides (so that the helix winds around in only one 'screw' direction), then this is the form that is amplified by replication and quickly comes to dominate.

THE RNA WORLD

In the chessboard analogy, you fail to complete the doubling task all the way to the 64th square because you have limited cash resources. Likewise, the ability of RNA to replicate will depend on the availability of free nucleotides from which it assembles copies of itself. As these molecular resources run out, the replication ceases.

The RNA that is produced may not start out as a single molecular structure, with a fixed number and sequence of base units in the chain. In fact, it's more than likely that different RNA strands are produced with a range of lengths and different sequences of base units. Also, the mechanics of replication are not fool-proof. If, for example, A is 'accidentally' paired with C instead of U when forming the negative, and C in the negative then pairs correctly with G, then in this 'replication' the base unit A has been mistakenly replaced with G. The resulting RNA is then a *mutated* version of the original with a slightly different sequence of base units.

What happens is that these different variations of RNA *compete* for the available resources. This is a competition only in the sense of chemical thermodynamics and kinetics. The RNA variant that is able to react with free nucleotides faster and more easily will replicate at the expense of those variants that can't. Over time, the more optimum (or more 'fit') RNA variant will come to dominate, with the other variants driven to 'extinction'.

This is Darwinian evolution at a molecular level.

Sounds incredible? I want to emphasize that this system of replicating RNA is not yet living. This is just complex chemistry which, according to the version of the story we're following, is occurring in the micropores of chimney structures at an alkaline hydrothermal vent system. But this is chemistry that can in principle happen anywhere, such as in a laboratory test tube.

In 1967, American molecular biologist Sol Spiegelman and his colleagues at the Department of Microbiology at the University of Illinois in Urbana demonstrated that RNA separated from a host virus would happily replicate when provided with the right chemical ingredients and a suitable enzyme. They allowed the replication to 'incubate' for 20 minutes before transferring a small sample to a fresh solution, repeating this process 75 times and reducing the incubation period to maintain a selection pressure in favour of the most rapidly replicating RNA.

These unusual circumstances encouraged the RNA to evolve into a fast-replicating form. What this meant was that, all other things being equal, the length of the RNA strand systematically reduced, as (quite logically) shorter RNA replicates faster than longer RNA. The starting RNA possessed something of the order of 4500 base units. After 74 generations, the evolved RNA consisted of just 218 base units. At the time this was the smallest-known self-replicating chemical entity. It became known as 'Spiegelman's monster'.

The fascinating chemistry of this 'RNA world' has been explored through some truly imaginative experimentation. We don't have an opportunity to dig deeper here, but perhaps there is one further experiment that's worth noting.

In 2009, Gerald Joyce and his co-worker Sarah Voytek at the Scripps Research Institute in California showed that when two different forms of RNA compete for the same chemical ingredients in order to replicate, then the thermodynamically and kinetically favoured RNA will eventually win out, driving the other to extinction. But when the range of ingredients is expanded, the two forms of RNA evolve in a way that allow them to co-exist, with each consuming a different subset of the available ingredients.

This is 'molecular ecology'. Each form of RNA adapts to survive in its own 'ecological niche', in which it can continue to replicate without competition.

There's one further dimension we need to consider. Random errors occurring in RNA replication may produce mutated varieties that are more thermodynamically and kinetically favoured and so replicate more efficiently. These mutants gain the upper hand in the competition for chemical ingredients. Over time, we might imagine that the system drives to a single individual form of RNA with a specific sequence of base units that is best adapted for the set of conditions that prevail. All other forms become extinct.

But computer simulations of molecular evolution performed in the 1970s by German Nobel laureate Manfred Eigen and Austrian theoretical chemist Peter Schuster suggested something rather different. Rather than driving towards a single individual form of RNA, the result of chemical selection pressures is instead a range or *population* of RNA forms with base unit sequences that share close similarities with the most successful sequence.

THE GENETIC CODE

As we've seen, the exogenous delivery and primordial soup hypotheses are focused on the production of amino acids in the prebiotic conditions of the primitive Earth. In the 'metabolism first' scenario, we start by constructing a primitive metabolism from which the RNA world of replication and molecular evolution emerges. So far, amino acids haven't featured.

This is all about to change.

In my description of the RNA world I have paid little attention to the importance of the *sequence* of base units. To all intents and purposes the sequence is random, but a further set of chemical interactions is going to lend a significance to the sequence that will have profound consequences when selection pressures are applied. This is because the sequence of base units *codes* for different amino acids in the production of proteins.

The process by which RNA does this in living cells is now well understood. But this mechanism is likely the result of several million years of chemical evolution, and does not necessarily reflect how this process originated in the prebiotic chemistry that led to the origin of the first life forms.

So, we speculate some more, based on what we know. And what we know is that each of the four RNA base units A, U, G, and C has a chemical affinity for (which translates as 'codes for') specific amino acid *precursors*. As the name implies, these are not amino acids—they are the molecules from which amino acids are made. Remember that the reverse citric acid cycle, which is also spinning away inside the chimney pores, produces—among other things—amino acid precursors.

A single precursor can be used to make a number of different amino acids. A second, adjacent base unit then specifies the precise nature of the conversion. For example, in the combination GC, the base unit G specifies the precursor pyruvate ($CH_3COCO_2^-$, produced in the reverse citric acid cycle) and C specifies its conversion to alanine through a process of 'reductive amination', in which the middle carbonyl group is modified by the addition of an amine group.

But, hold on. There are only eight different possible combinations of two-letter codes formed from an 'alphabet' of four letters. Forming pairs from four base units can only code for the production of eight amino acids, yet there are 20 involved in the biochemistry we know today. What's going on?

The short answer is that the code is actually formed from *three* base units. In principle, a three-letter code would yield 64 possible combinations, but there's a lot of built-in redundancy in the third letter. For example, the combinations GCU, GCC, GCA, and GCG all code for alanine. Such combinations are called *codons*, and form the basis of the genetic information at the heart of all life forms. This *genetic code* is summarized in Figure 65.

It's not at all clear why the code was expanded to a third letter when most of the useful chemical information is carried in the first two. We surmise that selection pressures will likely have favoured a system capable of producing a broader range of amino acids. Thermodynamics, kinetics, and molecular geometry may also have been important factors.

In present-day biological systems, an amino acid connects with an RNA codon through an adaptor molecule. This is another, much shorter length of RNA, called transfer RNA (or tRNA), typically consisting of 73–94 base units. To avoid confusion, we call the longer RNA strand which carries the sequence

<table>
<tr><th>First base</th><th colspan="4">Second base</th><th>Third base</th></tr>
<tr><td></td><td>U</td><td>C</td><td>A</td><td>G</td><td></td></tr>
<tr><td rowspan="4">U</td><td rowspan="2">Phenyl-alanine</td><td rowspan="4">Serine</td><td rowspan="2">Tyrosine</td><td rowspan="2">Cysteine</td><td>U</td></tr>
<tr><td>C</td></tr>
<tr><td rowspan="6">Leucine</td><td>STOP</td><td>STOP</td><td>A</td></tr>
<tr><td>STOP</td><td>Tryptophan</td><td>G</td></tr>
<tr><td rowspan="4">C</td><td rowspan="4">Proline</td><td rowspan="2">Histidine</td><td rowspan="4">Arginine</td><td>U</td></tr>
<tr><td>C</td></tr>
<tr><td rowspan="2">Glutamine</td><td>A</td></tr>
<tr><td>G</td></tr>
<tr><td rowspan="4">A</td><td rowspan="3">Isoleucine</td><td rowspan="4">Threonine</td><td rowspan="2">Asparagine</td><td rowspan="2">Serine</td><td>U</td></tr>
<tr><td>C</td></tr>
<tr><td rowspan="2">Lysine</td><td rowspan="2">Arginine</td><td>A</td></tr>
<tr><td>Methionine</td><td>G</td></tr>
<tr><td rowspan="4">G</td><td rowspan="4">Valine</td><td rowspan="4">Alanine</td><td rowspan="2">Aspartic Acid</td><td rowspan="4">Glycine</td><td>U</td></tr>
<tr><td>C</td></tr>
<tr><td rowspan="2">Glutamic Acid</td><td>A</td></tr>
<tr><td>G</td></tr>
</table>

FIGURE 65 The genetic code. Sequences of three base units in RNA, called *codons*, specify the sequence of amino acids that will be assembled to make proteins. There is considerable redundancy in the code, with many different three-letter codons specifying the same amino acids.

of codons messenger RNA (or mRNA). The reason for this particular choice of name will become apparent shortly.

Transfer RNA has a couple of loops, an acceptor stem, which binds to the amino acid, and an *anti-codon* arm, which contains a sequence of three base units that matches the mRNA codon. For example, the anti-codon for GCU is CGA. The tRNA with this anti-codon will selectively bind with alanine.

What happens now is that the sequence of three-letter codons in the mRNA strand draws together a sequence of amino acids that sit on the acceptor stems of the tRNA adaptors (Figure 66). The amino acids may now polymerize, forming so-called peptide bonds (—CO-NH—) between them. When first

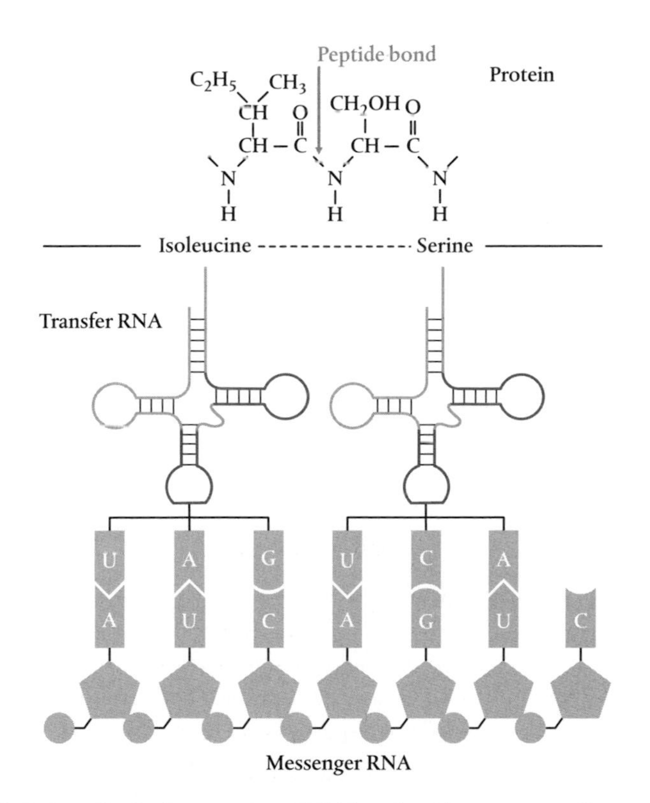

FIGURE 66 Each codon in the messenger RNA bonds with a corresponding anti-codon on the anti-codon arm of the transfer RNA (not drawn to scale in this figure). For example, the codon AUC bonds with the anti-codon UAG. From the genetic code in Figure 65 we know that AUC codes for the amino acid isoleucine, which bonds to the acceptor stem of the tRNA. The next codon AGU codes for serine. As the amino acids line up, they polymerize, forming peptide bonds between them. The sequence of amino acids then determines the properties and molecular geometry of the resulting protein.

encountered, the codon AUG (which also codes for methionine) indicates the *start* of this polymerization process. The codons UAA, UAG, and UGA indicate when the process should *stop*. The end result is a protein molecule, the sequence of mRNA codons between start and stop specifying the sequence of amino acid units in the protein. This sequence in turn determines the properties and molecular geometry of the protein, the way it will 'fold' into a unique three-dimensional structure.

It may not have worked quite this way in the chemical systems evolving in the chimney pores but, make no mistake, this is a game-changer. Proteins are ubiquitous life molecules with a broad variety of biochemical functions. Among other things, they act as enzymes, and they transport other molecules around the cell. The production of proteins now opens up a truly vast array of possibilities, many of which can make the process of RNA replication so much more efficient.

The chemical system now really kicks into overdrive. The deployment of proteins moves it firmly into the territory of biochemistry. Selection pressures now shape the evolution not only of the RNA but of the whole chemical system within the chimney pores. The game is much the same. The population of RNA sequences that replicate most successfully will come to dominate, but now success is also measured in terms of the range of proteins that the sequence can produce.

One further thought. If, by chance or chemical necessity, the D-enantiomers of the nucleotides came to dominate RNA replication, then the resulting geometry of the RNA may in turn demand that only the L-enantiomers of amino acids can be assembled into proteins.

THE INVENTION OF DNA

The sequence of base units in the RNA has now gained a tremendous significance. But the RNA molecule is rather exposed and vulnerable, prone to mutations which may lead to an under-optimized sequence, impairing its ability to compete in the replication stakes. Once again, there's an evolutionary advantage for the chemical system that finds a way to preserve its sequence of RNA base units.

The answer, of course, is DNA. Making a strand of DNA is, broadly speaking, no more difficult than making RNA. Deoxyribose replaces ribose, which makes the resulting nucleic acid a little less prone to reactions with water. And thymine replaces uracil as a pyrimidine base, structurally identical but for an extra —CH_3 group (look back at Figure 60).

Although this step is unusual, there is a modern precedent. Retroviruses such as human immunodeficiency virus (HIV) use an enzyme called reverse transcriptase to assemble DNA from RNA. Once the optimum sequence of base units has been translated into one of the helical strands of DNA, it pairs with its base unit counterpart to form the famous double helix structure (Figure 67). The sequence of codons previously held in the RNA is now much better protected.

The DNA replicates quite happily on its own (with the aid of suitable enzymes). The two strands of the double helix separate, forming a complementary 'positive' and a 'negative'. The positive assembles a matching negative and the negative assembles a matching positive. From one DNA molecule we now have two. Then four. Then eight. And so on.

Making proteins is now a little more involved. The DNA transcribes its base unit sequence into a strand of mRNA through a series of biochemical steps. The

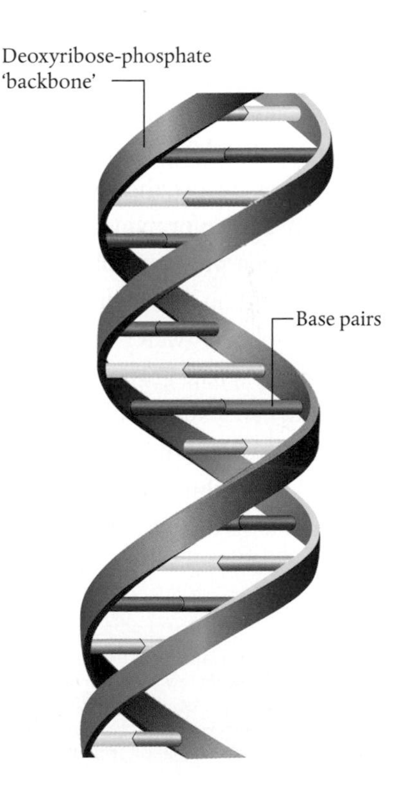

FIGURE 67 The iconic double helix structure of DNA was first elucidated in 1953 by James D. Watson, Francis Crick, Maurice Wilkins, and Rosalind Franklin. It consists of two spirals formed by two deoxyribose-phosphate 'backbones' tied together by hydrogen bonds between the base pairs adenine-thymine and guanine-cytosine.

mRNA then provides the template for assembling proteins, as we have seen. This is the reason it's called 'messenger' RNA: it carries the 'message' of the genetic code from DNA into the chemical system used to synthesize proteins. This, by the way, is the 'central dogma' of molecular biology.* Genetic information generally flows from DNA to RNA to proteins.

We're almost there. Systems developing in the chimney pores that are particularly successful may expand into or 'infect' neighbouring pores. We can imagine that a thriving 'community' develops.

Have we now crossed a threshold? Is this community of replicating chemical systems 'alive'? Without a clear understanding of what life is, it's obviously hard to be definitive. But almost all of the components of the chemical apparatus of what we will come to regard as living systems are now up and running.

THE BIRTH OF LUCA

Almost. This is a community that remains firmly (though comfortably) confined to its womb of chimney micropores. To become, as *Star Trek*'s Bones would have it, 'life as we know it', it must find a way to be 'born'.

Let's continue to exercise our imaginations. As the pores start to become clogged with chemicals that are busy metabolizing and replicating, it's not hard to imagine that these chemical systems start to get squeezed out. Perhaps whole populations are lost, the essential components of their chemical apparatus dispersed into an unforgiving sea.

Long-chain protein molecules or lipids now come to the rescue. The long chains of carbon and hydrogen in phospholipids behave much like oil—they are 'hydrophobic' (water-hating) and don't mix well with water. In contrast, the charged phosphate groups are 'hydrophilic' (water-loving). Once again, simple thermodynamics drives phospholipids to form *bi-layers*, a sandwich structure in which the hydrophobic chains are installed in the middle, shielded from water, with the hydrophilic groups installed on the outside, in contact with water.

The result is a *membrane* which serves to hold the chemical system in place, forming a cell.

---

* First stated by Francis Crick in 1958, although he misused the word 'dogma' (a belief that is beyond doubt). At the time there was little evidence for Crick's assertion and, as Crick himself acknowledged in his autobiography, 'central hypothesis' would have been a better term. However, we know today that this hypothesis is broadly correct although there are exceptions, such as retroviruses which can transcribe genetic information from their RNA into the DNA of the host they have infected.

There's still much to do, however. Molecular selection pressures have driven the evolution of a population of systems that has come to depend on several chemical 'tricks' in order to survive. Some of these tricks are no longer accessible in the environment of a cell. But, rather than start over, the changing environment instead selects cells able to reproduce or mimic these tricks.

For example, the natural electrochemical gradient between ocean and vent fluid that is thought to have been available in the pores to drive the conversion of ADP back to ATP will be lost as the cell wall becomes impermeable to the movement of H^+ ions. Starved of energy, such cells would quickly die out.

But this doesn't happen straight away, and cells that can evolve an alternative solution will greatly increase their chances of survival. In a paper published in 2012, Nick Lane and William Martin suggested that these fledgling proto-cells adapted as their cell walls became increasingly impermeable, drawing on analogies with the biochemistry of methanogens and acetogens.[8] They argued that the walls of the proto-cells would have first become impermeable to larger positive ions, such as sodium (Na^+).

This offered selection advantages to cells that had evolved a simple protein carrier—called an anti-porter—which carries Na^+ ions out of the cell as H^+ ions continue to leak in. This device couples the different types of ions together and creates electrochemical gradients based on Na^+ inside the cell. As the cell walls eventually seal off and became impermeable to H^+ ions, those cells that had evolved this trick could survive on their Na^+ gradients, continuing to drive the production of ATP and becoming self-reliant. These cells survive and replicate, and soon come to dominate. And they are now free to leave the confines of the vent pores.*

There are plenty more challenges to come. As the cells drift away from the vent system that gave birth to them, they lose access to the precious source of molecular hydrogen they need in order to fix carbon from carbon dioxide, metabolize, and grow. Once again, the changing environment will favour a population of cells that are adapted to extract hydrogen from other sources.

---

* Whilst the demand for an electrochemical gradient is universal to all cellular life, in principle any chemical trick used to replicate this inside the fledgling cells will be satisfactory provided it works reliably. It is not difficult to imagine that different populations of systems within the vent pores settle on different alternative solutions, leading to the evolution of distinct types of proto-organisms that we might recognize as primitive bacteria and archaea. In their modern form these organisms share many of the same essential biochemical processes, yet differ in their membrane biochemistry, possibly linked to different electrochemical gradient fixed. See Sojo et al., *PLoS Biology*, **12** (2014), p. e1001926.

Although we can't be specific about the precise nature of its biochemistry, we can nevertheless now welcome LUCA into the world. This primitive *prokaryote* (an organism formed of a single simple cell) will go on to shape all life on Earth.

It doesn't really matter whether you are inclined towards a theory based on exogenous delivery, primordial soup, or hydrothermal vents (or a combination of these—they are not mutually exclusive) or alternative scenarios that I've not had the opportunity to consider here. The one conclusion we can draw from all these approaches is that production of the building blocks of life appears to be a natural and inevitable consequence of chemistry on a rocky planet with liquid water. Life on Earth is deeply rooted in the composition and properties of, and the conditions on, planet Earth itself.

The particular sequence of events I've given here—primitive metabolism to RNA world to protein synthesis to DNA and cellular structures—is hotly debated among scientists specializing in the origin of life. In truth, events may not have unfolded in quite this way or even in this order. But, however it happened, the transition from building blocks to biologically significant molecules—ATP, DNA, RNA, and proteins—also appears rather irresistible. There really aren't that many tricks, and there's plenty of time to find them through trial and error. And, it seems, the principles of evolution through natural selection were already at work at a molecular level in the emerging systems chemistry.

There is plenty of uncertainty. There are lots of gaps and much still to learn. But we can craft a half-way plausible story. We are therefore just left with the assumption that, about four billion years ago, a complex chemical system evolving in this way at some stage crosses a threshold and becomes a living thing.

The world will never be the same again.

9

SYMBIOSIS

The Origins of Complex Cells and Multicellular Organisms

However it happened, abiogenesis changed all the rules. The sequence of base units carried within DNA, the transcription of this sequence via RNA and its role in the construction of proteins mean that we must now start to reach for a different language. At its root, this is a language that is still firmly grounded in chemical thermodynamics and kinetics—the energies and speeds of complex chemical reactions. But its words tell of a level of structure and organization that lies higher than chemistry.

This is the language of genes and heredity. If we're going to follow the next episode in our scientific creation story, we'll need to be familiar with at least a few of its key words.

The full sequence of bases in the DNA of an organism is called the *genome*. A gene is then defined as a section of the genome that contributes to a specific heritable biological property or trait, which we can think of as a kind of 'unit of inheritance'. Genes are not, as sometimes described in the popular press, like smartphone apps ('we've got a gene for that'). Life is much more complicated. Biological traits are typically traced to the operation of many genes,* and some genes may play a multiplicity of roles.

* For example, there may be up to 16 different genes responsible for eye colour in humans, but the two main genes are thought to be oculocutaneous albinism II (OCA2), a sequence of almost 350 000

An organism's specific collection of genes, or its gene sequence, is responsible for determining what set of proteins is to be constructed, and therefore what kind of organism it is, how it is to be put together, how it will metabolize and grow, and how it will replicate.

Now, we know that a strand of DNA has a double helix structure, as Watson, Crick, Wilkins, and Franklin discovered in 1953. But in living organisms the DNA adopts a higher-level geometric shape, referred to as a *chromosome*. In prokaryotes, it coils into a circle, the two 'ends' of the molecule joining together, making it a little less vulnerable to chemical attack.*

We're now in a position to give a rough outline of the structure of a prokaryotic cell. The circular chromosome forms what is known as a *nucleoid* ('nucleus-like') and floats freely in the cell cytoplasm, which consists of about 80% water with some inorganic salts and organic molecules. Dotted here and there in the cytoplasm are *ribosomes*, small molecular 'factories' which turn the genetic code mapped onto mRNA into proteins. The cell is enclosed by a membrane of phospholipid bi-layers which also contain some embedded proteins.

The genes are passed on to successive generations of the organism through the replication of its chromosomal DNA. In prokaryotic cells, this happens through the process of *cell division*. This starts with the replication of the circular chromosome to produce an identical copy. The two identical chromosomes separate within the cell and move apart. The cell expands, and the membrane pinches together and seals off the two chromosomes, each with a complement of ribosomes, to form two genetically identical 'daughter' cells.

If all life on Earth is indeed descended from the population of prokaryotic proto-organisms that first appeared about four billion years ago, and which we call LUCA, then in theory we should find some evidence of this ancestry in all forms of life on Earth today.

Of course, random mutations and evolution operating over billions of years has transformed the nature and, most importantly, the diversity of life forms on Earth, and we will explore some of the earliest of these evolutionary developments in this chapter. But beneath all of this glorious diversity there lies a

---

nucleotides which codes for a protein involved in the production of the pigment melanin, and a gene called HERC2. It is believed that a mutation in HERC2 occurred between 6000 to 10 000 years ago, affecting the expression of OCA2 and turning brown eyes blue.

* Apart from egg and sperm (sex) cells, human cells have 46 chromosomes, formed from chromosomes inherited from the parents. Human sex cells have 23 chromosomes.

foundation, a 'core' set of genes that can in principle be traced all the way back to LUCA.

The evidence for this heritage comes from the application of techniques used to determine the genomes of different life forms. In the 1970s, American microbiologist Carl Woese pioneered the use of sequencing techniques applied to ribosomal RNA (rRNA), which constitutes about 60% of the ribosome (the rest is proteins). The ribosomal RNA catalyses the formation of peptide bonds between amino acids attached to the acceptor stems of neighbouring tRNA molecules described in the last chapter, thereby stitching the protein together like a sewing machine.

These sequences allowed Woese and his colleagues to determine the 'relatedness' of different forms of life and identify a kind of 'lowest common denominator' collection of genes that are shared across all life forms. This is not simply a question of tracking genetic inheritance 'vertically' back down the generations to some point of origin. Single-celled organisms can also transfer genes 'horizontally', by inserting foreign genes into a cell's DNA or RNA.*

As I described in the last chapter, we can plausibly construct LUCA from the bottom up through the development of a primitive metabolism, the emergence of RNA and a genetic code, protein synthesis, and the building of cell membranes (though not necessarily in this order). The 'top-down' genetic profile that emerges from the sequencing of rRNA in present-day organisms describes a proto-organism that more or less matches this bottom-up synthesis. This is no coincidence, of course. Any origin-of-life scenario must be developed with more than half an eye on where we were trying to get to.

I'm calling LUCA a 'proto-organism' because the evidence doesn't suggest that it can be considered to be a fully fledged, separately identifiable organism. In *On the Origin of Species*, first published in 1859, Charles Darwin wrote:[1]

> Therefore I should infer from analogy that probably all the organic beings which have ever lived on this earth have descended from some one primordial form, into which life was first breathed.

It may not have been quite as simple as this. Writing 139 years later, in 1998, Woese commented:[2]

* Horizontal gene transfer is primarily responsible for the evolution of strains of bacteria that are resistant to antibiotics.

> The ancestor cannot have been a particular organism, a single organismal lineage. It was communal, a loosely knit, diverse conglomeration of primitive cells that evolved as a unit, and it eventually developed to a stage where it broke into several distinct communities, which in their turn become the three primary lines of descent.

There are arguments either way, but there's little doubting that, however it was constituted, LUCA gave rise to three fundamentally important lines of descent. We will go on to consider these shortly.

THE EVOLUTION OF PHOTOSYNTHESIS

Of one thing we can be reasonably certain. It seems that LUCA was born in virtual darkness in an environment with little or no free oxygen, lending support to the notion that life began in the deep, dark ocean. However they were formed, these first organisms would have faced the challenge of finding readily available sources of hydrogen with which to fix carbon from carbon dioxide and—even more importantly—sufficient energy to do so. Even if we accept the idea that LUCA was born in the environment of an alkaline hydrothermal vent, it would have lost access to its source of molecular hydrogen just as soon as it started to move away.

It would have worked something like this. Primitive organisms carrying bits and pieces of chemical machinery that could be re-purposed to strip hydrogen atoms from some other available source (such as hydrogen sulphide) would have survived and passed the relevant genes for this machinery to successive generations. Organisms without this advantage would be starved of hydrogen. They would become unable to fix carbon and metabolize, and would quickly die out.

I said 'virtual darkness'. In fact, hot magma seeping from a hydrothermal vent at a mid-ocean ridge glows in the cold sea and produces a very faint light. Today this is enough, it seems, for a species of green sulphur bacteria, which use the energy available in this faint light to fix carbon from carbon dioxide using hydrogen sulphide as a source of hydrogen.[3] It's not difficult to suppose that, as the primitive descendants of LUCA drifted around the bottom of the ocean, those with the chemical machinery to *photosynthesize*, to harness light energy in order to metabolize, would possess an evolutionary advantage.

The existence of such bacteria today provides important clues about these early life forms. Although the individual biochemical mechanisms are complex (as we will see soon enough), the general principles are relatively straightforward.

Photosynthesis works by using light energy to pull electrons from a suitable donor (such as hydrogen sulphide, stripping out the hydrogen in the process), and using these electrons to drive a sequence of reactions that either produce ATP or an intermediate called nicotinamide adenine dinucleotide phosphate (NADPH), which can be used to fix carbon and feed a metabolic cycle.

The light is captured by a protein structure called a photo-active centre, a variant of the familiar pigment *chlorophyll*. Green sulphur bacteria utilize a form of bacteriochlorophyll which is somewhat different compared with the more familiar green plant pigment, and which is sensitive to infrared light with wavelengths between about 700 and 750 nanometres.

The photosynthetic mechanism used by green sulphur bacteria is called Photosystem I. Other forms of bacteria use a second mechanism called Photosystem II. All that we really need to remember is that the photo active centre absorbs light, pulls in electrons, and either stores this energy (Photosystem II) or fixes carbon (Photosystem I). The photosystems are closely related and likely evolved from a single ancestral system. Unfortunately, horizontal gene transfer between prokaryotes tends to jumble the genes between different types of organism and makes it impossible to trace the genealogy.

This kind of photosynthesis is called *anoxygenic*, meaning that it does not involve the production of oxygen. The first photosynthetic organisms would likely have made use of this form, using one or other of the two photosystems.

But these organisms are still tied to environments that provide ready sources of hydrogen and electrons, such as hydrogen sulphide. This seems curious, as they now have access to a new fund of light-energy and they are surrounded by water which, of course, contains hydrogen atoms and can act as an electron-donor. Unfortunately, the cost in energy terms of stripping the hydrogen and electrons from water is just too high.

Impasse.

Evolution isn't easily beaten, however. The photo-active systems used in anoxygenic photosynthesis can't individually produce enough energy to strip hydrogen and electrons from water, but what if we now string a couple of these systems together in series? John Allen, a biochemist at Queen Mary College, University of London, has argued that early prokaryotic organisms may have evolved both photosystems in parallel (rather than evolving them individually in separate organisms), turning one or other on or off depending on the external environment. If this is right, then it would appear to be a relatively small step to harness the two photosystems in series.

This is another game-changer.

The resulting mechanism runs something like this (Figure 68). A couple of water molecules bond to an enzyme consisting of a core of four manganese atoms and a calcium oxide unit surrounded by proteins which, according to Russell, was likely 'borrowed' from useful bits of minerals such as ranciéite ($CaMn_4O_9$). For reasons that will become clear quite soon, this is called the 'oxygen-evolving complex'. It sits at the exposed edge of a larger protein structure that contains the more familiar form of chlorophyll, which absorbs predominantly blue and red light (reflecting green), with a peak absorption around 680 nanometres.

The photo-active centre in chlorophyll absorbs a photon of light, promoting an electron from a lower to a higher-energy molecular orbital. The excited electron doesn't come back. Instead, it goes on to trigger a series of chemical reactions that we can think of as 'transporting' the electron along a chain, producing a molecule of ATP through chemiosmosis. This is Photosystem II.

The electron is then passed on to the active centre of a second chlorophyll system, Photosystem I, with a peak absorption around 700 nanometres.

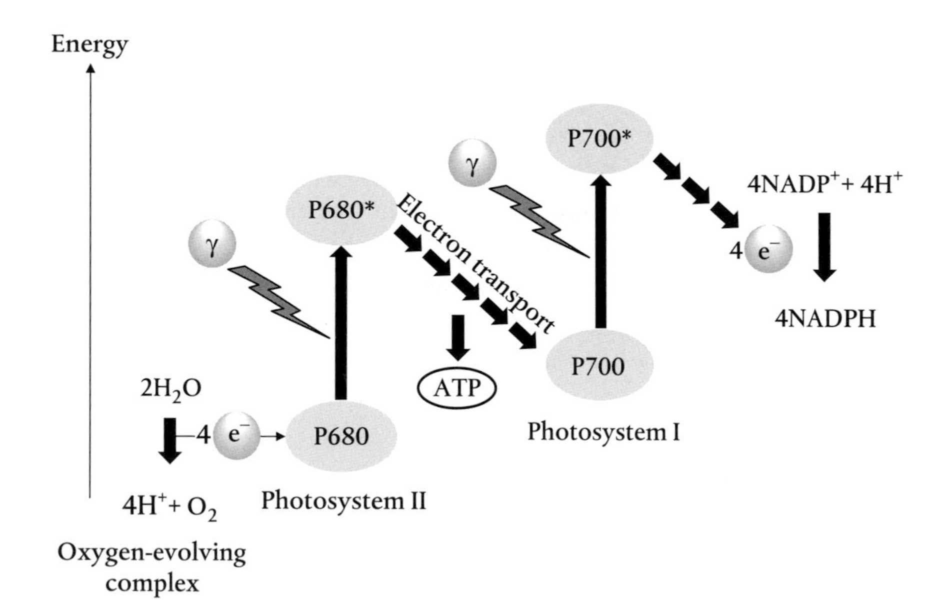

FIGURE 68 The 'Z-scheme' of photosynthesis, which uses light to fix carbon from CO_2 using water (H_2O) as the source of hydrogen. Molecular oxygen (O_2) is released as a waste by-product.

Meanwhile, the positively charged 'hole' left behind by the excited electron in Photosystem II is filled by ripping an electron from an O—H bond in a molecule of water bound in the oxygen-evolving complex, liberating a proton (H^+) in the process.

The chlorophyll in Photosystem I absorbs another photon, promoting an electron to an even higher energy molecular orbital. This in turn triggers another cascade of reactions involving an iron-sulphur protein before pushing the electron onto a positively charged form of nicotinamide adenine dinucleotide phosphate ($NADP^+$). The electron is added to $NAPD^+$, and it goes on to form a hydrogenated version, NADPH.

We're almost there. NADPH now provides the source of hydrogen used to fix carbon from carbon dioxide, with the aid of another enzyme and some ATP in a catalytic cycle called the Calvin cycle, named for American chemist Melvin Calvin. The carbon atom is inserted into a molecule called ribulose biphosphate. The resulting product is unstable, and immediately falls apart to produce two molecules of triose phosphate ($CHO—CH(OH)—CH_2OPO_3^{2-}$), which can then go on to form a molecule of glucose. This is called the 'dark reaction', as it doesn't involve the absorption of light.

Let's now imagine that we go around this process another three times, making four times in total. In doing so, Photosystem II pulls four electrons from two molecules of water in the oxygen-evolving complex, releasing four protons. The two oxygen atoms left behind combine to form molecular oxygen, O_2, in a reaction we can write overall as $2H_2O + \text{photons} \rightarrow O_2 + 4H^+ + 4e^-$ (and now we understand why it's called the oxygen-evolving complex). At the other end of Photosystem I, four electrons (and four protons) produce four molecules of NADPH, in a reaction we can write as $4NADP^+ + 4H^+ + 4e^- + \text{photons} \rightarrow 4NADPH$.

It takes two molecules of NADPH to help fix a single carbon atom from carbon dioxide in the dark reaction, so we can write a crude overall equation as $2CO_2 + 2H_2O + \text{photons} \rightarrow$ 'carbohydrate' $+ O_2$. I've put carbohydrate in inverted commas because the end product of the dark reaction is actually a precursor containing three carbon atoms, not two, so the equation as written doesn't quite balance. You get the idea though.

It doesn't matter that the hydrogen that's used to fix carbon doesn't come from the hydrogen liberated from the water molecules. It's the net result that counts. This is *oxygenic photosynthesis*, so named for rather obvious reasons.

Accessing the hydrogen atoms from water for biological purposes represents a singular moment in the evolutionary history of life on Earth. But this comes at a high cost. The waste by-product—molecular oxygen—is highly reactive and quite toxic. It begins to seep into the oceans and the atmosphere.

The terraforming of Earth has begun.

EVIDENCE FOR EARLY LIFE FORMS

When asked how long it might have taken for life to emerge on Earth, Stanley Miller speculated that: 'a decade is probably too short, and so is a century. But ten or a hundred thousand years seems okay, and if you can't do it in a million years, you probably can't do it at all.'[4]

So, what is the nature of the evidence for the oldest forms of life on Earth? Such a search inevitably relies on the curious processes of *fossilization*. These occur when, soon after its death, an organism is buried in sediment precipitated and accumulated beneath rivers and oceans, and which settles to form layers of rock. Spaces within the organism that are filled with liquid (such as cytoplasm) are now filled instead with water rich in minerals. The minerals precipitate, producing rock patterned by the shapes of the spaces in the organism. Most of the organic material decays, leaving a record of the organism's existence in the sedimentary rock.

This is a chance event, of a kind that doesn't happen very often. To make matters worse, the Earth's crust is constantly being renewed at the spreading centres, with older crust (and any sedimentary rock it might be carrying) disappearing at convergent plate boundaries. Consequently, much of the Earth's continental crust is much younger than the Earth itself, its slate regularly wiped clean.

But there are parts of the world where there is exposed rock that is less than a billion years younger than the Earth. For example, the Kaapvaal Craton in Southern Africa and the Pilbara Craton in Western Australia are thought to have once been joined together in Earth's first supercontinent, called 'Vaalbara' (a name derived by joining the second syllables of Kaapvaal and Pilbara). This is believed to have formed about 3.6 billion years ago, and started to break up about 2.8 billion years ago. The word 'craton' indicates that these are particularly stable rock masses, surviving for billions of years.

Within the Pilbara Craton lies the Warrawoona Group, a 12 kilometre thick formation consisting mostly of volcanic rock with some fine-grained sedimentary rock (called 'chert'). Analysis of zircons found in the volcanic rock at the base of this formation has allowed geologists to determine that it is 3471 million years old. Zircons found in volcanic rock at the top of the formation are a slightly younger at 3458 million years old.

Hints of structures in Warrawoona chert first emerged in 1980, and were tentatively identified as *stromatolites*. These are formations produced in sedimentary rock by communities of bacterial cells that come together in films or 'mats', the most common example being cyanobacteria (sometimes—and wrongly—called 'blue-green algae'). These are photosynthesizing bacteria, ubiquitous today and one of the most successful microorganisms on the planet.

In 1986, palaeontologists J. William (Bill) Schopf and Bonnie Packer from the University of California at Los Angeles reported striking evidence for the existence of ~3465 million year old photosynthesizing bacteria in the Warrawoona chert. In 1993, Schopf followed through with evidence for 11 different types of ancient cyanobacteria-like organisms (see Figure 69 for some examples).

In 2002, Oxford palaeontologist Martin Brasier challenged Schopf's findings. He argued that the structures observed in the Warrawoona chert had been misinterpreted. They were caused by purely geological, not bio-geological processes, he claimed. He accused Schopf and his colleagues of seeing what they wanted to see, arguing, in a subsequent paper, that: 'We have tended to ask "what do these structures remind us of" rather than "what are these structures"?'[5]

Schopf countered with evidence of remnant organic material characteristic of photosynthesizing organisms found at the margins of these structures, suggestive of the decayed remains of the organisms that had created them. Brasier dismissed this as carbon from inorganic sources.

Their dispute was quite incendiary.

In a 2006 paper, Schopf summarized the evidence from 48 examples of Archean-eon structures dated between 3496 and 2548 million years old from Australia, South Africa, Canada, India, and Zimbabwe. He had by this time accepted that these could not be the fossils of ancient photosynthesizing cyanobacteria, but closed with the claim: 'These compilations support the view of most workers in the field of Precambrian palaeobiology, worldwide, that the "true consensus for life's existence" dates from more than or equal to 3500 million years ago.'[6]

A 2009 study of stromatolites in the Strelley Pool Formation (part of the Warrawoona Group) by a team of American and Australian palaeontologists rejected the possibility of microfossils, but went on to conclude that '... evidence preserved in the Strelley Pool Formation suggests that microbial mat communities probably existed 3.43 billion years ago in the Pilbara

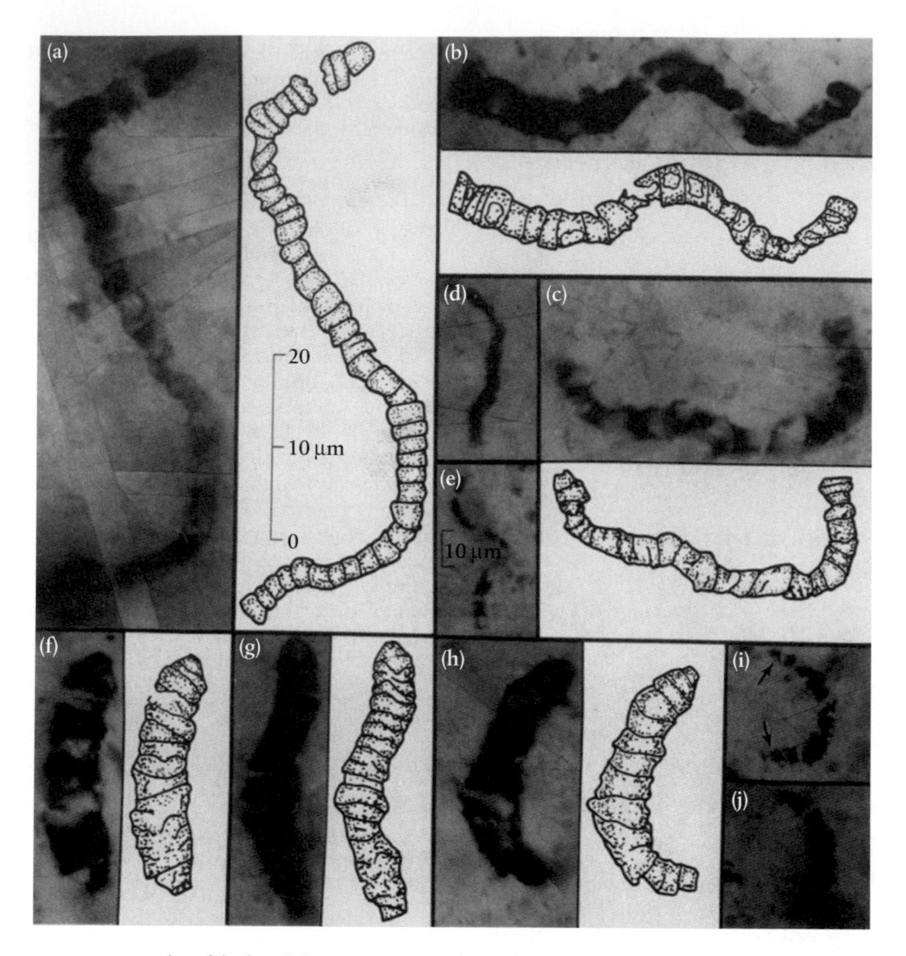

FIGURE 69 Schopf declared these structures, found in the Warrawoona chert, to be examples of *microfossils*, the fossilized remains of some of the earliest microscopic organisms. The drawing to the right of image (a) provides a scale, in which μm means micrometres, 10^{-6} metres. Schopf referred to these structures as 'cyanobacteria-like', but later accepted that there is insufficient evidence for this assertion. Today, there appears to be no consensus that these are genuine microfossils (some of the angles in the drawings appear more characteristic of crystal growth), although other evidence for early Archean organisms is still sound. Reproduced, with permission, from J. William Schopf, *Science*, **260** (1993), p. 643.

sea, flourishing under shifting environmental conditions'.[7] This was followed a couple of years later by a detailed analysis of microstructures from the Strelley Pool Formation which concluded that these are microfossils of early sulphur-based organisms. Brasier was one of the co-authors.[8] Similar evidence for 3.48 billion-year-old microbially induced sedimentary structures have since been found in the Dresser Formation, which is also part of the Pilbara Craton.[9]

These results suggest that relatively complex 'cyanobacteria-like' and 'sulphur-bacteria-like' life was already flourishing just a few hundred million years after the Late Heavy Bombardment. But, surely, for life to have been flourishing it must have originated even earlier than this? A 1996 paper detailing the carbon isotope ratios of carbonaceous inclusions in grains of apatite (calcium phosphate) from the Isua Supracrustal Belt in Western Greenland concluded that these were indicative of biological activity, 3.8 billion years ago.[10] More recent (2014) studies of carbon isotope ratios in graphite from 3.7 billion-year-old rocks from the same formation arrived at the same conclusion.[11]

Could it be that primitive life predates, and therefore survived, the Late Heavy Bombardment? The results of computer simulations published in 2009 provide some strong hints that this might have been the case. These simulations show that although the bombardment would have been pretty devastating, parts of the Earth's crust would have remained habitable, and any life that had emerged could have survived.[12]

The slate might not have been completely wiped clean, after all.

The palaeontologists have learned to be cautious. Getting any kind of definitive evidence from the oldest accessible rock structures on Earth has proved to be extremely difficult. But there is now a fairly strong consensus that bacterial life was firmly established on Earth at least 3.5 billion years ago.

THE DOMAINS OF BACTERIA AND ARCHAEA

We should now qualify the use of the term 'bacterial'. Until 1977, biologists believed that there was only one 'kingdom' of simple, single-celled prokaryotic organisms—bacteria. But when Carl Woese started to trace the ancestry of living organisms by sequencing their ribosomal RNA, he discovered that there was not one kingdom, but two. In 1990, Woese and his colleagues suggested that these should be called 'domains'. The term 'kingdom' is now reserved for the next level of biological taxonomy.

In addition to the domain of bacteria, there is also a domain of 'archaebacteria', or just 'archaea'. These life forms are very similar in size and shape to bacteria but they possess distinctly different sets of genes, metabolic cycles, and cell membranes. They were initially thought to persist only in extreme environments, such as hot springs and hydrothermal vents, but are now known to be

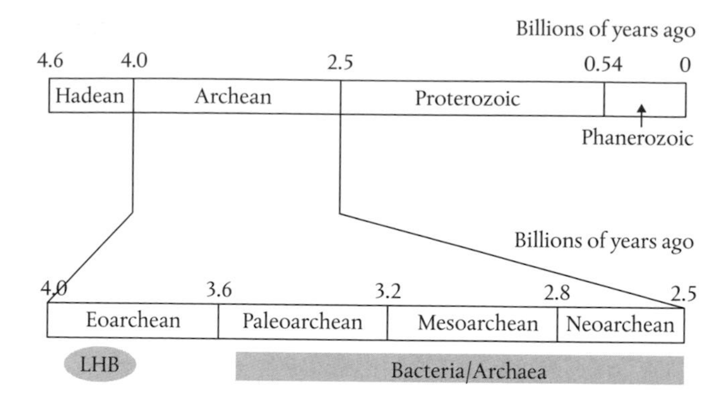

FIGURE 70 The Earth's geological timescale is split into a series of four eons—the Hadean, Archean, Proterozoic, and Phanerozoic. The Archean eon spans the time period from 4.0 to 2.5 billion years ago, and is further divided into a series of four 'eras'. These are the Eoarchean, Paleoarchean, Mesoarchean, and Neoarchean. The Late Heavy Bombardment (LHB) occurred in the Eoarchean era, about 3.8 billion years ago. The evidence from microbially induced sedimentary structures suggests that bacteria and archaea were already thriving by the early Paleoarchean era.

quite common. The methanogens that live in your gut (and aid digestion while producing a sometimes socially unfortunate waste product—methane) are archaea, not bacteria.

It's not clear when and how these two forms parted company. It may be that they evolved separately from LUCA, 'branching' off in different genetic directions. Some biologists have argued that this is much too unlikely, and that one form must have evolved from the other. However it happened, we can be confident that these different forms of life evolved on Earth over a one-billion-year period, between 3.5 and 2.5 billion years ago (Figure 70).[13]

Within each of these domains, the processes of natural selection have resulted in the evolution of a variety of different *phyla*. Each phylum is genetically distinct and identifiable, but shares characteristics common to the domain. In 1990, Woese and his colleagues mapped a selection of bacteria and archaea to a phylogenetic 'tree of life', in which the order of branching and the length of each branch is derived from the 'relatedness' of the phyla based on rRNA sequencing (Figure 71). Few scientists today believe that sequencing the genes in such a small sub-unit accurately reflects the 'true' nature of the genetic relationships between phyla, and more recent versions of this tree include more phyla and different branch structures. But the basic principle is unchanged.

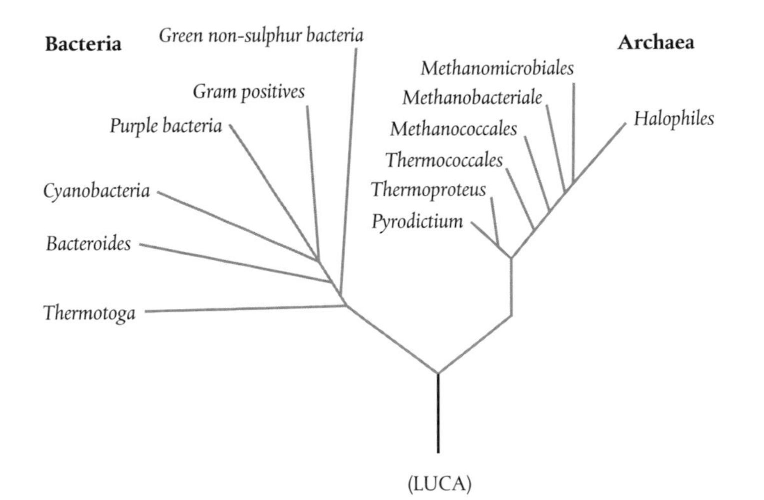

FIGURE 71 The first phylogenetic 'tree of life' was published by Carl Woese and his colleagues in 1990. The order of branching and the lengths of the branches reflect the 'relatedness' of different life forms based on rRNA sequence comparisons. Note that only two domains are shown here (see Chapter 10 for the full story). Adapted from Carl R. Woese, Otto Kandler, and Mark R. Wheelis, *Proceedings of the National Academy of Sciences*, **87** (1990), p. 4578.

Likewise the notion of a branching tree of evolution is evocative but it can be potentially misleading, particularly for prokaryotes. We tend to think of evolution operating through genetic inheritance, with genes passed vertically from parent to offspring. But prokaryotes can pass genes horizontally, and genes can be transferred between bacteria and archaea. A tree based on the genes in rRNA is not necessarily replicated in other gene sequences. Consequently, some molecular biologists prefer to picture prokaryote evolution in terms of a 'circle' or 'ring' of life.

Now, we know from the simple fact of our own existence that there's much more to evolution on Earth than bacteria and archaea. But it's a mistake to think that evolution irresistibly drives life to greater and greater complexity. Bacteria and archaea are highly successful life forms, and have persisted largely unchanged in evolutionary terms for something like three and a half billion years. If we were to bet on the possibility of discovering extra-terrestrial life, my money would be on some form of prokaryote.

And, before you start getting delusions of grandeur, just remember that there are ten times more bacterial cells in your body than there are human cells . . .

THE GREAT OXIDATION EVENT

We can now start to pull some threads together. By the end of the Archean, bacteria and archaea are flourishing, as we have seen. Although cyanobacteria may not have been present in their current form 3.5 billion years ago, there is evidence to suggest that they were here 2.7 billion years ago. We can imagine the surfaces of the ocean are now home to huge mats or 'blooms' of cyanobacteria which, along with other kinds of photosynthesizing organisms, have been slowly leaking oxygen into the atmosphere and oceans for hundreds of millions of years.

The scientific evidence for what happens next is in places very firm, in other places weaker and disputed but no less compelling. This is what we think we know.

As I mentioned, molecular oxygen is very reactive. As it leaks into the atmosphere and oceans it starts to oxidize any chemicals it encounters. The oceans contain dissolved iron produced by volcanic activity, and this is now oxidized to give ferric oxide (rust). The rust precipitates, forming layers of sedimentary rock characteristically red in colour, called *banded iron formations*, which are the principal sources of iron ore today.*

Of course, oxygen is a waste product of carbon fixation by oxygenic photosynthesis, and organic substances are also prone to oxidation (to produce—guess what?—carbon dioxide and water). It's not hard to imagine that in this situation oxygen might be simply cycled back and forth, produced by photosynthesis and consumed by dissolved iron and dead organic matter. But, as the Vaalbara supercontinent starts to break up, a substantial quantity of organic matter becomes buried in the shallow seas formed from new continental shelves, removing an important oxygen 'sink' and throwing the cycle off-kilter. Oxygen starts to accumulate in shallow parts of the oceans where sunlight is plentiful.

Any oxygen leaking into the atmosphere is quickly reacted with molecular hydrogen and methane. The latter is produced both by volcanic activity and methanogens, but methanogens are anaerobic organisms that struggle to survive in an oxygen-rich environment. As they die back, less methane is produced and the oxygen starts to overwhelm the atmospheric sinks, too.

A line has been crossed, and there's no going back.

---

* Note that banded iron formations can be produced in the absence of oxygen. Some photosynthetic bacteria use iron as a source of electrons, which results in the oxidation of the iron and its precipitation eventually to form haematite, Fe_2O_3.

Oxygen levels in the atmosphere start to build up. Any molecular oxygen reaching the stratosphere intercepts the Sun's ultraviolet light, breaking apart to form free oxygen atoms. The oxygen atoms react with molecular oxygen to form ozone (O_3), which in turn absorbs more ultraviolet light and breaks up to form molecular oxygen and free oxygen atoms again. The oxygen atoms shuttle between oxygen and ozone, all the time soaking up radiation. An ozone layer starts to form, which shields the Earth's surface from ultraviolet light.

Atmospheric sulphur oxides (such as SO_2), which are produced by volcanic activity, are now also shielded from ultraviolet light. Sulphur has four main isotopes, and a variety of geochemical processes cause variations in the distribution of these isotopes that depend on their relative masses. But photolysis by ultraviolet light causes variations that do *not* depend on the masses of the isotopes. This is called *mass-independent fractionation* (MIF). The MIF of a sulphur isotope such as ^{33}S is therefore a tell-tale signature of sulphur oxide photolysis by ultraviolet light.

The sulphur oxides rain back down to Earth and react with iron in the oceans, eventually precipitating to form a sedimentary rock called iron pyrite (FeS_2, also called fool's gold). So, any atmospheric sulphur MIF signature is translated into an equivalent MIF in iron pyrites.

Studies of banded iron formations reveal that these were formed periodically throughout the Archean, but halted around 2.4 billion years ago, at the beginning of the Proterozoic eon (Figure 72(a)). This coincides with an even more telling transition in the MIF of ^{33}S in iron pyrites (Figure 72(b)), which falls to zero. It seems that 2.4 billion years ago, the photolysis of atmospheric sulphur oxides simply ceased. The most logical explanation is that atmospheric oxygen levels had by this time built up sufficiently to provide a protective ozone layer.

There's more evidence that we don't have the space to consider here, and there can be little doubt about the overall scenario. It's called the Great Oxidation Event.

How much oxygen? This is really hard to tell. Today, molecular oxygen accounts for about 21% of the atmosphere by volume. Our best scientifically informed estimates suggest that the early Proterozoic atmosphere contained something of the order of only 5–18% of today's levels. It's not much, but it's more than enough to change the world.*

---

* Getting an accurate fix on atmospheric oxygen levels in the Proterozoic has proved to be very tricky, and these levels could well be over-stated.

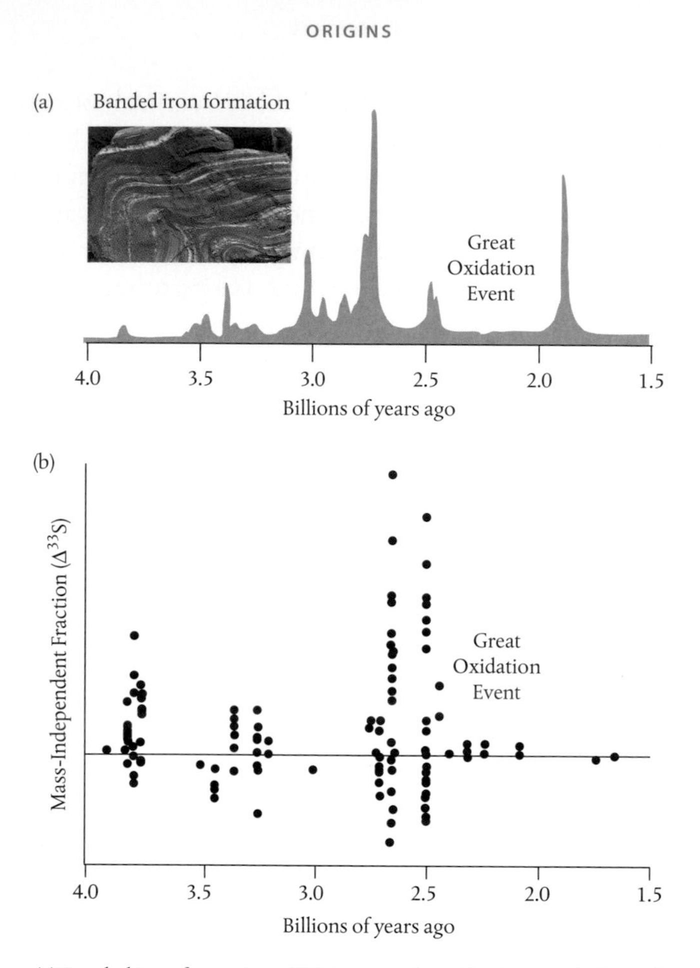

FIGURE 72 (a) Banded iron formations (BIFs) are produced by iron oxide (rust) formed in the reaction of dissolved iron and molecular oxygen. Studies of BIFs suggest that these occurred fairly regularly throughout the early history of the Earth, but stopped suddenly about 2.4 billion years ago. This is believed to have been due to the accelerated oxygenation of the atmosphere and oceans in the Great Oxidation Event. A number of explanations have been suggested for the deposition that occurred 1.8 billion years ago, and this likely reflects tectonic movements as much as anything. The mass-independent fraction (MIF) of ^{33}S varies through geological time until 2.4 million years ago, when it falls to zero (b). This is believed to signal the formation of a protective ozone layer in the stratosphere, shielding the Earth from the Sun's ultraviolet light.

Today we are rightly concerned that burning fossil fuels is impacting Earth's climate, producing rising average surface temperatures and unpredictable extreme weather. Burning (actually, oxidizing) fossil fuels simply returns organic material that was once buried in sedimentary rock back to where it came from—carbon dioxide, a 'greenhouse gas' which traps heat and raises the planet's surface temperature. Perhaps it's hard to imagine how lifeforms can have such planet-wide impacts.

But there's a precedent. At the beginning of the Proterozoic, oxygen released by photosynthesizing bacteria and archaea changes Earth's climate. It removes methane from the atmosphere, and methane is an even more powerful greenhouse gas. The Earth loses one of its insulating blankets. Heat escapes and the surface cools.

There are other contributing factors. Rain drags carbon dioxide out of the atmosphere as carbonic acid. It falls on silicate minerals uplifted and exposed by the shifting continents, 'weathering' them and carrying calcium and bicarbonate ions into the sea, never to return. The Earth starts to lose another of its insulating blankets. The shifting continents also bury more organic carbon, throwing the system further out of balance and driving more oxygen into the atmosphere.

The Earth is caught in a vicious cycle. This is not the first 'ice age' it has experienced, but it is by far the deepest. Russian climatologist Mikhail Budyko predicted that as glaciers creep to within 30° of the Earth's equator, a threshold is crossed. The ice reflects sunlight, including the Sun's heat, cooling the planet still further (this is called the ice-albedo effect) and the glaciers continue to creep to the equator. These are the Huronian and Makganyene glaciations, which will last an astonishing 300 million years.

Welcome to 'Snowball Earth'.

ENDOSYMBIOSIS AND THE RISE OF THE EUKARYOTES

The Great Oxidation Event is a double-edged sword. It means curtains for some forms of anaerobic organism, either because they can't handle the oxygen or because they lose access to key nutrients (such as sulphides) which the oxygen now cruelly snatches away from them.

But then there's the other edge. Every time we turn the key in our car's ignition, we appreciate the value of dead organic matter as a source of energy.* The proliferation of bacteria and archaea at the beginning of the Proterozoic inevitably means that there's a lot of dead organic matter to go around. There's some evidence to suggest that bacteria able to scavenge this matter and 'burn' it

* Unless you drive an electric car, of course, although the chances are high that the source of electricity used to charge your car battery can still be traced back to fossil fuels.

using oxygen to provide a source of energy (ATP) evolved long before the Great Oxidation Event, perhaps relying on inorganic oxygen sources. If this is the case, then for these bacteria the Great Oxidation Event is one big long party.

Such organisms don't fix carbon. They make use of readily available sources of 'dissolved organic carbon' that was previously fixed by other organisms.* *Pelagibacter ubique*, a species of unusually small, so-called α-proteobacteria (a form of purple bacteria, Figure 71), is possibly the most common bacterium on Earth today and accounts for a substantial proportion of all marine plankton. It scavenges dissolved organic carbon and uses oxygen to convert this into carbon dioxide, water, and ATP. This is called *cellular respiration*. Although the cells are not 'breathing' in a physiological sense, they are nevertheless 'inhaling' oxygen through their cell walls and 'exhaling' carbon dioxide.†

We need to pay attention to what happens next, as this opens an evolutionary pathway that will lead eventually to us.

Let's imagine a mixed community of bacteria and archaea, living quite happily in close proximity, side-by-side. As the oxygen levels rise, natural selection forces a shift in the bacterial population towards organisms much like the α-proteobacteria. But now suppose that their supply of dissolved organic carbon is running out. Meanwhile, the anaerobic archaea are struggling. Perhaps deprived of a ready source of essential nutrients, they can no longer make ATP. They are starving to death.

This is a fateful encounter. The archaeon needs ATP. The α-proteobacterium needs organic carbon to make ATP. It's not clear how their relationship begins, but it develops into an elaborate biological version of 'I'll scratch your back if you'll scratch mine'. In fact, the organisms become strongly co-dependent—they cannot survive without each other. The small α-proteobacterium is sooner-or-later absorbed into the larger archaeon, and the cell swells in size. The result is *endosymbiosis*, the endosymbiont bacterium living within its archaeon host.

The bacterium is not digested but it does not survive in its original form. It remains enclosed in its own membrane, but holding on to a duplicate set of

* I must admit that 'making use of dissolved organic carbon' sounds rather better than 'feeding on the decayed remains of dead organisms'.

† Again, this is not the first time in life's history that organisms have respired. We tend to think of respiration from our uniquely human, aerobic, perspective but microscopic organisms have been likely respiring a variety of inorganic substances throughout the Archean, depending on the nature of their metabolic cycles. Anaerobic respiration (without oxygen) is possible.

genes and ribosomes within the larger cell is just too inefficient. Some of the bacterial genes are inserted into the archaeon genome, though some remain in a small, circular chromosome. And the former bacterium retains a set of ribosomes needed for the manufacture of a more limited collection of proteins.

We're on our way to making a *complex cell*.

The absorbed bacteria become *mitochondria*, so-called 'organelles' (sub-units) within the larger complex cell which function as cellular 'power plants'. They oxidize organic compounds made available in the cell cytoplasm, such as glucose, producing carbon dioxide and water, passing back energy in the form of ATP. By holding on to a chromosome and the chemical apparatus needed to make proteins, the mitochondrion retains some local operational autonomy, thought to be necessary to ensure that energy is delivered to the larger host cell precisely when it is needed.

This is a powerful evolutionary strategy, one that is repeated with archaea and photosynthesizing cyanobacteria. Instead of forming mitochondria inside the larger cell, the absorption of the endosymbiont cyanobacteria (complete with their photo-active systems for converting light into ATP and fixing carbon) produce a different set of organelles called *chloroplasts*.

Of course, we know that there's no such thing as a free lunch. The evolutionary success of bacteria and archaea has to this point depended in part on horizontal gene transfer, the ability of these organisms to mix things up a little and through random mutations and trial-and-error to discover gene sequences that offer selection advantages. There are several mechanisms, but some genes are very mobile, forming 'transposable elements' (or *transposons*) whose behaviour is illustrated by a more prosaic name—*jumping genes*, discovered in 1944 by American geneticist Barbara McClintock.

Jumping genes are extremely 'selfish', and behave rather like a virus. A jumping gene installed within a strand of DNA is not transcribed into RNA. Instead, during transcription it folds back on itself and cuts (splices) itself out of the sequence. The ejected gene provides a template for rapid (we might even say aggressive) replication. The duplicated genes then insert themselves back into the DNA, at essentially random points. Any attempt to transcribe a jumping gene triggers the gene to spread in its host DNA.

This is starting to get messy, but it is in principle okay for as long as the jumping genes can't be transcribed and used to make proteins. The problem is that when these genes become dormant, they no longer splice themselves out of the sequence and consequently do get transcribed into RNA. Such dormant jumping genes are called *introns*. Because the introns are installed randomly in the genome, they don't code for a functional protein and can actually turn other genes on or off. The end result is chaos: the ribosomes end up making a lot of non-functioning or 'junk' protein.

The α-proteobacteria evolved a mechanism to cope with all this, although it's not clear how they do it. These organisms manage to eliminate nearly all the jumping genes and the introns, although a few do persist.

But although they may be few, the host archaeon has no mechanism to cope with them. As its genome starts to gather up lots of new genes from its endosymbionts, it becomes 'infected' with jumping genes and introns. This 'free' lunch comes with a heavy price, it seems, or at least a very bad case of indigestion.

Substantial numbers of these fledgling complex cells were likely wiped out by this 'intron catastrophe'. Those that survived evolved a way of dealing with the introns by splicing them out before they could be transcribed. They achieved this simply by taking the active splicing centres from the jumping genes themselves (each consisting of about 150 nucleotides) and wrapping these in protein to make *spliceosomes*.

Slight problem. In a prokaryote cell the DNA in the nucleoid and the ribosomes—the chemical factories where proteins are made—sit quite close together. It takes no time at all to transcribe the gene sequence from DNA into mRNA and then transport this to a ribosome to produce proteins. In modern cells, proteins are assembled at a rate of about one amino acid per second. But the spliceosome works rather more slowly, taking on average 10 to 200 seconds to remove an intron. This isn't good enough. If the introns can't be spliced out quickly, many will get transcribed, and junk protein will still be produced.

We're building up to an ingenious solution proposed in 2006 by American botanist William Martin and Russian-American biologist Eugene Koonin. They argued that the only way to resolve this problem of timing is to physically separate the cell genome from the ribosomes, by tucking the genome away in a *cell nucleus*.

The nucleus is the complex cell's command and control centre. It consists of an outer double membrane, dotted with pores made from large protein complexes through which large molecules can be moved in and out using carrier proteins. It holds the cell's DNA which, because of the acquisition of new genes from its endosymbionts (and its accumulation of jumping genes and introns), now forms a series of long linear strands embedded in a matrix of protein structures.* It also contains the *nucleolus,* where ribosomes are manufactured.

Now the spliceosome can quite happily go to work on the 'precursor' mRNA in the nucleus, taking as long as it likes to splice out all the introns before the 'mature' mRNA is carried out through the pores to the ribosomes. Order is restored, but only by dealing with the symptoms, not the cause. The host DNA remains infested with introns, and these get passed on to succeeding generations.

The evolution of a cell nucleus signals the origin of *eukaryotic* cells. The name is derived from the Greek *eu* and *karyon,* meaning 'true nucleus'. Now at last I can admit that the name *prokaryote* indicates cells that do not have a true nucleus. In other words prokaryotic cells are named for what they're not (see Figure 73 for a comparison).

The endosymbiosis theory of the origin of eukaryotic cells has a long history. It was first proposed in 1905, by Konstantin Merezhkovsky, a botanist at

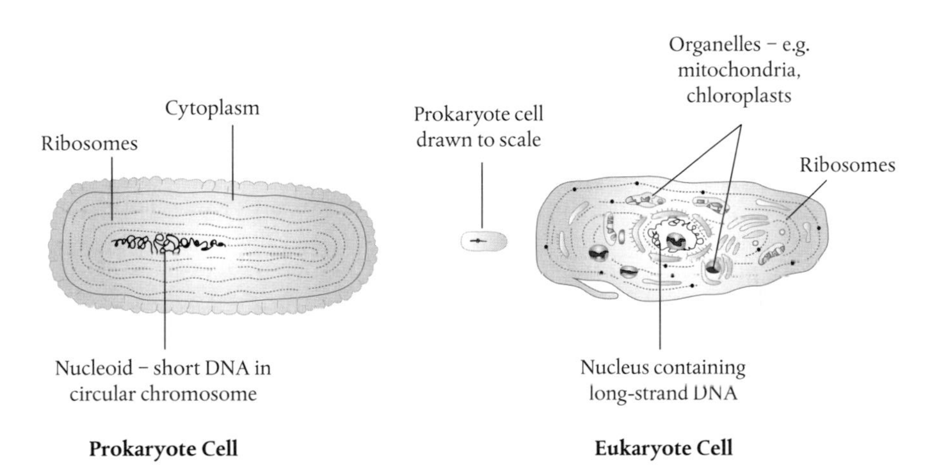

FIGURE 73 Comparison of prokaryote and eukaryote cells. Prokaryotes are typically 0.2–1.0 microns in length and possess a simple cell structure consisting of a short length of DNA, typically formed into a circular chromosome and ribosomes—small protein manufacturing complexes in a cytoplasm which consists of about 80% water. Eukaryotes are much larger, being 1–10 microns in length with a much more complex internal structure consisting of a nucleus, organelles (such as mitochondria and chloroplasts), ribosomes, cytoplasm, and more.

* To put this in perspective, note that every human cell contains roughly two metres of DNA.

Kazan University in Russia. It was independently rediscovered in 1967 by Lynn Margulis (married to American cosmologist Carl Sagan and writing as Lynn Sagan). It was a radical idea, and her paper was rejected 15 times before it was eventually published in the *Journal of Theoretical Biology*.

Although endosymbiosis of the kind I've described here isn't the only candidate scientific explanation for the origin of eukaryotic cells, it is now broadly accepted. The evidence that mitochondria, the power plants of eukaryotes, share a common ancestor with α-proteobacteria seems pretty firm.[14] There is also something of a consensus that the host organism must have been an archaeon, and there is some evidence from genetic analysis regarding its identity.[15] The Martin–Koonin hypothesis for the origin of the cell nucleus appears much less secure, but is nevertheless still rather compelling. In 2009, British biochemist Nick Lane wrote: 'This is the sort of idea that adds magic to science, and I hope it is right.'[16]

SIZE MATTERS

When is this happening? Good question. The fossil record for single-celled eukaryotes is murky, to say the least. This may seem surprising. After all, eukaryotes are surely larger than prokaryotes and, although a lot of the details of structure within the cell are lost in the process of fossilization, surely size is a good indicator? Alas, it's not this simple. While there are no bacteria larger than the largest eukaryotes, and no eukaryotes smaller than the smallest bacteria, there's actually a lot of overlap in between.

There is some tentative evidence for the existence of 2.7 billion year old eukaryotes in the Pilbara Craton, first reported in a study published in 1999. This comes from the observation of so-called biomarkers, or 'molecular fossils', traces of ancient chemicals associated with eukaryote biology. Interpretation of such evidence is fraught with concern over the possibility of sample contamination, although further studies with improved laboratory procedures appear to support the original findings. A recent review declares the evidence tantalizing, but goes on to add that it '. . . should be viewed with caution and awaits further verification'.[17]

The evidence for the presence of eukaryote life forms about 1.8–2 billion years ago is rather less contentious, and fits more comfortably with the idea that endosymbiosis may have been driven (or was at least accelerated) by the Great

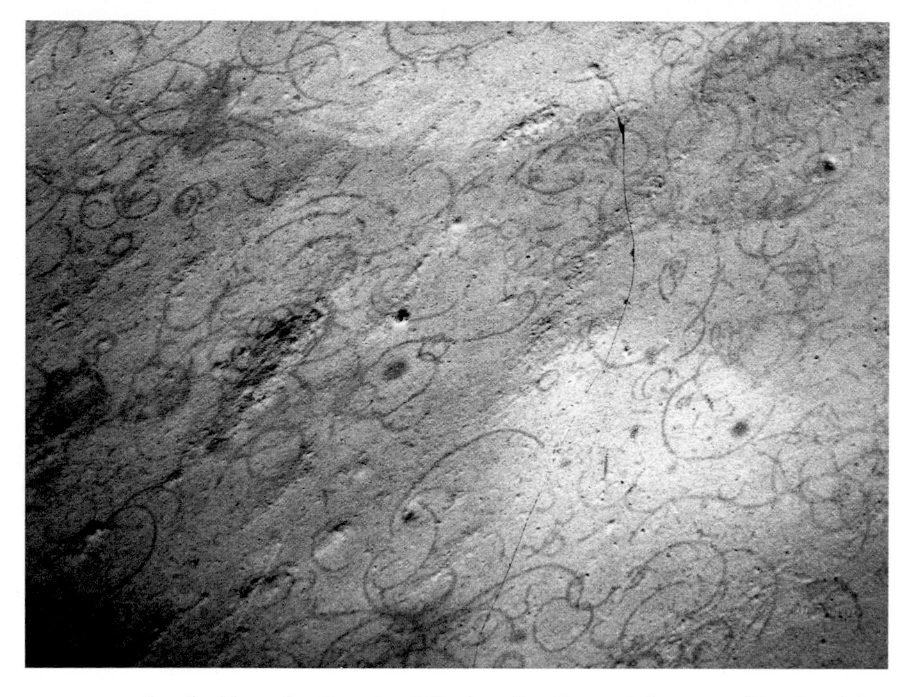

FIGURE 74 Fossil evidence for *Grypania spiralis*, thought to be a 1.9 billion year old multicellular alga, from the Negaunee Iron Formation in Michigan's Upper Peninsula. These ribbon-like fossils measure a few millimetres in width and up to 10 centimentres in length.

Oxidation Event.* It really doesn't matter all that much. We know that within a few hundred million years of the end of the Huronian and Makganyene glaciations, eukaryotes had evolved rather dramatically. The oldest known eukaryote fossils are those of *Grypania spiralis*, a spiral, tube-shaped photosynthetic alga, found in the Negaunee Iron Formation, a banded iron formation which outcrops in Michigan's Upper Peninsula. These fossils were initially dated at 2.1 billion years old, subsequently revised to 1.9 billion years.

Consider this. *Thiomargarita namibiensis*, a sulphur bacterium found in ocean sediments of Namibia's continental shelf, is one of the largest bacteria ever discovered. It can achieve sizes approaching 0.75 millimetres, big enough to be visible to the naked eye. In contrast, the fossils of *Grypania spiralis* measure several *centimetres* in length with widths of the order of 0.5–0.6 millimetres (Figure 74).

* We should note that William Martin doesn't share this obsession with oxygen. Together with Miklós Müller, he has argued that the seemingly vast variety of mitochondrial structures seen in modern eukaryotes can only be explained if eukaryotes emerged under predominantly *anaerobic* conditions.

The volume of *Grypania spiralis* is about a million times larger than the average volume of prokaryote organisms.

Size (and, for sure, cellular organization) matters. Bacteria and archaea are constrained in terms of size and complexity by limits imposed by physics and chemistry. While they are highly successful, these prokaryotes will always be 'simple' life forms, which is why they have remained largely unchanged over three and a half billion years of Earth's history.

The eukaryotes, by contrast, represent a fundamentally different proposition. However it happened, the birth of complex eukaryotic cells is an utterly transformative development. It's surely no coincidence that the dramatic increase in the size of organisms is observed in the fossil record shortly after the Great Oxidation Event. We can only speculate on the connection, but gaining access to oxygen and oxidizing dissolved organic carbon to provide a source of energy would appear to have liberated life from its physical and chemical constraints.

The eukaryotes are, literally, a new lease of life.

FROM ONE TO MANY

The evolution of complex eukaryote cells would not have been so dramatic had it not been for other 'inventions' that evolved before, during, or shortly after this singular development.

A good example is multicellularity (we will go on to consider other inventions shortly). By this I simply mean single-celled organisms coming together to form more complex multicellular organisms. Now, I should confess that the word 'multicellular' has many meanings. Communities (or 'colonies') of bacteria and archaea that form large-scale films or mats and which enjoy symbiotic relationships among themselves can be considered to be a multicellular form of life. In this sense, multicellular life was already present in the 'cyanobacterial-like' mats that are believed to have left chemical traces in the fossil record dating back 3.5 billion years. *Grypania spiralis*, and other algae dating back more than a billion years, are thought to be multicellular life forms.

This kind of multicellularity is obviously quite primitive. At the other end of the scale are complex plants and animals (including ourselves), which have evolved a highly sophisticated form of multicellularity, with cells specialized to different functions (brain cells, blood cells, liver cells, skin cells, and so on) and elaborate chemical and electrochemical mechanisms for transmitting information between them. These cells possess the same genetic information (they are part of a single organism), their specialization is rather determined by their biochemical environment.

It helps to draw a line somewhere between these extremes. For sure, simple single-celled organisms form multicellular communities, with a modest degree of cell differentiation. For example, spherical communities of algae position cells with flagellae—small whip-like structures that are used to propel the community around the ocean—on the outside of the sphere, which is logically where these are needed. The cells on the inside are spores, which hold the genetic information and reproductive capability necessary for the organism's ability to pass its genes to future generations.

But the simple truth is that the cells on the inside and the outside are the same cells at different stages of their life-cycle. In this case the cell differentiation is one of *time*, not type.

This is not to say that more sophisticated cell differentiation is impossible, although it is very unusual among prokaryotes. As we have seen, cyanobacteria fix carbon using photosynthesis, producing oxygen as a waste product. But these bacteria also need to fix molecular nitrogen, N_2, in order to build proteins. Slight problem. The enzyme involved in fixing nitrogen is inhibited in the presence of oxygen. Single-celled cyanobacteria can't photosynthesize and fix nitrogen at the same time.

Some forms of the bacteria opt to photosynthesize by day and fix nitrogen at night. In other, multicellular, forms such as *Nostoc punctiforme*, nitrogen starvation transforms about 1 in every 9–15 cells. These cells, called heterocysts, build extra cell walls, creating an anaerobic environment by insulating the interior from oxygen. They also switch off Photosystem II (but retain Photosystem I). In this way, the organism divides the labour between different cells, with cells dedicated to photosynthesis sitting comfortably alongside cells dedicated to fixing nitrogen.

The survival advantages of such communities are obvious. But if the environment changes and becomes more conducive then some cells may revert to type, leaving the community to go back to their single-celled way of life. This is a symbiosis of convenience, without commitment.

'True' multicellularity is the result of symbiosis on a much grander scale. It involves the evolution of cells that simply cannot survive on their own. Leaving the 'community' to go their own way is simply not an option, under any circumstances. Such cells are committed to the greater good of the larger organism (with some caveats, which we will look at shortly).

How did this happen? This is a question without a straightforward answer. It has been argued that multicellular organisms have evolved from single-celled ancestors at least 25 times during Earth's life history, and a transition to multicellularity is a gambit that continues to be used today by single-celled organisms experiencing selection pressures.

For organisms that evolved in rivers and oceans, multicellularity was likely the result of incomplete cell division, a consequence perhaps of a genetic mutation. As I explained above, primitive, single-celled organisms reproduce by cloning themselves, making genetically identical daughter cells which then separate. Incomplete separation leaves the daughter cells 'glued' together, forming an extended array. There may have been no immediate evolutionary advantages, the mutated organisms drifting along quite happily. But the introduction of another invention of evolution, coupled with rising oxygen levels, may have made size even more important.

LET US PREY

The invention in question is *phagocytosis*. Put simply, this is the ability of one cell to absorb another into its interior and use it as a source of food.

Hang on. Isn't this the same as endosymbiosis, the mechanism that gave rise to the first eukaryotes?

Yes, and no. The origin of eukaryote cells involved the absorption of bacterial cells into an archaeon host, to form organelles such as mitochondria and chloroplasts. But this, it is argued, was a fateful encounter, a rare (but possibly quite predictable) event. This was the result of a neo-Darwinian roll of the dice. The usual Darwinian mechanism of experimentation through countless trial-and-error genetic variations and natural selection didn't apply. This was a sudden jump (hence 'neo'), but one that was no less subject to selection pressures (still 'Darwinian').

Phagocytosis, from a combination of Greek words meaning 'to devour', 'cell', and 'process', evolved by a more traditional Darwinian mechanism, probably in response to a growing shortage of readily available dissolved organic carbon. Instead of feeding on the decayed corpses of dead organisms, the evolving eukaryotes began to target the living.

Now there are predators. And prey.

There are lines of argument in favour of the idea that at this stage in life's evolutionary history, phagocytosis was not possible without mitochondria. If this is the case, then phagocytosis was an invention of eukaryote cells. Phagocytosis of a certain type may have been important in providing selection pressures for multicellularity, based on the simple logic that single-celled eukaryote organisms that clung together following cell division would have been harder to prey on.

The pieces of life's jigsaw puzzle are starting to assemble, and the picture is beginning to emerge. Rising oxygen levels, the evolution of complex eukaryote cells, their coming together to form multicellular organisms, and the invention of a primitive form of eating (phagocytosis) combine to drive evolution in the direction of larger and larger organisms. There are obvious survival advantages in being the largest organism around, sitting at the top of the food chain.

But increasing size and complexity bring new challenges. What happens if a mutation in one of the differentiated cells changes it from loyal citizen to revolutionary agitator, threatening the viability of the whole organism? And what should be done about the growing threat of jumping genes from absorbed bacterial cells? These invade the host cell chromosome. Some of the incorporated genes may be useful. Most fill the host DNA with junk, which can be dealt with through the deployment of spliceosomes. Some genes may be quite harmful.

The bacterial method of reproduction by cell division simply replicates the chromosome, producing identical clones complete with any harmful genes. It's not hard to guess where this leads. In a relatively small population successive generations may accumulate more and more harmful genes, degenerating to extinction. This irresistible decline in fitness is called *Muller's ratchet*, for American Nobel laureate Hermann Muller.

Not good.

SEX AND DEATH

The response to the first of these challenges is relatively straightforward. Chemical signals from any defecting cells—mutant cells that no longer play nicely for the greater good of the organism—are picked up by mitochondria. These then release a sequence of cysteine-aspartic protease (or caspase) enzymes, also known as 'executioner proteins', to kill off the defectors. This is *apoptosis*, one type of 'programmed cell death', a handy mechanism likely imported into eukaryotes lock, stock, and barrel from bacterial endosymbionts.

The caspase enzymes invade the offending cells in a cascade, one enzyme activating the next. They act to chop up the cell from the inside. Their remains are then removed by phagocyte cells.

Let's put this in perspective. It is estimated that in an adult human between 50 and 70 *billion* cells are exterminated by apoptosis every day. In this way any defecting cells are sacrificed in the interests of the organism, the greater good here being the ability of the organism to reproduce and pass its genes on to successive generations.

Which brings us neatly to the second challenge.

For the evolving eukaryotes, reproduction by cloning is no longer viable. That way lies genetic degradation, an irresistible journey to extinction. Eukaryotes increase their chances of survival if they evolve a different mechanism to propagate their genes, one that mixes things up rather more and at a faster rate than is possible with horizontal gene transfer, offering prospects of 'shuffling the deck' and randomly assembling genomes with a higher proportion of 'good' genes.

The solution is sex.

Now, let's be clear. This is not the only reason that can be found for the evolutionary invention of sex (though it is a rather compelling one). Some bacteria practise a very rudimentary form of sexual reproduction and we know that sexual life-cycles had already evolved in single-celled eukaryote organisms. But sex comes into its own in their multicellular descendants. This may not have been a new invention but, as selection pressures start to drive multicellular organisms to greater size and complexity, the variation afforded by sex becomes an essential method of maintaining and occasionally increasing genetic 'fitness', repairing DNA and masking the effects of damaging mutations.

It works like this. Multicellular organisms differentiate some cells into sex cells—sperm and egg. In sexual reproduction these different cell types fuse together. The chromosomes from both cells are first duplicated. This is no different in principle to the first step in bacterial reproduction, except that we now have chromosomes from each of two genetically distinct 'parents', so duplication makes four (2 × 2) chromosomes in total.

What happens next is an extraordinary dance of the genes. Genes from the four chromosomes intermingle, splicing themselves randomly into each other and recombining to produce a set of four new chromosomes, each a distinctly different collection of genes assembled from the originals. The fused cell then divides to form two daughter cells, and each of these divides again. The end result is four distinct 'granddaughter' cells, each with a chromosome bearing a shuffled set of genes inherited from both parents.

Suppose a situation arises in the external environment in which two favourable genetic mutations offer clear survival advantages. In a population of bacteria reproducing clonally, simple probability dictates that the chances of one or the other mutation arising might be relatively high, but the chances of both arising together are likely to be very low. Organisms carrying one or the other mutation do not interact, except through occasional (and random) horizontal gene transfer. If the organisms cannot respond quickly enough through successive generations, then selection pressures may drive them to extinction.

The chances of acquiring both of the favourable mutations, one inherited from each parent, are much higher in a population that reproduces sexually. Of course, the converse is also true. Offspring can also accumulate unfavourable genes. But such offspring are less likely to survive and therefore less likely to reproduce. By concentrating 'bad' genes in offspring that then die out, the 'bad' genes are eliminated, improving the overall genetic fitness of the population.

But it's wrong to think of this in black-and-white terms, only 'good' genes accumulating at the expense of 'bad' genes. Genetic fitness is about statistics. Organisms are 'rewarded' in the lottery of evolution by surviving long enough to reproduce. 'Good' genes that help make this happen may become 'bad' genes later in the organism's life-cycle, resulting in diseases associated with ageing and leading eventually to death (if the organism survives that long and doesn't fall prey to the multiplicity of things out in the external environment that are trying to kill it). Evolution is indifferent to their ultimate fate. Clearly, organisms engaging in sexual reproduction early in their life-cycles in an environment in which nutrients are plentiful are more likely to pass their genes to successive generations.

This is what evolution works with and is all it cares about.

THE GROWTH OF DIVERSITY

Nailing these inventions of evolution to the mast of geological time is no straightforward task. But we can be confident that eukaryotes, multicellularity, phagocytosis, programmed cell death, and sexual reproduction (and other inventions I've chosen not to consider here) were assembled in life's armoury during the Proterozoic eon (Figure 75).

It might seem that there was a lot going on, but there was an awful long time for all this to be happening in. The Proterozoic spans nearly two billion years, from 2.5 billion years ago to 540 million years ago. Palaeontologists sometimes

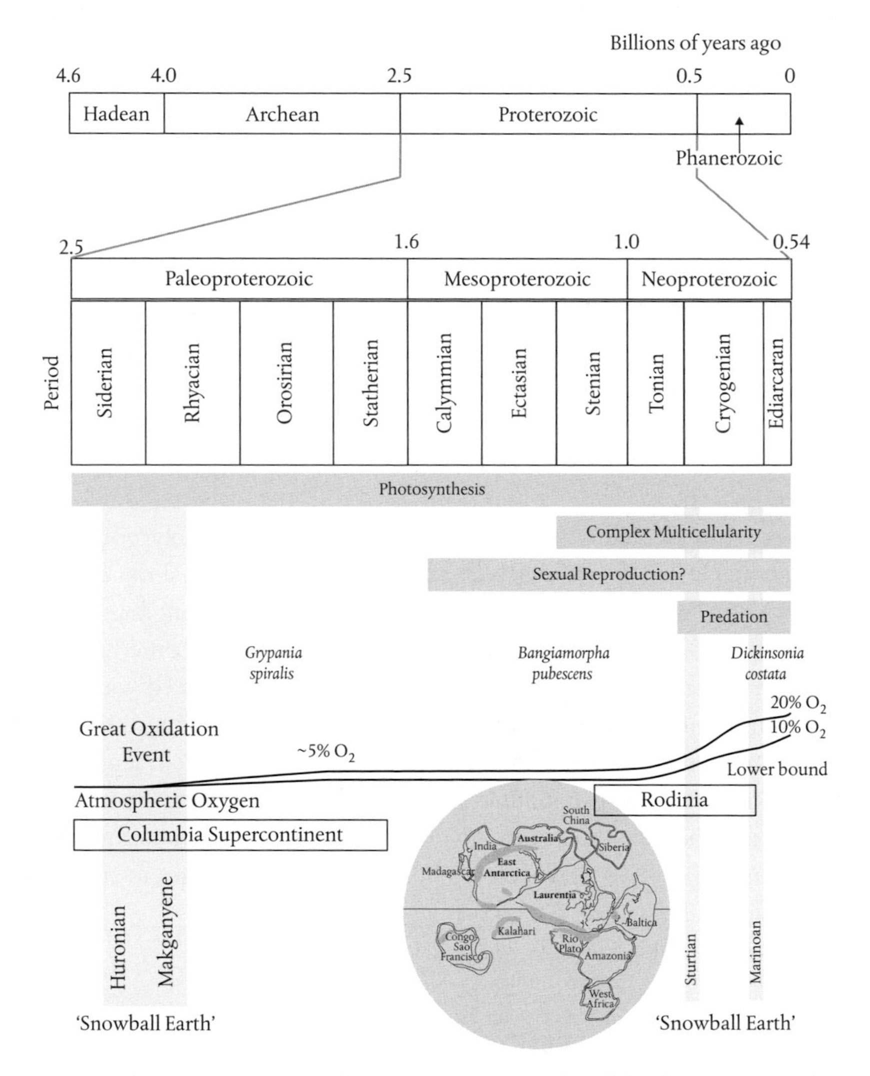

FIGURE 75 An attempt to map evolutionary inventions, selected fossil organisms, and geological events against the geological timescale of the Proterozoic eon. Evolution accelerates the development of diversity, leading eventually to the first animals, as a result of changes in the assembly of Earth's continents, the rise of oxygen in the atmosphere and oceans, and 'Snowball Earth' episodes.

refer to the period from about 1.8 to about 0.8 billion years ago as the 'boring billion'.*

It's certainly true that there's not much to see in the fossil record, at least in comparison with what is about to happen next. But although evolutionary biologists are still debating many of the finer points (of the origins of multicellularity and sex, for example), there are more than enough grounds for confidence in the basic correctness of this broad-brush evolutionary history. I'll try to illustrate these grounds with a few choice examples.

So, from fossils of *Grypania spiralis*, dated at 1.9 billion years old and mentioned above, let's fast-forward 700 million years to the late Mesoproterozoic era, and fossils of *Bangiamorpha pubescens*, discovered in the Hunting Formation above the Arctic Circle in Canada. These provide robust evidence for multicellular organisms similar in many ways to modern photosynthetic red algae, whose various species are common in marine environments and which account for many different types of seaweed. They possessed reproductive cells and differentiated 'holdfasts', root-like adhesive cells that allowed the organism to fix itself to rock. Red algae belong to the taxonomic 'supergroup' *plants*, which also includes green algae and the more familiar land plants (which come later).

Other examples include *Paleovaucheria clavata*, a relative of brown algae discovered in the 1 billion-year-old Lakhanda Suite in Siberia and the 750 million-year-old Svanbergfjellet Formation in Spitsbergen, part of the Svalbard archipelago in Norway. The Svanbergfjellet Formation has also yielded *Proterocladus major*, a green alga.

Vase-shaped microfossils believed to be different species of so-called testate amoebae† have been discovered in the 750 million-year-old Chuar Group, in the Grand Canyon, and several other locations aged 800–740 million years old. These are believed to have possessed rigid shells formed from organic materials.

By the early Neoproterozoic era, the eukaryotes have evolved and diversified into fungi, several species of amoebae, red and green algae, and many other forms.

There follows not one but two more multi-million-year-long 'Snowball Earth' episodes, the Sturtian glaciation (from about 720 to 660 million years ago) and the Marinoan glaciation, which comes later and which recent

* A term that seems to have been coined by Martin Brasier—see *Secret Chambers*, p. 190. Brasier explained that this period is not without interest, but it is a period of extraordinary evolutionary stability.

† 'Testate' means organisms with a shell.

studies suggest lasted 4–5 million years or so.* These episodes fall within the Neoproterozoic era and define a period aptly named the Cryogenian.

It's not clear what triggered these glaciations. It's possible that a continued accumulation of atmospheric oxygen, and its effect on greenhouse gases such as methane, conspired with the formation and subsequent breakup of the Rodinia supercontinent, which gathered around Earth's equator. As the planet cooled once again, glaciers may have formed at the equator, where the Sun's radiation is most intense (but still 6% weaker than it is today), the ice-albedo effect cooling the planet still further.

The fossil record informs us that life on Earth hunkered down, hoping and waiting for better conditions. For example, extraordinarily well-preserved seaweeds can be seen, pressed like flowers between the pages of a book, in the black shales of the Doushantuo Formation in Guizhou Province, China, which is dated at 635–551 million years old.

Every cloud, as they say, has a silver lining. Again, we can't really be sure what happened but a plausible scenario runs something like this. The icy conditions of the Sturtian glaciation allow an accumulation of atmospheric carbon dioxide and methane from volcanic activity over 30 million years, trapping heat and warming the Earth once more. As the glaciers recede, more land is exposed and the albedo effect is quickly reversed.

Sediments rich in phosphorus and other essential nutrients for life are released by the glaciers and flow into the oceans, and an atmosphere now once again rich in carbon dioxide triggers an explosion of photosynthesizing cyanobacteria. Oxygen levels are now propelled to giddy heights, reaching their present-day atmospheric concentrations and seeping ever more deeply into the ocean.

The price is another cycle of cooling, glaciation, and warming, but by the end of the Cryogenian period, oxygen levels are higher than ever and a new, frantic, round of eukaryote evolution begins.

THE EDIACARAN RADIATION AND THE FIRST ANIMALS

There may have been another rather short-lived supercontinent—called Pannotia—formed around 600 million years ago, after the breakup of Rodinia.

* I should point out that not everyone agrees. Some geologists argue that the Earth was not completely covered by ice during these glaciations, and that large surface areas of sea were left open and accessible. According to this view, the Earth was a 'slushball' rather than a 'snowball'.

If it did form, then this was an 'accidental' supercontinent, assembled through glancing collisions between the main tectonic plates. It had broken up again within about 60 million years, just at the border of the Ediacaran and Cambrian periods, forming four separated large continents, called Laurentia, Baltica, Siberia, and Gondwana, and some smaller bits and pieces.

Such movements can wreak havoc, lowering and then raising sea levels, releasing oxygen-starved waters from the depths of the oceans and filling the atmosphere with carbon dioxide and noxious volcanic gases. During this time the Earth experiences another ice age, called the Gaskiers glaciation, though this is not a full-on Snowball Earth episode. It 'only' lasts a couple of million years, about 582–580 million years ago.

The balance between oxygen and carbon in the oceans and the atmosphere is extraordinarily delicate. Living organisms prefer to fix the more ubiquitous ^{12}C isotope, and this selectivity is transferred to the organisms' remains. Oxidizing these remains simply restores the ratio of ^{12}C to ^{13}C in carbon dioxide in the atmosphere. But when these remains become buried in sediments, the ratio can't be restored and the atmosphere's carbon dioxide is enriched in ^{13}C, eventually to rain back down to Earth to form deposits of calcium carbonate. As we know, burying the carbon allows oxygen to accumulate (as it no longer has quite so much dead organic matter to react with). The $^{13}C/^{12}C$ ratio in carbonate rocks therefore provides a clue to the extent of carbon burial and oxygen levels in the atmosphere and oceans.

The isotope ratio suggests that the Ediacaran was a period of massive upheavals in the Earth's carbon cycle. The Shuram 'excursion', named for the Shuram Formation in Oman, is one of the most significant negative excursions in the $^{13}C/^{12}C$ isotope ratio on record. It occurred in the middle of the Ediacaran, and suggests that this was a time of substantially *lower* carbon burial or, for some reason, an enhanced release of ^{12}C-enriched material. Explanations are hard to come by, but it is possible that as the oxygen soaks into the sea a huge reservoir of dissolved organic carbon is oxidized, releasing carbon enriched in ^{12}C.[18]

High carbon burial rates were subsequently restored, and atmospheric oxygen levels increased once again. Higher oxygen levels mean access to more energy, encouraging a further increase in size.[19] It may also have triggered what evolutionary biologists refer to as a 'radiation'—a frantic period of evolution and species diversification.

Particularly interesting from our human perspective is the emergence in the fossil record of the first *animals*. These include fossils interpreted to be relatives of sponges and simple cnidarians, the group that includes sea anemones, jellyfish, and corals (though these affinities are still being debated). Fossils recovered from the Doushantuo Formation have been interpreted as the *embryos* of some unknown animal species. Curiously shaped fossils from the 550–543 million year old Nama Group in Namibia cause no little head-scratching, as American palaeontologist Andrew Knoll, writing in 2003, explains:[20]

> If we examine Nama sandstones carefully—preferably in late afternoon when the sun, set low in the sky, throws surface features into high relief—we see fossils that are almost shockingly different from anything found in older rocks . . . We see the impressions of large, complicated organisms, as well as simple tracks and trails unambiguously made by animals. In truth, the Nama fossils are shocking whether we approach them from above or below, for if they have no counterparts in older beds, Nama impressions bear equally little resemblance to most fossils found in Cambrian or younger rocks. Thus, the debate: do the remarkable fossils in Nama and other latest Proterozoic rocks record the ancestors of modern animals or a failed evolutionary experiment at the dawn of animal evolution?

One of the first fossil discoveries from a period earlier than the Cambrian that had palaeontologists sitting up and taking notice was made in 1957 by a young schoolboy called Roger Mason, at a quarry near the village of Woodhouse Eaves in Charnwood Forest in Leicestershire, England. This was named *Charnia masoni*, a frond-like specimen with segmented ridges that branch alternately left and right. It looks like it could be a form of algae or a plant, but the rock in which it was found indicates that it lived in deep water, well below the reach of light from the Sun. This organism couldn't photosynthesize, and it might therefore have been a primitive animal, some kind of fungus, or perhaps an evolutionary experiment that didn't make it beyond the Ediacaran. Further examples of *Charnia* have since been discovered in Australia, Newfoundland, and Russia, as have many other types of 'frondose' fossils (such as *Fractofusus misrai*, discovered in 1967 at Mistaken Point, Newfoundland—Figure 76).

The archetypal (and rather iconic) Ediacaran fossil is *Dickinsonia costata*. This was discovered ten years earlier by Reginald Sprigg in the Ediacara Hills, a range of hills in the northern part of the Flinders Range in South Australia (Figure 77). Sprigg named the fossil organism for the Director of Mines for South Australia, Ben Dickinson. He had thought the rocks he was examining were dated to the

FIGURE 76 *Fractofusus misrai* is named for the Indian geologist Shiva Balak (SB) Misra who was a graduate student when he discovered a rich assemblage of Ediacaran fossils at Mistaken Point in Newfoundland, Canada, in June 1967. The area where this picture was taken is now carefully protected, and forms part of the Mistaken Point Ecological Reserve. This fossil is about 10 centimetres in length, and 565 million years old.

FIGURE 77 *Dickinsonia costata* was discovered in the Ediacara Hills in Southern Australia in 1947. It is thought that these were primitive cnidarians, a phylum that includes jellyfish, corals, and sea anemones that lived in colonies. Whatever they were, they didn't survive much beyond the end of the Ediacaran period.

early Cambrian period, which follows the Ediacaran, and it was only on the discovery of *Charnia* that attention returned to *Dickinsonia*.

Other *Dickinsonia* examples have been found, measuring anything from a few millimetres up to a metre and a half in length, each a few millimetres thick. What were they? Some palaeobiologists have argued that they were annelid (ringed) worms, with a bilateral, segmented body and a central gut. Others have contested this, and the evidence—such as it is—suggests that the creatures sat motionless on the sediment surface, which is not very worm-like behaviour. Knoll suspects that they were primitive cnidarians, with an outer layer of cell tissue (epithelia) lying above and below relatively inert jelly like material. Possible modern equivalents are the placozoa, simple multicellular animals that creep along the seafloor.

As these organisms start to prey on one another, various defence strategies evolve. Worm-like creatures burrow their way to safety, revealed through a dramatic rise in *bioturbation*, a churning of sediment which changes its texture and leaves fossil traces. The appearance of *Cloudina* and *Namacalathus* towards the end of the Ediacaran signals a more widespread use of *biomineralization*, the ability to produce tough, mineralized tissue formed from calcium carbonate and calcium phosphate to protect the soft tissue beneath. This was not a new invention, but it now comes in very handy as a survival mechanism.

Whatever they were, the creatures of the Ediacara 'biota' signal a unique moment in the evolution of life on Earth. It's taken three and a half billion years to get from LUCA to the first plants and animals, by way of photosynthesis, bacteria, archaea, endosymbiosis, multicellularity, phagocytosis, sexual reproduction, programmed cell death, biomineralization, and many, many other inventions of evolution. The Earth has played a crucial role in this story, and it is abundantly clear that life and planet are inextricably bound to one another.

Life on Earth is now pushing at an open door. It's quite possible that the animals of the Ediacaran period were not a failed evolutionary experiment, although it is certainly true that they don't survive—with very few exceptions no Ediacaran fossils are found in the Cambrian period that follows.

It's not clear what happens next, but there is evidence from another negative excursion in the carbon isotope ratio and studies of various trace elements to suggest that, at the end of the Ediacaran, the oceans are now *depleted* of oxygen.[21] Nobody can be sure what might have caused this, but there are a few geological mechanisms which can dredge oxygen-poor and carbon-dioxide-rich waters from the depths of the ocean and bring it to the surface. Shifting continents and seas may have played a part and, indeed, there may have been multiple causes. But whatever happened there can be no doubting the consequence—mass extinction.

We can't be sure what, if any, effect this had on the animals of the Ediacaran period. But we do know that almost all of them die out. Their 'short' 20 million year reign is at an end.

Life may appear tenuous at times, but the history of life on Earth suggests that it is impossible to kill off completely. Some lifeforms survive. And, in a scenario that we will see repeated again and again in the next chapter, those lifeforms that do make it through difficult times find themselves in an almost empty ecology. As the oxygen is restored, evolution *really* starts to go to work.

10

A SONG OF ICE AND FIRE

The Origin of Species

Okay, the subtitle is a bit rich. Perhaps we can call it poetic licence.

This chapter deals with the cycle of successive ice ages, fiery volcanic eruptions and at least one (possibly two) deadly asteroid impacts spanning some 475 million years, beginning in the Cambrian. This song of ice and fire changed the face of the Earth many times. It caused a number of mass extinctions and subsequent evolutionary radiations which resulted in a great expansion in diversity of specifically animal lifeforms.

So, the chapter does *not* deal with the 'origin of species', as such. By this time, different species are of course already established in the oceans of the four-billion-year-old Earth and we examined some of these in the previous chapter. Also, my focus in this chapter will be firmly on *animals*, but 'The Origin and Diversification of Animal Phyla' somehow didn't seem quite so resonant.

Of course, my title makes more than a nod in the direction of the series of epic fantasy novels by American author George R. R. Martin, of which the first is titled *Game of Thrones*, and my chosen subtitle alludes to Darwin's famous work. Darwin was well aware of a very notable and very visible challenge to his theory of evolution by natural selection. To Darwin, evolution is a mechanism that works rather gradually, by slow degrees over many generations. At the time he was writing his most famous work, however, the fossil record told a rather different story. There simply were *no* fossils predating the beginning of the Cambrian period.[1]

Darwin was at a loss, writing: 'To the question why we do not find rich fossiliferous deposits belonging to these assumed earliest periods prior to the Cambrian System, I can give no satisfactory answer.'[2]

In fact, David Attenborough, Britain's favourite naturalist, attended Wyggeston Grammar School in Leicester a few years before Roger Mason and had walked in Charnwood Forest before him. Attenborough was also a passionate fossil hunter, but he hadn't thought to look for fossils in Charnwood because he knew the rock there was Precambrian. 'And every geologist knew or at least was convinced that rocks of such extreme age couldn't possibly contain fossils of any kind.'[3]

In the Cambrian period, organisms tumbled into mud and sediment, died, and became fossilized like there was no tomorrow. It looked to palaeontologists as though life had somehow just 'switched on', seemingly wholly inconsistent with evolution by slow degrees. This is the *Cambrian radiation*, a period of evolutionary experimentation so frantic in geological terms that, even though we now know it follows a long history of evolutionary development that did leave traces in the fossil record, it arguably still deserves to be called an 'explosion'.

We're going to follow these developments by focusing our attention on animals, which the evidence suggests made their first appearance in the Ediacaran. So let's start by locating animals on the tree of life. Figure 78 shows the full picture of the phylogenetic relationships established in 1990 by Woese and his colleagues. We've seen part of this picture already (Figure 71), but we now complete it with Eukarya, the domain of all eukaryote organisms.*[4]

Recall all the caveats from the last chapter about representing genetic lineage and diversity in terms of a branching tree. Figure 78 shows Eukarya to be more closely related to Archaea than Bacteria and, indeed, if endosymbiosis happened the way we think it did, then this is entirely reasonable. On this basis eukaryote organisms may have inherited much of their rRNA sequence from their archaeon host. But some genes from the bacterial endosymbionts would also have become incorporated, a relationship that is hard to depict in a diagram such as this.

---

* Domains? Kingdoms? Phyla? Species? If you're puzzled by these taxonomic groups and would like to understand something of the nature of the relationships between them, then please consult this chapter's endnote 4.

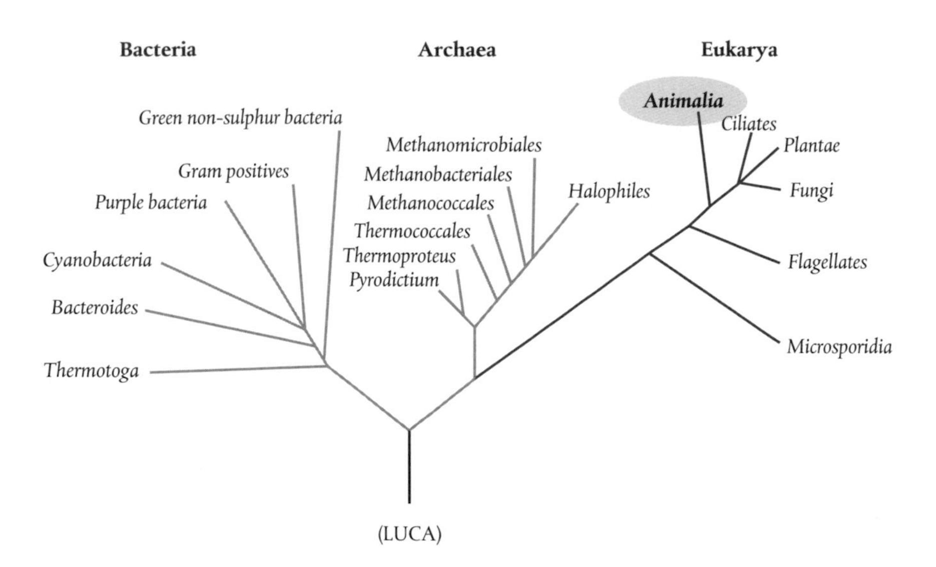

FIGURE 78 The full phylogenetic 'tree of life' published by Carl Woese and his colleagues in 1990 includes the Eukarya, shown here on the right. Animalia (animals), Plantae (plants), and Fungi are kingdoms in the domain of Eukarya and are reasonably self-explanatory. However, you may be unfamiliar with Ciliates, a group of single-celled eukaryotes that possess hair-like organelles called cilia. Flagellates are eukaryote cells possessing a whip-like structure called a flagellum which the organism uses to move around (think of male sperm cells). The Microsporidia are single-celled fungal parasites. Recall from Figure 71 that the order of branching and the lengths of the branches reflect the genetic 'relatedness' of different lifeforms based on rRNA sequence comparisons. This diagram was the first of its kind but has now been superseded by studies based on many more organisms and more gene sequences, and the details of eukaryote evolution are now accepted to be rather different. However, the late appearance of plants and animals holds firm. Adapted from Carl R. Woese, Otto Kandler, and Mark R. Wheelis, *Proceedings of the National Academy of Sciences*, **87** (1990), p. 4578.

THE CAMBRIAN EXPLOSION

It goes without saying that the journey from LUCA to the Ediacaran biota is simply marvellous, a testament to the elegant interplay between Earth's geology, geochemistry, and biological organisms striving to survive long enough to pass their genes to successive generations. And yet, if we're in a bad mood and want to be picky, the truth is that all we have to show for 3.5 billion years of evolution are some jellyfish and few small shelled creatures, or 'small shellies'.[5]

Hindsight is a wonderful thing. You know by virtue of the fact that you're sitting here reading this book that at some stage animal life must have evolved a diversity and complexity that can hardly be compared with the rather simple soft-bodied organisms and small shellies of the Ediacaran radiation. The clock is ticking, and we're now only 540 million years or so from the present.

Evolution needs to get a move on.

And move it does. But where, how and in what direction it moves is, once again, determined by the geological environment created by a restless Earth and solar system. Of course it's no coincidence that we're witnessing a key event in life's history at a key geological time boundary. We move from the Proterozoic eon (Greek for 'earlier life') into the Phanerozoic (meaning 'visible life'). The Phanerozoic is in turn divided into three eras, the Paleozoic ('ancient life'), Mesozoic ('middle life'), and Cenozoic ('new life'). The reason is simple. Until fossil evidence for earlier lifeforms was discovered, geologists and palaeontologists believed that the Cambrian was the period of the Paleozoic era in the eon in which ancient life first became visible.

So, at the start of this first period in the era of ancient life, precisely *what* did evolution do?

Any answer to such a question is necessarily speculative, but the fossil evidence tells us that this was a truly remarkable period of evolutionary experimentation. Within a few million years life moved on from simple forms to complex animals with many different body parts, such as segments, legs, antennae, eyes, sex organs, and so on. These evolutionary inventions were augmented by experimentation in their 'body plans'.

A simple consequence of movement (through the ocean or along the ocean floor) is that the body of an animal necessarily 'polarizes' into head and tail. The head is simply that part of the body that is first to encounter a new environment, and therefore the most logical place to put all the sensory organs. Forward movement defines 'left' and 'right', establishing a plane of symmetry. Such a design, with head end and tail end and a plane of symmetry along the length, is said to be *bilateral*.

A 'body plan' then simply means the genetic blueprint for the size and placement of body parts. Precisely how these parts are assembled is regulated by an astonishingly small number of genes, called Homeobox (or 'Hox') genes. Humans have just 39 Hox genes, organized in four clusters on four different chromosomes.

Why so few? Because these are command-and-control genes. They act like switches, turning on or off hundreds of other genes which then do all the work, specifying how cells in the animal embryo should be differentiated and where they should go. The embryo is first divided into front and rear and a body axis is formed, mirroring the pattern in which the genes are physically organized on the chromosome. The embryo is then segmented through the actions of Hox proteins manufactured by RNA, according to the nucleotide sequences specified by the Hox genes. Further genes then differentiate the cells and so elaborate the structures to be formed on each segment.

Much of our knowledge of Hox genes comes from studies of *Drosophila melanogaster*, the fruit fly. The discovery and classification of these genes led to the award of the 1995 Nobel Prize in Physiology or Medicine to American geneticist Edward Lewis, German biologist Christiane Nüsslein-Volhard, and American developmental biologist Eric Weischaus. Generally speaking, the simpler the organism the simpler its set of Hox genes (the fruit fly has eight, clustered on two segments of a single chromosome).* But the same or very similar genes are found in all animals, even in jellyfish, strongly suggesting that this developmental toolkit was already present in the DNA of the last Ediacaran common ancestor of the animals now ready to emerge in the Cambrian explosion.

The toolkit comes in very handy, as the mass extinction at the end of the Ediacaran creates what Knoll calls a 'permissive ecology':[6]

> By 'permissive' I mean an ecological landscape in which competition for resources is rare or weak. (You don't have to be good to win the game of evolution; you only have to be better than the other players.) Permissive ecologies may arise because environmental change makes new physiologies possible, or because an evolutionary novelty allows organisms to exploit resources in a new way, however poorly. Environmental catastrophe provides still another route—populations that survive mass extinction may radiate in the ecological emptiness that follows . . .

Mutations in the Hox genes can lead to some really bizarre body structures. For example, in one fairly common mutation in fruit flies, the antennae become legs.† Mutations in the Hox genes make it possible for evolution to ask some searching questions concerning the adaptability and survivability of the organisms that result. Does it make sense to put a leg there, or an eye? It does not take a great leap of imagination to suppose that, as evolution now experiments with

---

* But note that there are exceptions to this general 'rule of thumb'.

† I'm reminded of some of the fantastical creatures that feature in Spanish film director Guillermo del Toro's *Pan's Labyrinth* and the subsequent *Hellboy* films. In these movies Del Toro appeared to have a particular fascination with creatures with eyes in lots of unusual places.

its developmental toolkit, creatures will be born with some weird and wonderful mutations that would perhaps not survive in a less permissive ecology.

Whatever the circumstances, such a friendly environment cannot last forever. The shallow seas now contain more oxygen, and access to this energy source offers a whole new set of opportunities. The environment quickly becomes a lot *meaner*.

Evolution responds by selecting organisms better able to prey on their victims, and better able to protect themselves from predators. The result is a kind of arms race, as animals start to 'tool up'. Legs are deployed for advancing on unwary prey or scuttling away in a hurry. Eyes and antennae develop better to sense the location of prey or predator. Biomineralization now comes into its own, as organisms hide away their soft and vulnerable body parts beneath a hard carapace.

This is a world of *arthropods*, with segmented bodies, jointed legs, eyes and antennae, and an external skeleton. There are other creatures, too, and not all of them have shells. But because of their nature and composition, shells are much more likely to be preserved in the fossil record (they are, after all, already mineralized). The proliferation of organisms with shells is one of the reasons why, to earlier palaeontologists, life just seemed to switch on in the Cambrian.

A good example of a Cambrian marine animal is the *trilobite*, which first appears about 20 million years into the period. Trilobites are near-ubiquitous (and so rather familiar) fossils, which span some 270 million years of geological history (Figure 79). The carapace, made of calcium carbonate on a lattice of chitin (a long-chain polymer derived from glucose), has three lobes, which give trilobites their name. These are illustrated in Figure 79(b).

Beneath the axial lobe sits the organism's primitive brain and its digestive tract. Sitting on its head are a set of antennae (not shown) and two eyes formed from crystals of calcium carbonate orientated along their optical axes. Below the head there extends a thorax consisting of a series of segments, each with a pair of jointed legs overlain by a set of gills, tucked away beneath the carapace. Finally, the tail is formed from a series of fused segments.

Note that although only the hard parts tend to become fossilized, there are some (though rare) examples of trilobites whose soft body parts have become fossilized as well, making it possible to elucidate the body plans of these creatures in considerable detail.

The trilobites look superficially like a common woodlouse (for example, *Oniscus asellus*), but it helps to know that they varied in size from one

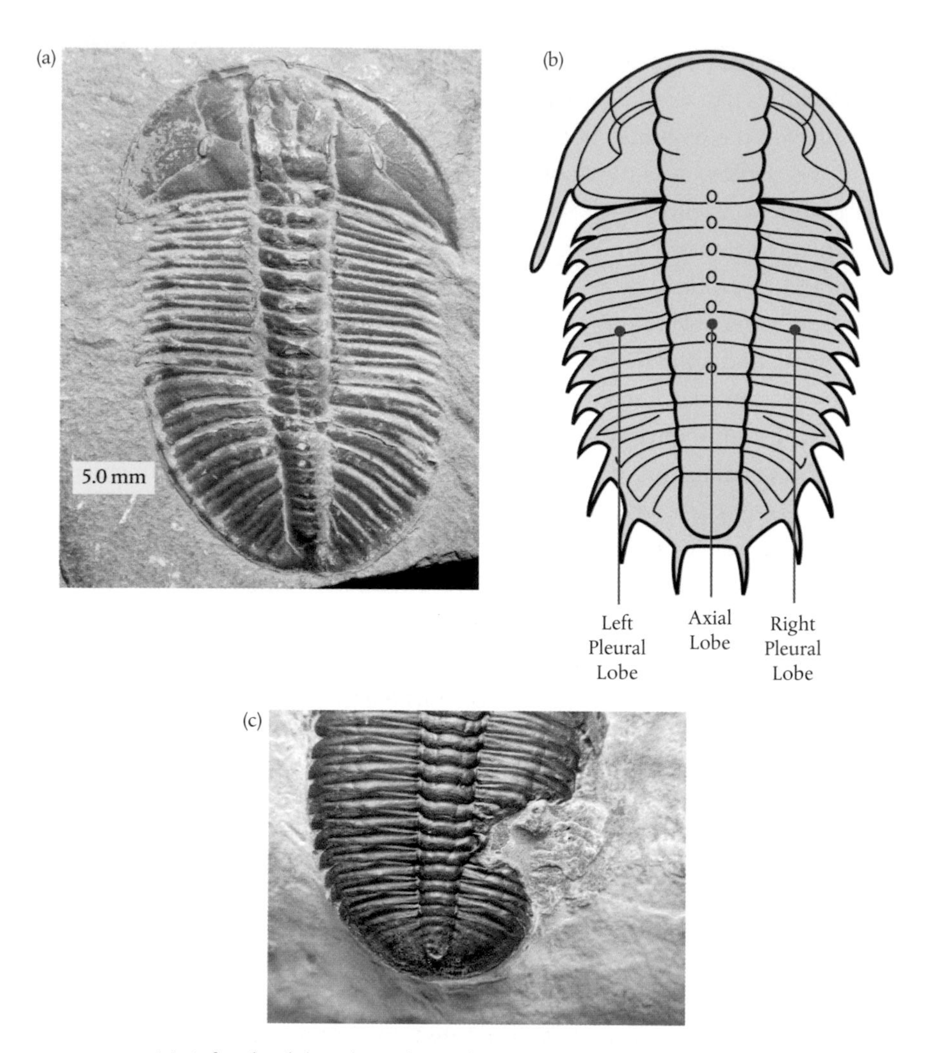

FIGURE 79 (a) A fossil trilobite from the mid-Cambrian. (b) Schematic showing the main exterior parts. (c) This trilobite carries a prominent bite mark.

millimetre up to more than 70 centimetres in length, with typical sizes of 3–10 centimetres.

However it has happened, and by whatever means, it is clear that in just 20 million years we have come an awfully long way from jellyfish and small shellies.

Figure 79(c) provides a hint as to why the evolution of such designs has accelerated. Despite its protective carapace, this trilobite has nevertheless suffered an attack from a larger predator. It carries a prominent bite mark on its right pleural lobe. The fossil evidence suggests that trilobites could survive having such chunks bitten out of them, provided these were below the head.

If the trilobites are being hunted, what is hunting them?

THE WEIRD (BUT WONDERFUL) DENIZENS OF THE BURGESS SHALE

Although the trilobites have left no surviving descendants, they nevertheless look superficially familiar. Their body plans are recognizable (or, at least, not wholly alien), with parts that we can see in creatures living today. To get a real sense of the nature of the experiments that evolution is now conducting, we need to move forward another few tens of millions of years, to the middle Cambrian about 505 million years ago.

Picture a well-oxygenated shallow sea, a few kilometres from the northern coastline of Laurentia.[7] A short distance from the coast in the Panthalassic Ocean, surf is breaking over the top of a near-vertical submarine cliff called the Cathedral Escarpment. The cliff is a reef, formed by cyanobacteria, and its base lies about 100 metres beneath the ocean's surface. Here the seafloor consists of mud and silt, eroded from the land and washed down into the sea. Sunlight struggles to penetrate to this depth. There is light, but it is rather gloomy here.

The seafloor is alive with marine animals. Some live burrowed in the mud, others are held fast to the seafloor, and they are filtering microscopic organisms that are pushed towards them by the current. Others, like the trilobites, crawl across the seafloor in search of food. Yet others swim above, eyeing the crawlers as their hunger grows.

Except for the trilobites, you could be fooled into thinking that these are marine animals pretty much as they appear today. But when we look more closely, we see that many of these animals have weird and wonderful designs that we simply cannot recognize in any living creature.

Look over there. What on Earth are those? Here is a small group of worm-like creatures, aptly named *Hallucigenia sparsa*, with bulbous head (or maybe that's its tail?) walking on seven pairs of tentacle-like legs, with an array of discouraging spines protruding from its back (Figure 80(a)). It is a scavenger, walking around on the muddy seafloor in search of carrion. Moving slowly across the seafloor nearby is *Wiwaxia corrugata*, possibly a distant relative of the molluscs, with soft parts like those of a slug or snail. But it is covered in armour plate and

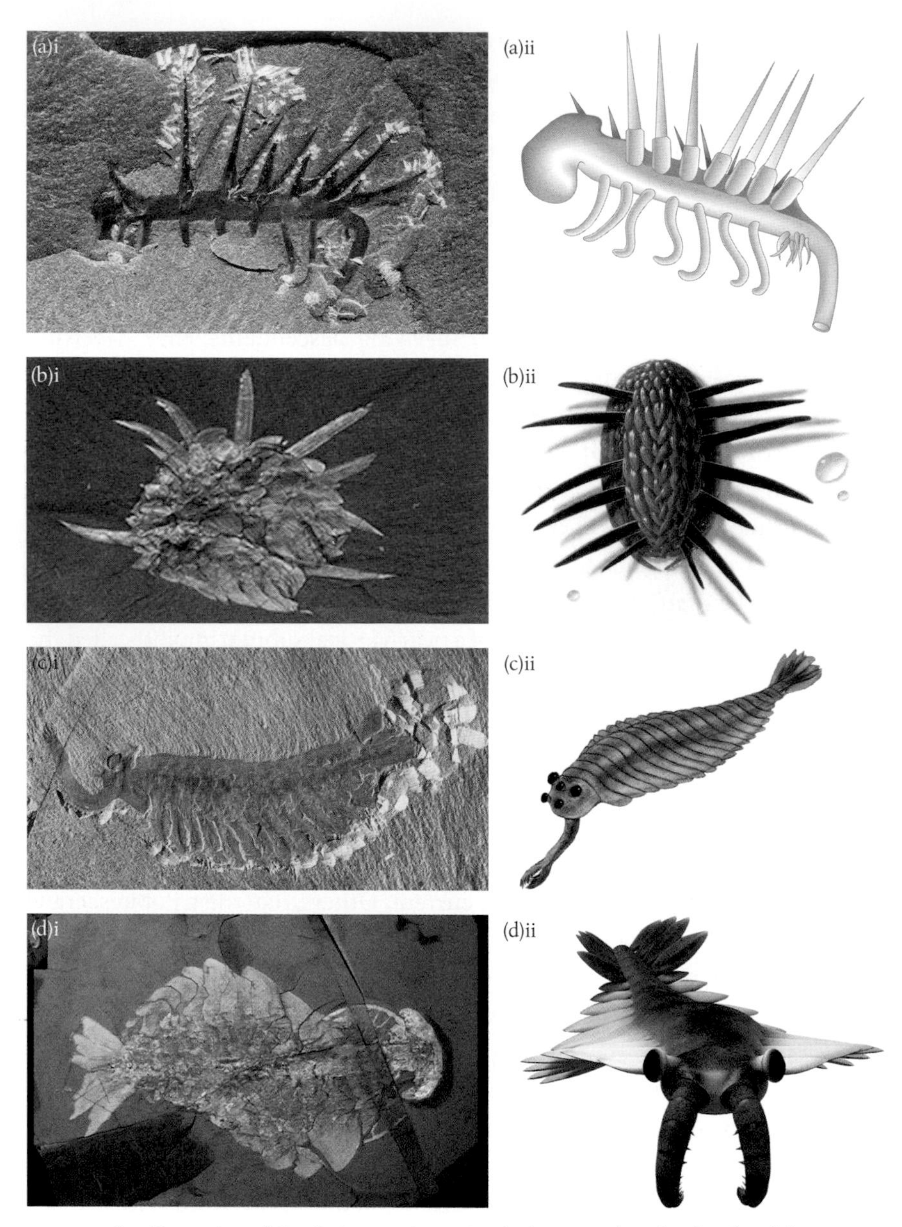

FIGURE 80 Examples of Cambrian marine animals discovered in the fossils of the Burgess Shale. (a) *Hallucigenia sparsa*, showing an example fossil specimen on the left and an artist's reconstruction on the right. (b) *Wiwaxia corrugata*. (c) *Opabinia regalis*. (d) *Anomalocaris canadensis*.

carries its own set of spikes designed to make potential predators think twice (Figure 80(b)).

These are weird enough, but over here we see *Opabinia regalis* (Figure 80(c)). When a diagram of *Opabinia* was shown at a meeting of the Palaeontological Society in Oxford, it was greeted with loud laughter, 'presumably a tribute to the strangeness of this animal'.[8] It too walks along the seafloor, looking for worms that it captures with its long trunk. The hapless worm is then quickly whisked into the creature's mouth, on the underside of a head that sports a distinctly alien collection of *five* large eyes. *Opabinia* is clearly very wary. Perhaps it is vulnerable to attack from above, and the eyes are needed to provide something of an early warning.

And then we see why. Our attention is drawn by a trilobite scuttling across the seafloor, and we jump involuntarily when a large dark shape suddenly swoops from above. There's a flurry of disturbed mud, from which emerges *Anomalocaris canadensis,* a six-foot-long 'strange shrimp' (Figure 80(d)). It has the trilobite trapped in two large appendages which protrude from its head and is already pushing the victim towards a vicious-looking circular armoured mouth, ready to take a bite. The *Anomalocaris* swims away with its prize.

These animals continue their daily struggle for survival, unaware of imminent danger. The seafloor is dotted here and there with fissures and hollows, evidence of an instability perhaps related to the shifting North American plate on which Laurentia sits.

Without warning, the seafloor suddenly falls away. Mud piled against the wall of the escarpment becomes unstable, and starts to slip. Thousands of animals caught in the dense swirl of mud are dragged to the bottom and are rapidly buried, at all kinds of crazy angles. Ordinarily, dead animals will quickly decay, a process that is greatly accelerated in the presence of oxygen. Some of the animals do start to rot, but the covering of mud soon cuts off the supply of oxygen and the carbon in the dead animal tissue instead forms a robust polymer called kerogen, preserving the detail of the soft body parts which it once formed. The processes of fossilization then begin.

Hundreds of millions of years pass. Laurentia has now become North America. The sea bed carrying its prize of exquisitely preserved Cambrian fauna now sits high up in the Canadian Rocky Mountains, about 8000 feet above sea level, in Yoho National Park close to Banff and Lake Louise. To find it you need to drive or take the train to the small town of Field, population now less than 200, lying 27 kilometres west of Lake Louise, and climb about 3000 feet towards Burgess Pass.

It was here that Charles Doolittle Walcott, an American palaeontologist and Secretary of the Smithsonian Institution, noticed a stray slab brought down by a snow slide which bore a fascinating collection of fossil crustaceans. Fossil hunting was a family affair, and in August 1909, together with his wife, daughter, son, and an assistant, he traced the stray slab to its source, a shale bed he later named the Burgess Shale. This is what had become of the mud that had engulfed our Cambrian animals, 505 million years before. The compressed mud had not only preserved the hard, shelly parts of these animals, but the soft body parts, too.

The story of Walcott's discovery of the fossils of the Burgess Shale was told by American palaeontologist Stephen Jay Gould in his deservedly best-selling book *Wonderful Life*, published in 1989. Walcott 'shoehorned' (Gould's phrase) the Burgess fossils into known taxonomic groups, so arguably concealing their marvelous diversity, and their weirdness. Something of their true nature came to light on their re-examination in the 1970s and 1980s by Harry Whittington, Derek Briggs, and Simon Conway Morris, a team of palaeontologists from Cambridge University in England.

Gould argued that the great diversity of Burgess Shale animals demanded revisions to the way we think about evolution. But the Cambridge team didn't share this conviction and, despite the apparent weirdness of the Burgess animals, further work revealed more similarities with familiar arthropods than differences. Some errors were also corrected on the way: Conway Morris had initially pictured *Hallucigenia* upside-down, walking on its spikes with a single row of tentacles pointing upwards from its back. Subsequent fossils from Changiang, in Yunnan Province, China, showed that *Hallucigenia* had two sets of tentacles, not one, though this doesn't really make it look any less weird.

The consensus is that these strange Cambrian animals form so-called phylogenetic *stem groups*. They represent evolutionary developments or experiments that ended in extinction. This may have been because they were evolutionary 'blind alleys'. Or they might simply have been the unfortunate victims of a series of natural geological extinction events. *Opabinia* and *Anomalocaris*, for example, are considered to be stem arthropods. They share some, though not all, of the features of modern arthropods, which includes crustaceans such as crabs, lobsters, and shrimp.

A stem group forms a single branch on the tree of life that doesn't go any further and doesn't sprout any new branches. Contrast this with a *crown group*. By definition, this includes the last common ancestor and all its surviving (and extinct) descendants, a part of the tree of life that radiates many branches. In this way, the animals of the Burgess Shale and similar fossil deposits in Greenland and China are seen, despite their weirdness, to be closely related to existing phyla rather than representing something completely different.

MASS EXTINCTIONS: THE 'BIG FIVE'

Darwin took a dim view of claims of species extinction resulting from natural catastrophe. Although examples of mass extinctions appearing in the fossil record were known at the time he was writing *The Origin of Species,* he tended to dismiss these as a tendency to over-interpret an evidential record that could well be fragmentary and incomplete. He wrote:[9]

> . . . so profound is our ignorance, and so high is our presumption, that we marvel when we hear of the extinction of an organic being; and as we do not see the cause, we invoke cataclysms to desolate the world, or invent laws on the duration of the forms of life!

As we know, Darwin was a gradualist. He perceived extinction as the natural result of competition between lifeforms within a sometimes precariously shifting ecology. Those forms best adapted to, or which 'fit' their changing environment are more likely to survive, procreate, and pass their genes to successive generations. Those less well-adapted are more likely to be driven to extinction.

But we now know that life on Earth has been profoundly shaped by a series of five mass extinctions over the past 400 million years, all of which can be tied to major geological upheavals.

Figure 81 shows the pattern in the number of marine animal genera from the Cambrian period to the present. Genera (singular 'genus') lie one rank up from species in the taxonomy (for example *Anomalocaris canadensis* is a species within the genus *Anomalocaris; Homo sapiens* is a species within the genus *Homo*).[4] This picture results from literally counting the number of different marine animal genera found in the fossil record at different moments in geological time, and was first put together from a database built by American palaeontologist J. J. Sepkoski. This version is adapted from Sepkoski's Online Genus Database, which is maintained by the University of Wisconsin-Madison.[10]

The result is rather startling. Of course, there are fluctuations in the number, which may be due to the possibility that some genera simply never left any evidence in the fossil record, or palaeontologists may have 'missed' some. But

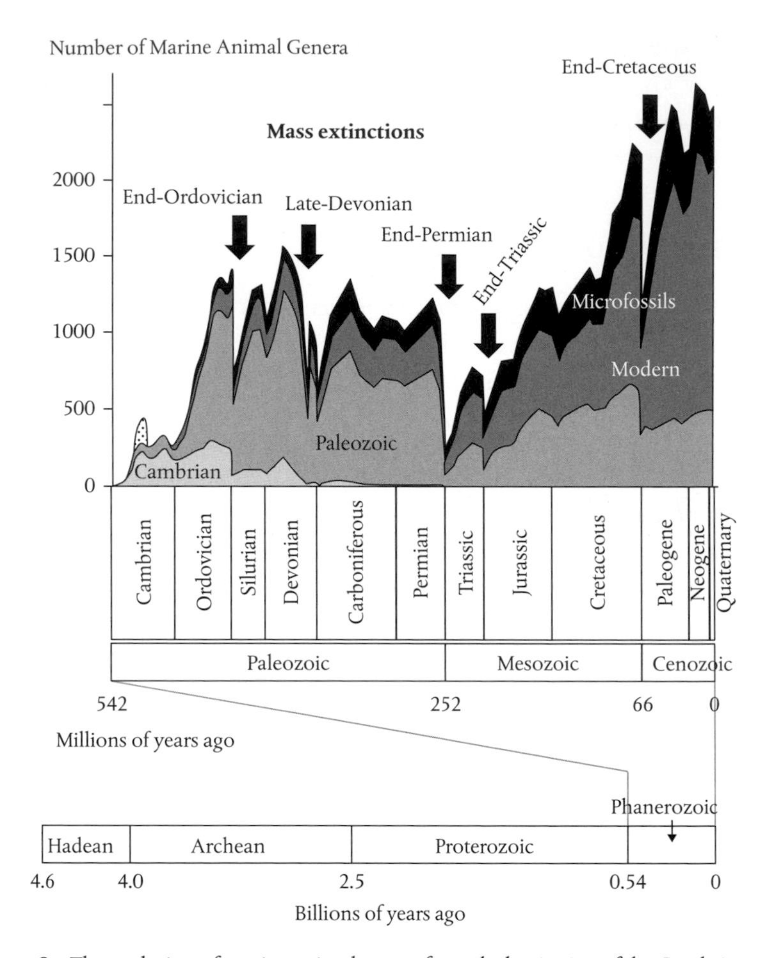

FIGURE 81 The evolution of marine animal genera from the beginning of the Cambrian period to the present, deduced from the fossil record. While marine life has become more diverse over time (the number of genera has increased) the pattern is punctuated by a series of dramatic and rapid declines, corresponding to mass extinctions. The 'big five' mass extinctions are marked with arrows. Adapted from Sepkoski's Online Genus Database: http://strata.geology.wisc.edu/jack/

there's no mistaking the dramatic and very rapid decline in numbers at five key moments in time (marked by arrows in the figure). These mass extinctions come at the end of the Ordovician, towards the end of the Devonian, and the ends of the Permian, Triassic and Cretaceous periods.

The mass extinction at the end of the Permian is particularly stark. It seems that about 252 million years ago 90–95% of all marine animal life was wiped out. Life had by this time also migrated to land (I'll describe this later in this chapter), and Figure 81 tells us nothing about what was happening simultaneously with land animal genera. There's little doubt that land animals suffered mass extinction too, but it seems that there's no evidence from the fossil record to support the figure of 70% extinction which is often quoted.*

The last mass extinction occurred about 66 million years ago, and marks the boundary between the Cretaceous and Paleogene (formerly Tertiary) periods. Marine life suffered once again, but we know this as the extinction event that erased the dinosaurs from the face of the Earth. Again, more on this later.

The pattern revealed in Figure 81 is rather sobering, and leads us to ponder on the fragility of life on an Earth that can be homely and welcoming, in a 'Goldilocks' sense, but which can also sometimes be very unforgiving and rather deadly. But notice how, after each extinction episode, life 'bounces back'. It radiates. The 'permissive ecology' created by clearing out innumerable life-forms offers the space for evolution to go to work once again and, within a few million years, life is again thriving and diverse.

But as we will see, the kind of life that thrives following a mass extinction is likely to be very different from what went before.

THE GREENING OF THE WORLD

I remember being dragged sleepily from my bed at almost 3 a.m. on the morning of 21 July 1969. I was 12 years old, and my parents thought that, despite the hour, I wouldn't want to miss history in the making. Grainy black-and-white flickering television images were showing Neil Armstrong slowly climbing down the ladder of Apollo 11's lunar module, pausing to explain: 'That's one small step for [a] man; a giant leap for mankind,' before stepping onto the surface of the Moon.[11] I still have my copy of Patrick Moore's *Moon Flight Atlas*, a beautifully illustrated commemorative volume published later that year.

This volume is full of evocative, full-colour photographs and drawings (I'm looking at it now). But, for me, some of the most compelling pictures show views of the Earth—Carl Sagan's 'pale blue dot'—as seen from space and from

* Knoll (personal communication, 25 October 2014) bet me a beer that I couldn't find any data to back up the assertion that 70% of land animal genera became extinct. I confess I didn't even try, so his bet is safe.

the surface of the Moon. Now, of course, we can just google images of 'Earth from space'.

Yes, such pictures are dominated by the blue of Earth's vast oceans, which cover about two-thirds of its surface, and which contrasts with the white candy floss of shifting cloud patterns above it. But it's not all pale-blue-and-white. Beneath the clouds we can discern a palette of brown hues betraying the arid landscapes of Northern Africa and the Middle East. And we also see a narrow spectrum of chlorophyll greens, perhaps most prominent in the Amazon region of South America. Blues, whites, and browns are what we get when we assemble a rocky planet at the right orbital distance from a third-generation Population I star. But the greens can only come from life.

Here's the thing. The greening of the world had to wait until life was ready to spread from the oceans to the land. By current reckoning, this didn't happen for almost 4.2 billion years after Earth was first formed.*

It's really not that hard to understand why. Even after all this time the land is a bleak and inhospitable place, where weather is free to do its worst, eroding the rock, scouring the plains, lashing rain producing flash floods and howling winds pulling sand dunes into stark shapes. There is no soil here, as soil is produced by living things.

Colonizing this 'empty' ecology might seem like a perfectly logical thing for life in a rather crowded marine environment to do. But this is life that has, thus far, evolved in the oceans, and all its biophysical systems and biochemical mechanisms are geared to survival in water. Removing life from water presents a raft of challenges that life on land must adapt to, or die.

It seems that the first stumbling steps towards colonization of the land were taken by photosynthetic green algae, of a kind that today is found in shallow fresh water. Natural selection forces the pace. Life depends on water, and evolving structures that help organisms to retain water and find alternative sources is much simpler than trying to evolve a form of life that is somehow less dependent on this substance.

So, the organisms evolve a waxy coating containing holes (called stomata) which admit air, and the carbon dioxide it contains, but which can close up to prevent water loss. They draw energy needed to fix carbon and split water

* Although there is some evidence for 1 billion-year-old non-marine eukaryotes from the Torridonian Supergroup, in the Northwest Highlands of Scotland. See Strother *et al.*, *Nature*, **473** (2011), pp. 505–9.

from sunlight, using photosynthetic pads. They extend roots into the ground in search of more water and other nutrients. They go on to evolve multicellular structures that offer some tensile strength, allowing them to grow shoots and eventually green leaves that reach upwards into the air. These structures are vascular—they circulate the sap (mostly water) around the organism. The sap is drawn upwards against the downward pull of gravity by the evaporation of water from its leaves, supported by water's natural surface tension and capillary action. This process is called *transpiration*.

The reproductive systems of some species of algae rely on the production and dispersal of spores. This mechanism is retained in the early land plants (and is used in many plant species today), although the spores—suitably coated to survive out of water—are now dispersed through the air, borne on the wind.

These first plant species did not survive to the present day, but we can get a sense of what they might have been like by studying modern plants such as liverworts, mosses, and ferns. Microfossils of plant spores similar to those of the modern liverwort suggest that plants had begun to colonize the land by the mid-Ordovician, about 460 million years ago. They were well established by the late Devonian, about 100 million years later. The Rhynie chert, near the village of Rhynie in Aberdeenshire, Scotland, is an abundant source of fossils of early Devonian land plants.

By the end of the Devonian, plant life had proliferated, and competition encouraged selection of taller and taller structures, to ensure that sufficient sunlight could be gathered, to increase the chances of wide dispersal of their spores, and to put competing plants literally in the shade. *Wattieza*, a relative of the fern, evolved a strong supporting and water-transport structure composed of a matrix of cellulose and lignin, which today we call wood. These prehistoric trees grew taller than eight metres, and formed forests that ranged far and wide across the mid-Devonian landscape.

All this greenery uses oxygenic photosynthesis to split water, fix carbon, and produce oxygen. If the carbon from dead and decayed organic matter becomes buried, then oxygen levels in the atmosphere further increase. But now certain chemical processes of combustion become possible that are simply impossible in the ocean. The forests are susceptible. Bolts of lightning during an electrical storm spark wildfires, reducing the plant life to charcoal. Such charcoal deposits are visible in the fossil record dating back to the early Silurian.

One note of caution. The fossil record is all we have but, as Darwin himself emphasized, we shouldn't be misled into thinking that this is somehow 'impartial'. Fossilization involves complex chemical processes and depends not only on the accidental circumstances under which lifeforms become buried, but also on the geological and atmospheric conditions that prevail at the time. Fossil beds tend to be found in one or two specific locations on Earth where the rock that preserved them is accessible to palaeontologists. It's all we have and it is imperfect.

THE RISE OF THE TETRAPODS

The plants don't have the land to themselves for very long. The earliest fossil evidence for land animals comes in the form of tracks left behind by arthropods, found in rocks from the late-Ordovician and early Silurian. Among these arthropods are mites, millipedes, spiders, and water scorpions. Their body plans are already suited to movement on land with external skeletons that help to prevent them losing water and drying out. Breathing air requires the evolution of *trachaea*, small tubes that—rather like stomata in plants—allow air in so that oxygen can be absorbed and transported around the body.

For a short time the arthropods reign supreme. The greening land provides a marvellous—and rather empty—ecosystem in which they can thrive. Mites crawl over the decaying organic matter that falls to the floor from above, feeding on fungi and helping to form the first soils.

But the ecosystem is quickly occupied, and competition between prey and predator once again favours size and armoury. The evolutionary possibilities are limited only by simple physical mechanics. The arthropod body plan calls for jointed legs with soft tissues and muscles hidden inside tubes of calcium carbonate exoskeleton. The larger the animal, the more unwieldy the joints and the greater its difficulty breathing. Nevertheless, evolution pushes the limit. *Slimonia* is a genus of Silurian water scorpion that grew to a rather frightening two metres in length, with primitive lungs in the form of folds of soft tissue on the underside of its body. It would lie close to freshwater, waiting for sight of its prey before pouncing.

Although our attention has now turned to the land, evolution continues in the seas and estuaries. The earliest fossil fish are ostracoderms, jawless, armoured creatures which originate in the Cambrian. These are *vertebrates*, with an internal skeleton consisting of a backbone. The ostracoderms possess a bony shield covering the head, thought to share a common genetic origin with rows of teeth. Now extinct, their nearest surviving relatives are lampreys.

By the Silurian, placoderms have appeared, also armoured but now with a jaw that likely shared a common genetic origin with the first gill arch. We would probably leap to the conclusion that this development created survival advantages because it aided feeding (which, aside from talking, is what we use our jaws for today). However, it seems that the immediate advantage was in fact in breathing more efficiently, the jaw being used to pump water over the gills.

Whatever the reason, the development of a jaw led to the diversification of jawed fishes through the Silurian and the Devonian. Thus we are led to the *Sarcopterygii*, or lobe-finned fish, whose living members include coelacanths and lungfish.

Let's focus on a late-Devonian lobe-finned genus called *Eusthenopteron*. On the surface, this fish appears relatively unremarkable. But it has some rather intriguing anatomical features. The skeleton of its fore-fin has a distinct humerus, ulna, and radius and its pelvic fin a femur, tibia, and fibula. These are bone structures we associate with land-living *tetrapods*, a biological super-class which includes all four-limbed creatures: amphibians, reptiles, birds, and mammals (including us).

For a time, *Eusthenopteron* was pictured as the archetypal 'fish out of water', using its fins to crawl onto dry land and so spawning whole new generations of land-borne tetrapods. Today, we recognize that this was a fish, albeit a 'tetrapodomorph', sharing some common anatomical features with tetrapods. In fact, further fossil evidence gathered over the past 20 years or so has allowed us to define a spectrum of morphologies, ranging from tetrapod-like fish to fish-like tetrapods.

Figure 82 summarizes our current understanding. *Eusthenopteron* and *Pandericthys* are tetrapod-like fish, their left forelimbs possessing humerus, ulna, and radius with *Pandericthys* showing a slightly more tetrapod-like structure, including short, unjointed digits. *Tiktaalik* is even more tetrapod-like, with a longer snout and larger eyes and a joint construction that could support the front of the body, like a wrist. These genera are roughly contemporary, dating to the mid- to late-Devonian.

By the end of the Devonian, fish-like tetrapods appear, such as *Acanthostega*, with eight digits on each forelimb and at least eight (possibly more) on its hind limbs. Despite its much more amphibian-like appearance, this is likely a creature that prefers to live in shallow waters (it has a tail fin), perhaps emerging to snatch prey that wanders a little too close to the water's edge. *Ichthyostega* appears to have combined a number of evolutionary specializations,

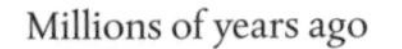

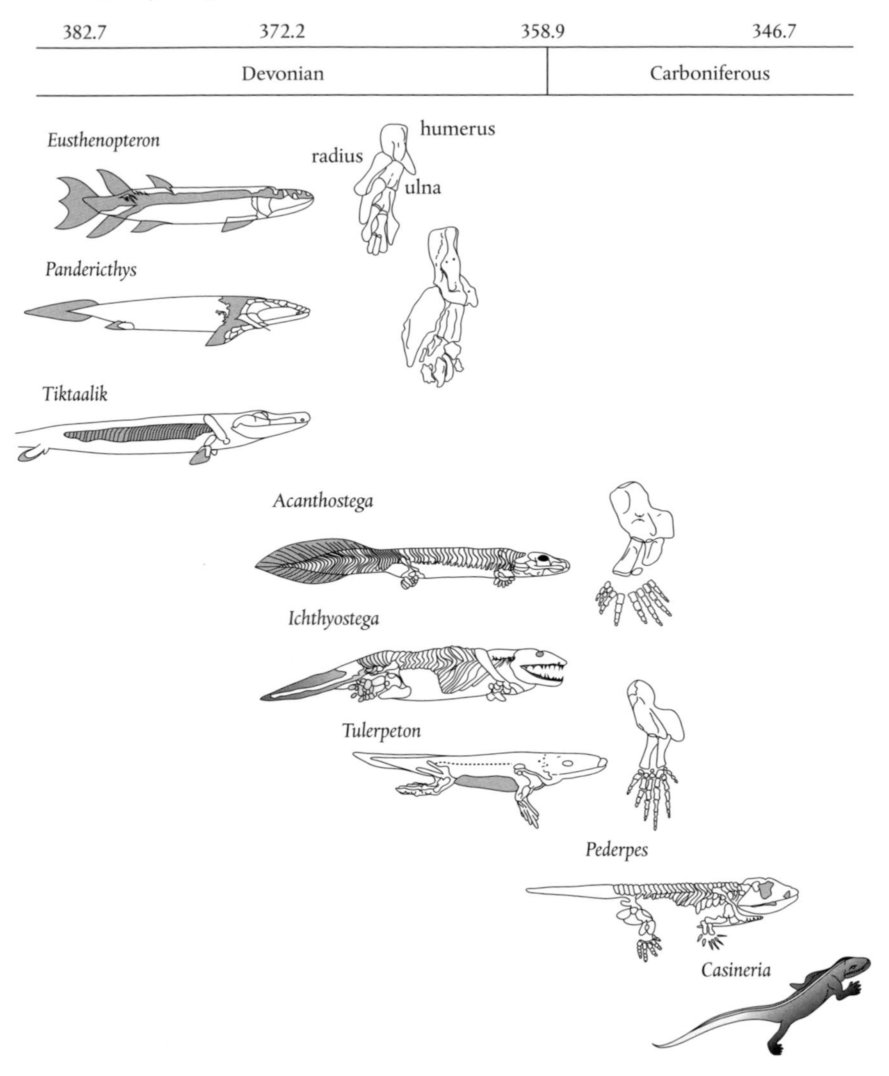

FIGURE 82 The transition from lobe-finned fish to tetrapod was likely complex, involving the evolution of fins into limbs with digits, and the development of anatomical structures adapted to walking. The fossil record reveals a range of creatures, called tetrapodomorphs, with morphologies in a spectrum ranging from tetrapod-like fish to fish-like tetrapods. *Casineria* was a fully fledged land-living tetrapod, dated to 340 million years ago. This figure has been put together from Figures 1, 5, and 12 in Jennifer A. Clack, *Evolution: Education & Outreach*, **2** (2009), pp. 213–223.

including an auditory apparatus for hearing underwater, now with seven digits to each limb.

Tulerpeton looks even more like a terrestrial animal, with six digits on its forelimbs and likely six on its hind limbs. *Pederpes* is an early Carboniferous creature with hind limbs bearing five digits that is more typical of a tetrapod, with feet pointing forward rather than to the side.

Finally, *Casineria,* discovered in 1992 by an amateur fossil collector on the shore of Cheese Bay near Edinburgh, Scotland, is a firmly terrestrial tetrapod from the early/mid-Carboniferous, about 340 million years old. It has acquired all of the evolutionary adaptations needed for life on land, such as lungs (already available in lungfish) and the ability to lay eggs out of water. It has five fingers with claws on each forelimb. Although this is likely to belong to a stem group, it has many features in common with the earliest reptiles.

The picture of a gasping *Eusthenopteron,* crawling from the water to take its first steps on land is firmly bedded in the public consciousness and likely to remain an evocative image. But the most recent fossil evidence suggests that the transition from fish to tetrapods was not quite so simple. Jenny Clack, professor and curator of vertebrate palaeontology at the University of Cambridge, UK, is an international expert on this transition. In 2009, she wrote:[12]

> The boundary between 'fish' and 'tetrapods' is becoming progressively more difficult to draw, and a more complex story is emerging in which, for example, the origin of limbs with digits, the origin of walking and terrestriality, and the origin of tetrapods in a strict sense, may be three different things.

PANGAEA: THE LAST SUPERCONTINENT

This is not all plain sailing. Weathering of rock produced by volcanic activity in the mid-Ordovician may have drawn carbon dioxide out of the atmosphere and, as Gondwana drifted over the South Pole, and sea levels declined towards the end of the Ordovician, a period of glaciation began. These events may have contributed to a mass extinction (Figure 81).

The reasons for the pulses of mass extinction in the late-Devonian are unknown, but these impacted largely marine animals living in warm, shallow water, while other species were relatively unaffected. Continental movements may again have had a role to play, perhaps in dredging up oxygen-starved waters from the depths of the oceans. By this time the continents of Laurentia and Baltica had collided to form Euramerica (sometimes called Laurussia), pushing up the northern Appalachians in the process.

In the Carboniferous, in contrast, life is positively bacchanalian. The spread of greenery across the land means that organic carbon becomes buried on an unprecedented scale, pumping more oxygen into the atmosphere. Much of the world's coal deposits are formed during this period (Carboniferous means 'coal bearing'). Carbon burial prevents the return of carbon to the atmosphere as carbon dioxide and so shifts the balance, and there are arguments to suggest that this allows atmospheric oxygen levels to increase to something like a giddy 35%. The atmosphere is now noticeably thicker.

Evolution enjoys an orgy of size, pushing species to the very limits allowed by their physiologies. Continental Earth in the Carboniferous is a true Land of the Giants.

Giant lycopods, trees growing 30 metres tall with trunks one-and-a-half metres in diameter, now dominate the Carboniferous swamps (to conjure an image, think of the rainforests and jungles of present-day Indonesia). These swamps are home to a variety of alarming giant arthropods, insects, and tetrapods. It is eerily silent.

We have already encountered *Slimonia,* a giant water scorpion. There are also giant millipedes, two metres long. The ground is crawling with insects, including large cockroaches, and insects have also taken to the air. What might have started as membranous outgrowths that allowed the insect to glide through the air from one plant to another have now evolved into wings. The silence in the fetid air above the swamp is occasionally broken by the beating wings of giant insects, such as *Meganeura,* the 'Bolsover dragonfly', with a wingspan of 65 centimetres. We keep a watchful eye on the carnivorous *Cochleosaurus,* a primitive amphibian and powerful predator that grows up to three metres in length, feeding on fish and small tetrapods much like modern-day crocodiles, as it lies in wait for unwary prey.

Arguments have raged over the circumstances and conditions that made this evolutionary excursion to gigantism possible, but something of a consensus is building in favour of oxygen.

Another (and, from our perspective, the final) round of supercontinent-building starts to reach its conclusion. Gondwana, holding the landmasses that we will later come to recognize as the southern hemisphere continents of South America, Africa, India, Australia, and Antarctica, drifts into Euramerica, itself an amalgam of Laurentia (North America) and Baltica (Europe). As north-west Africa grinds into south-eastern Euramerica, the southern Appalachians are formed. Siberia collides first with the microcontinent of Khazakstania, forming the Altai Mountains, before drifting into Baltica and forming the Urals.

The formation of Pangaea completes about 300 million years ago, towards the end of the Carboniferous. It straddles the equator and extends to the poles (Figure 83). It was famously hypothesized by German geophysicist Alfred Wegener, part of his (at the time, outrageous) theory of continental drift, which he first advanced in 1912. He referred to it initially as the 'Urkontinent', meaning 'primal continent'. The Greek Pangaea means 'all-lands'.

As always, there are consequences. The formation of a supercontinent causes sea levels to decline. The single large land mass shapes Earth's climate, propelling it towards glaciation. The Karoo ice age begins, which lowers sea levels still further. Atmospheric oxygen levels fall back.

FIGURE 83 The supercontinent of Pangaea was formed about 300 million years ago, towards the end of the Carboniferous. It was assembled from Gondwana (South America, Africa, India, Australia, and Antarctica), Euramerica—consisting of Laurentia (North America) and Baltica (Europe), and Siberia. It stretched from pole to pole, with much of the landmass in the southern hemisphere.

The forests and swamps of the Carboniferous go into decline, and as the ice retreats, arid desert forms in Pangaea's central interior. Lycopods give way to seed-bearing ferns and early conifers. Cockroaches and (smaller) dragonflies continue to thrive, as do amphibians such as *Cochleosaurus*. The arid conditions especially favour reptiles, including archosauriforms, meaning 'ruling lizard', semi-aquatic, crocodile-like predators, the stocky *Pareiasaurus*, a large herbivore, and the herbivorous *Lystrosaurus*, a mammal-like reptile about the size of a pig.

Life continues much as it has done for the past 200 million years.

THE GREAT DYING

The mass extinction that now occurs at the Permian-Triassic boundary, 252 million years ago, is sometimes called the 'Great Dying'. We can't be sure what causes it but we are not short of possibilities. The line-up of 'usual suspects' include an asteroid impact, a prolonged series of massive volcanic eruptions, a destabilization of ecosystems caused by the formation of Pangaea and the consequent decline in sea levels, anoxia (low oxygen levels) caused by the churning of the deep ocean into shallow water by *rising* sea levels, and the release of vast amounts of methane into the atmosphere, causing rapid global warming.

For a time it was suspected that there may not be a single cause, the marine and land animals of the Permian falling victim to a gradual weakening of their survival capability followed by a double or even triple geological 'whammy', what the American palaeobiologist Douglas Erwin referred to as the *Murder on the Orient Express* hypothesis.[13] Few (if any) palaeontologists now hold to this view.

In fact, a recent (2014) study suggests that the Great Dying lasts just 60 000 years, plus or minus 48 000 years.[14] This simple fact points the finger of blame at one particular suspect in the line-up. A quick, execution-style death implies a relatively sharp catastrophic event or short sequence of events, such as an asteroid impact or volcanic activity on a massive scale.

Finding evidence for an asteroid impact dating from this time is fraught with difficulty. The cratered surface of the Moon today bears mute witness to the Late Heavy Bombardment because there is no weather there. The craters are not eroded, or erased by plate movements that on Earth tend to wipe much of the slate clean every 200 million years or so. There are, nevertheless, some tell-tale signatures in the form of trace chemical substances and subtle variations in local gravity around candidate impact sites, and these have been used to argue in favour of the impact hypothesis. The evidence is not wholly compelling, however, and the jury is still out.

The evidence for volcanism is much clearer. The Siberian Traps is a large area of stepped volcanic rock, or so-called 'flood basalt', the result of the biggest sequence of volcanic eruptions in 600 million years of Earth history, one that lasted a million years (still 'sharp' in geological terms). It occurred between 252 and 251 million years ago, and this can surely be no coincidence.

Now, it's possible that you think you can imagine this event because you've watched volcanic eruptions on television or seen fictional eruptions in movies. Be assured—you can't. The Volcano Explosivity Index (VEI) is the volcanic equivalent of the Richter scale for earthquakes, and correlates with the height of the plume of gas, smoke, and ash and the volume of material ejected during the eruption. The Mount St Helens eruption in 1980, which produced a plume greater than 25 kilometres and a cubic kilometre of ejecta, measured 5 on the VEI scale.* An eruption of the massive supervolcano at Lake Toba in Indonesia is thought to have occurred about 70 000 years ago, and produced 1000 cubic kilometres of ejecta, measuring the highest level (8) on the VEI scale.

The eruptions that created the Siberian Traps are believed to have produced 4 *million* cubic kilometres of ejecta, covering an area the size of the continental United States to depths between 100 and 6000 kilometres or more. For sure this didn't happen all at once. But rough estimates suggest massive supervolcano-sized eruptions occurring every 50 years or so, for a million years.

Under such an assault, the atmosphere becomes filled with dust and smoke, shielding the Sun's radiation and choking off all photosynthesis. Atmospheric carbon dioxide levels increase dramatically, sufficient to raise global temperatures by 5°C and—possibly—triggering the release of frozen methane locked up in marine reservoirs, warming the world still further. The world has gone dark, and the heat is rising. Sulphurous volcanic gases return from the atmosphere in the form of deadly acid rain. The shallow seas become more acidic and anoxic, because oxygen is less soluble in the warmer water.

* I worked as a post-doctoral researcher at Stanford University, near San Francisco in California, in 1981–82, and remember vivid purple sunsets caused by residual atmospheric dust from the Mount St Helens eruption.

It's as though the world has returned for a time to the Hadean, to the vision of Hell on Earth.* The land is scoured, and the sea and air turn their backs on life.

It's not at all clear how an extended period of volcanic activity causes such a sharp extinction event. But we can imagine that ecosystems experience a prolonged period of stress, eventually reaching a tipping point beyond which few species can survive. Perhaps it starts at the bottom of the food chain, with the first dominoes falling. Either way, the consequences are utterly devastating.

The forests and the swamps disappear, and 90–95% of all marine animal genera are lost, including the hardy trilobites, which have survived for 270 million years. Many land organisms die out. This is the only known mass extinction of insects.

In *Life: An Unauthorised Biography*, first published in 1997, Fortey wrote:[15]

> If an average small town were affected to the same extent we can imagine looking around deserted streets afterwards and recognizing no familiar faces among the few survivors on the pavements. Nobody would man the checkouts in the supermarkets. There might be crying in the night from a lost child. But mostly there would be emptiness.

THE AGE OF THE DINOSAUR

The regeneration is protracted. It takes about ten million years for life to recover from the effects of the Permian-Triassic extinction. Not so much a bounce-back, then. More like a desperate crawl.

The land-based tetrapods have almost completely disappeared. *Lystrosaurus*, the pig-sized herbivore which we met in the late-Permian, is among the few to survive. It's not clear why, as it seems to have no features that would make it especially suited to the post-extinction ecology of the early Triassic. Its smaller size and heavy build, tusk-like canines, and horny beak may help it to cope with a new, meagre diet of scrub and weeds. But other tetrapods that seem no less fit have vanished, never to be seen on Earth again. Maybe *Lystrosaurus* burrowed its way to survival, or just got lucky.

---

* It's one thing to point the blame at volcanism, but quite another to explain precisely what caused volcanic activity on such a massive scale. The truth is we don't really know, but it has been argued that continental rifting may have been responsible. Some scientists have suggested that this volcanism was actually triggered by an asteroid impact.

Lystrosaurus is one of a number of 'disaster taxa', lifeforms that survive disaster and quickly take advantage of the emptiness, diversifying and spreading across Pangaea. Fossils of different species of *Lystrosaurus* have been discovered in South Africa, India, and Antarctica, providing evidence that these were once joined in a great continent.

One or more *Archosauriformes* also survive, evolving in the early Triassic to produce *Archosaurs*, the common ancestors of birds, crocodilians, and non-avian dinosaurs. The reptilian *Archosaurs* are sauropsids (meaning 'lizard-faces'). The mammal-like *Lystrosaurus* is a therapsid (meaning 'beast-face', now generally referred to as synapsids). In the evolutionary tussle for domination that follows, the dice is heavily weighted in favour of the sauropsids. The arid landscape that once again dominates Pangaea favours cold-blooded reptilian creatures able to conserve water.

The mid-Triassic signals the return of the forests and the beginning of the age of the dinosaur.

We have always been rather fascinated by monsters. They pervade myths ancient (think of *Hydra*, the many-headed serpent, the *Kraken*, and medieval dragons) and modern (J. R. R. Tolkien's *Smaug* and the *Balrog*). Fossil bones of dinosaurs were discovered and first identified as such in the 1820s, and the name ('terrible lizard') was coined by English palaeontologist Richard Owen in 1841. Owen went on to become the first Director of London's Natural History Museum, which was established in 1881.

The dinosaurs are monsters that are not mythological creatures born of a fertile imagination. They really existed. They have become firmly fixed in the public consciousness ever since the first discovery of their fossil remains, brought to life through ever more life-like illustrations and, most recently, through popular movies such as the *Jurassic Park* trilogy.* We already know many of the Latin names for their genera and species.

Anticipating popular demand, I thought I'd try to illustrate the evolution of dinosaurs through the Triassic, Jurassic, and Cretaceous periods with the aid of some of our favourite genera (Figure 84).

Dinosaurs are classified according to the structures of their hip bones. The *saurischians* are 'lizard-hipped', the *ornithischians* are 'bird-hipped'. When this scheme was introduced, the latter category was not meant to imply any

* A fourth film in the franchise, called *Jurassic World*, appeared in cinemas in June 2015 as this book was in production.

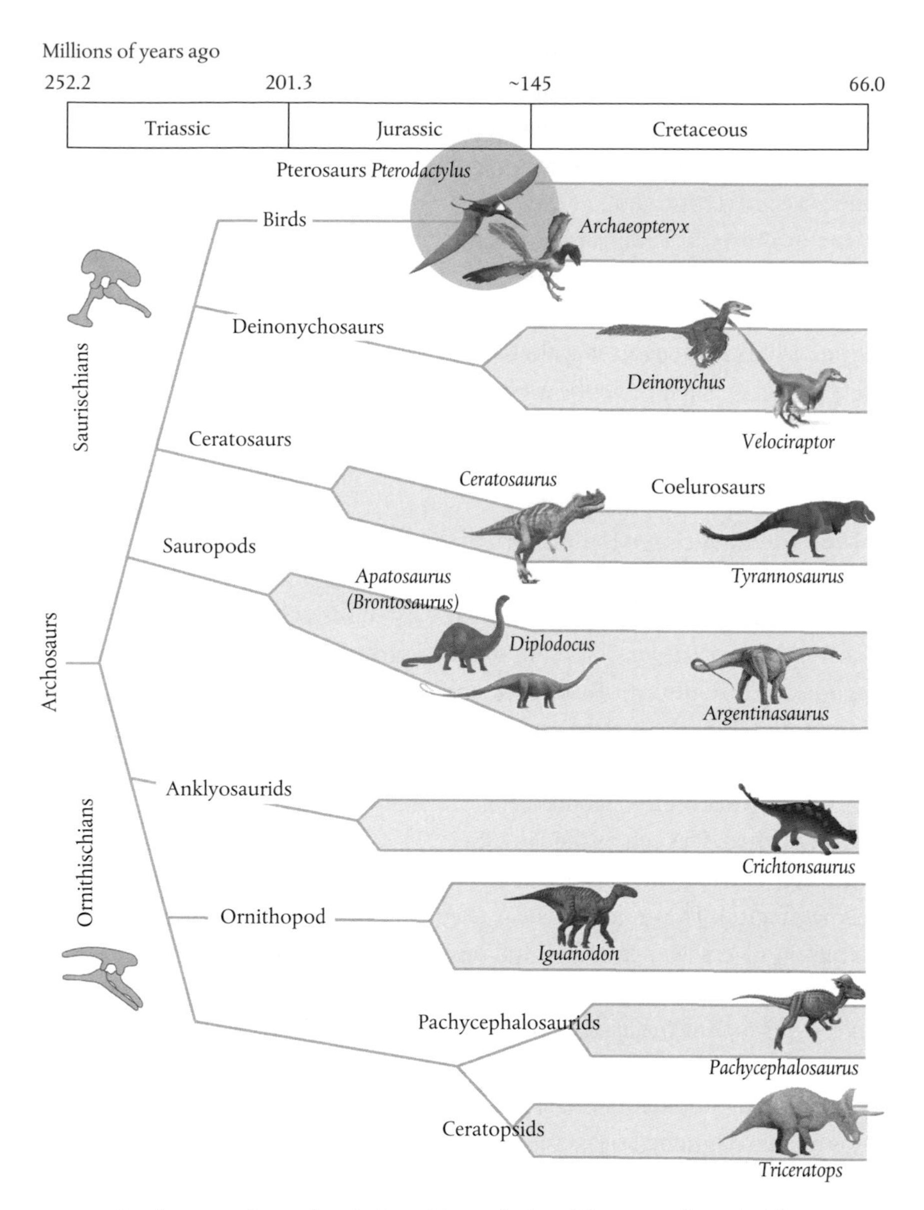

FIGURE 84 An approximate description of the evolution of dinosaurs, from their lowest common ancestor at the beginning of the Triassic to the end of the Cretaceous. Dinosaurs are categorized by their hip bones into saurischians ('lizard-hipped') and ornithischians ('bird-hipped'). Although many of the various orders of dinosaur may be unfamiliar, each is illustrated with a more familiar genus. One exception is *Pterodactylus*, which is not considered to be a dinosaur (and is thus marked by a grey circle) but is included here to illustrate the evolution of flight.

anatomical relationship to birds and, at first glance, it seems curious to find birds categorized as 'lizard-hipped' saurischians. In truth, their hip bones evolved over time and the bird-like structure was reinvented.

The first dinosaurs to evolve from their *Archosaur* ancestors are the sauropods, which make an appearance in the fossil record towards the end of the Triassic. These are the very familiar long-necked herbivores shown—correctly—in *Jurassic Park* to be moving in herds and grazing on the tops of trees. They include *Apatosaurus* (perhaps better known to older readers as *Brontosaurus*), a late-Jurassic sauropod which grew to over 20 metres in length, weighing about 16 tonnes, and *Diplodocus*, weighing about the same but with an extra-long neck that stretches its overall body length to 35 metres.*

Towards the end of the Cretaceous different sauropod genera vie for the title 'largest animal on Earth'. They are appropriately called *titanosaurs*. The *Argentinasaurus* currently claims top spot, although it will soon be deposed by another, as yet unnamed, sauropod discovered in May 2014, also in Argentina. *Argentinasaurus* measures 30–35 metres in length, is over seven metres tall and weighs an astonishing 70 tonnes. Based on the length and circumference of the thigh bone, the newly discovered titanosaur weighs in at 77 tonnes. This is about the weight of an average house.

The ceratosaurs ('horned lizards') evolve around the same time or a little later. The archetypal ceratosaur is obviously *Ceratosaurus*, a large predator which appears in the late-Jurassic, around the time of *Apatosaurus* and *Diplodocus*. It makes a guest appearance in the third *Jurassic Park* movie. This is a rather fierce predator, with large jaws and razor-sharp teeth. It is bipedal, its forelimbs short but powerful, and it grows to about six metres in length and weighing something between half a tonne and a tonne.

At first sight, *Ceratosaurus* looks rather familiar. It might remind you of everyone's favourite dinosaur, *Tyrannosaurus rex* ('tyrant lizard'), which puts in an appearance towards the end of the Cretaceous. I've included it in Figure 84 although *T. rex* is not a ceratosaur, it is a related coelurosaur, a taxonomic subgroup that has undergone many revisions over the years. *T. rex* is definitively tyrannical, measuring 12 metres in length and weighing almost seven tonnes, and with the strongest bite force of any terrestrial animal. Ever.

---

* The *Brontosaurus* was the inspiration for the theory of Anne Elk (Miss), memorably played by John Cleese in an episode of *Monty Python's Flying Circus* first broadcast by the BBC on 16 November 1972. For those unfamiliar with the sketch, Elk's theory is that: 'All brontosauruses are thin at one end, much much thicker in the middle and thin again at the far end.'

At the risk of eroding its image as the greatest terrorist that ever stalked prey on the Earth, I should nevertheless point out that it may have *feathers*. It is also not as quick as famously depicted, and would not have been able to pursue an accelerating jeep. But then we would have been denied one of the most memorable scenes in *Jurassic Park*, as Robert Muldoon (played by Bob Peck), glances nervously at the *T. rex* caught in the jeep's wing mirror, which advises him: OBJECTS IN MIRROR ARE CLOSER THAN THEY APPEAR.

The ceratosaurs are closely followed by ankylosaurids, armour-plated dinosaurs with a fearsome bony club on the end of their tails. They make their appearance in the early Jurassic and, in a nice twist, an ankylosaurid discovered in rock from the late-Cretaceous in 2002 by Chinese palaeontologists was named for Michael Crichton, the original author of *Jurassic Park* and *The Lost World*, on which the first two films in the franchise were based.

Ornithopods such as *Iguanodon* possess a horny beak but no armour, and are highly successful herbivores. The pachycephalosaurids ('thick-headed lizards') develop thick, dome-shaped skulls, believed to be useful for head-butting. The related ceratopsids, including *Triceratops*, possess elaborate frills and horns, possibly for head-butting fellow species members or defending themselves from larger predators.

This brings us to the birds. Strictly speaking, pterosaurs are not considered to be birds (they are not even considered to be dinosaurs) and it's not clear how they fit among the Archosauria. But I think it would be wrong to exclude them from this story. I've included them alongside birds in Figure 84, not because these creatures are related (birds are likely more closely related to crocodiles than they are to pterosaurs), but simply because they are both capable of flight.

The genus *Pterodactylus* ('winged finger') belongs to a sub-order of pterosaur, first appearing in the fossil record in the late Jurassic. It is a fearsome flying carnivorous reptile, with a long beak and wings of skin and muscle membrane, like webbing, spanning from an elongated finger to its hind legs. Its wingspan is about one metre: more like a large bat than the birds we know today. Later species of pterosaur evolve to larger sizes. *Quetzalcoatlus* appears towards the end of the Cretaceous with wingspans of more than ten metres.

The fossil remains of *Archaeopteryx* ('first bird') were first discovered in 1861 and represent a crucial link between feathered dinosaurs and modern birds. Early fossil discoveries were used by Darwin to argue in favour of evolution by natural selection in *The Origin of Species*. It is roughly the size of a raven. On balance, it is more dinosaur than modern bird, sharing many features in common with deinonychosaurs, which we will meet next. Although its feathers have by now evolved into recognizable bird feathers, it possesses small teeth and a long bony tail.

Among the many things that the *Jurassic Park* books and movies achieved, was to bring to prominence *Velociraptor,* a hitherto little-known dinosaur genus. These were envisioned in the movies as wily, vicious, bipedal lizard-like creatures that hunted in packs. The character Alan Grant, portrayed by Sam Neill in the first and third movies, was convinced that *Velociraptor* bore many close similarities with birds.* Discoveries made since the first movie was released bear this out.

Velociraptor has feathers, and is a lot smaller than depicted, growing to two metres in length and standing half a metre high. The raptors of popular imagining are more closely related to *Deinonychus,* which at the time Crichton was writing the first book had been named as a species within the genus *Velociraptor. Deinonychus* ('terrible claw') is somewhat larger, more than three metres long and almost a metre in height. It also bears feathers. Arguments have raged over whether *Deinonychus* pack hunted or pounced solitary on its prey and ripped it apart rather like an eagle.

This extraordinary bestiary does not include genera that evolve during this period but which don't survive to the end of the Cretaceous, such as stegosaurs. There can be little doubt that the dinosaurs are *hugely* successful. They survive another mass extinction event at the end of the Triassic (once again associated with massive volcanism, a kind of end-Permian 'lite'). They survive the breakup of Pangaea, which begins in the early Jurassic as Laurasia (Laurentia and Eurasia) rifts from Gondwana, and which continues in the early Cretaceous as Gondwana starts to break up into the southern hemisphere continents that we recognize today.

One last thing to note. The later dinosaurs roam in the Earth's first *gardens.* Flowering plants (called *Angiosperms*) radiate in the mid-Cretaceous, their evolution driven by symbioses established with the great varieties of insects and animals that now populate the land and the air. The plants become dependent on creatures carrying off or dispersing their pollen, which contain the

* The character of Grant was inspired, in part, by the American palaeontologist Jack Horner, who grew up in Montana and found his first dinosaur fossil at the age of eight. He is curator of palaeontology at the Museum of the Rockies and teaches at Montana State University, both in Bozeman, Montana. Horner acted as a consultant on all three movies.

precursors to male sperm cells. Some plants evolve tricks designed to appeal to a single species of insect.

To pick one modern example from a great many, *Ophrys* is a genus of orchid that produces flowers in the form of a female bee. The flower emits a sex pheromone to signal that it is ready to mate. Suitably deceived, amorous male bees mistakenly copulate with the flower, in the process either picking up pollen or pollinating it. Individual bees are only fooled once, it seems, but this is enough to ensure the plant's survival.

The world of chlorophyll greens now suddenly bursts forth with colour and scent.

THE CRATER AT CHICXULUB

When I was young, dinosaurs were depicted as lumbering, dim-witted beasts whose bodies greatly outgrew the capacities of their brains. They were doomed to extinction; a failed, if prolonged, evolutionary experiment. 'Dinosaur' is a term still used today to signify a backwards-looking failure of imagination, an incapacity or a wilful refusal to evolve with the changing times.

To a large extent, the reputation of dinosaurs has now been restored, thanks to the pioneering efforts of American palaeontologists such as Jack Horner and Robert Bakker.*

Bakker argued that, contrary to prevailing opinion, dinosaurs were warm-blooded creatures capable of much greater activity than cold-blooded reptiles. The birds, deinonychosaurs, and ceratosaurs are classed as *theropod*, or 'mammal-like' dinosaurs.† They engaged in much higher levels of social activity, herding, and nesting in ways that are common and familiar among modern mammals and birds.

Mammalian synapsid ancestors had co-evolved with their sauropsid counterparts from the mid-Carboniferous, but by the late Cretaceous, mammals had been driven to the margins of their ecosystems. They took the forms of small, nocturnal insectivores (such as *Megazostrodon*), much like tree-shews,

* Bakker was the inspiration for the character Robert Burke (played by Thomas F. Duffy) in the movie version of the *The Lost World*. Bakker and Horner had argued about whether *T. rex* was a predator or a scavenger and, as a favour to Horner, in one scene director Steven Spielberg shows Burke being eaten by *T. rex*. Bakker recognized himself, but was nevertheless delighted, declaring to Horner: 'See, I told you *T. rex* was a hunter!'

† Theropod means 'beast feet', not to be confused with *therapsids*.

and other creatures that we might compare with rats, armadillos, and flying squirrels and *Repenomamus*, a badger-sized mammal that might have preyed on dinosaur young.

There can be little doubt that dinosaurs held all the cards and there is no reason to suspect that they would not have continued to dominate the world.

Most readers will know how this chapter of the story ends. The evidence for an asteroid impact 66 million years ago is now largely unquestioned. It coincides with the disappearance of the dinosaurs at the end of the Cretaceous and the beginning of the rise of mammals in the following Paleogene period.

The meek shall indeed inherit the Earth.

The notion that an asteroid impact was responsible for the mass extinction event at the Cretaceous-Paleogene (sometimes abbreviated as K-Pg) boundary* was famously advanced in 1980 by Luis Alvarez, an American physicist who had worked on the Manhattan Project, and his son Walter, a geologist. The K-Pg boundary is marked by a thin layer of clay which separates lower Cretaceous rock from upper Paleogene rock. In searching for radioactive isotopes suitable

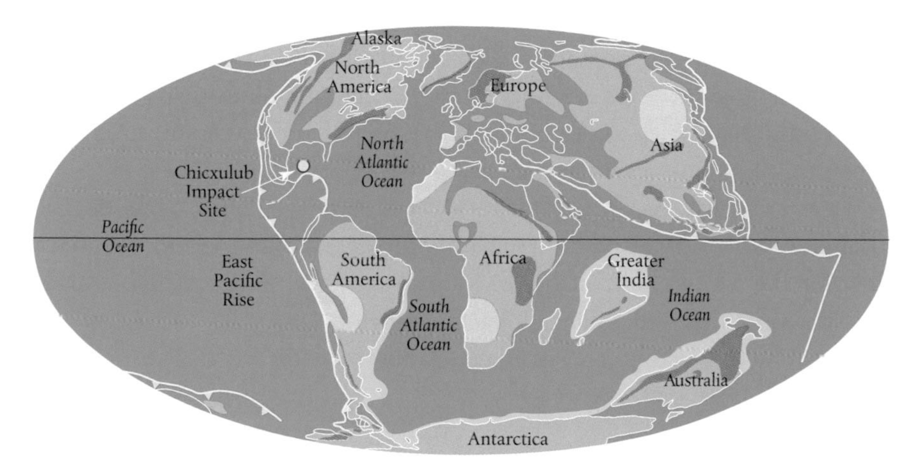

FIGURE 85 The Chicxulub impact site, as it would appear on the late-Cretaceous Earth. Laurentia has now rifted from Eurasia, opening up the North Atlantic Ocean. Gondwana has broken up, with South America rifting from Africa to form the South Atlantic Ocean. India is making its way northwards to collide with southern Asia. Australia is still connected with Antarctica but will soon break free and head northwards.

* At the time, this geological boundary was between the Cretaceous and Tertiary periods and referred to as the K-T boundary. The Tertiary is no longer recognized by the International Commission on Stratigraphy.

for use as a marker to determine the speed with which this layer was deposited, father and son settled on an isotope of iridium, ^{77}Ir.

In his 1987 autobiography, Luis Alvarez described what happened next:[16]

> Frank [Asaro] and Helen [Michel] analyzed our sample. They showed, to our very great surprise, that three hundred times as much iridium was concentrated in the clay layer as in the limestone layers above and below.

There's typically much more iridium in asteroids and chondritic meteorites than in the Earth's crust. Once they had worked out the effect of the impact of a 10-kilometre diameter asteroid on Earth's climate, they figured they had enough evidence on which to convict the prime suspect.

The idea remained highly controversial, however. More evidence was needed. Specifically, a 10-kilometre diameter asteroid impact would surely have left a tell-tale impact crater. In fact, an impact crater close to the town of Chicxulub, on the Yucatán Peninsula in Mexico, was discovered in 1978 by geophysicists working for Pemex, the Mexican state-owned oil company (Figure 85). Their discovery was not widely publicized until the early 1990s.

Although there's little or no real evidence to support it, volcanism may have also played a part, perhaps further amplifying the effects of the asteroid impact. The Deccan Traps is a flood basalt located on the Deccan Plateau, in west-central India, believed to be the result of a series of volcanic eruptions beginning 68 million years ago and lasting two million years.

We can speculate on the all too familiar consequences. Dust, darkness, the disappearance of photosynthesizing organisms (including all those pretty flowers), the widespread collapse of food chains, acid rain, the high atmospheric oxygen levels unleashing global firestorms.

Death and destruction cast their dark shadows over the Earth once more.

11

THE HUMAN STAIN

The Origin of Humanity

So, what are mammals, exactly? Mammals are distinguished from other types of animals by possession of a number of specific characteristics. They are warm-blooded, which means they maintain a constant body temperature as a result of an elevated metabolic rate. Unlike reptiles, mammals don't have to wait for the warmth of the Sun's radiation to raise their core body temperature before they can go off in search of breakfast. Mammals nevertheless need to regulate their body temperature, and the function of hair or fur (another defining mammalian feature) is to insulate when external temperatures fall, and to cool (through sweating) when temperatures rise. Reptiles do not shiver, or sweat.

The earliest mammals are also *amniotes*—they lay eggs containing a pouch which holds water but which allows the developing embryo to breathe—as indeed did all animals descended from the early land-living tetrapods. A few modern-day mammals, such as the platypus and spiny anteaters, have retained this reproductive strategy. But mammals adapted to giving birth to live young start to appear towards the end of the Cretaceous. These include marsupials, which carry their relatively undeveloped young in a pouch for a time, and placental mammals, which give birth to fully developed young after prolonged gestation in the womb.

Of course, one of the defining features of mammals (and the one that gives them their name) is that the females nurture their infant young by *suckling*. The timing and nature of the evolutionary development of mammary glands remains a topic of scientific debate, but one theory suggests that these were adapted from sweat glands.

The mass extinction at the K-Pg boundary is very deadly, but it is not as devastating as the Great Dying. Most archaic bird species become extinct but some do survive, and fossils of a few bird species have been found in rock strata either side of the boundary layer. The terrifying pterosaurs have disappeared from the sky. About half of all crocodilian species are lost. The impact on mammals is more regional. Although there are some losses, most mammalian species survive. But no marsupials survive in North America, and we only have to look at present-day Australian fauna to know that they not only survive but have continued to thrive there. Distance from the impact site may have been a factor.

For those readers looking for tips on surviving a mass extinction event, it would seem that being small is an important prerequisite. Smaller animals need less food, reproduce faster, and produce more offspring, which encourages greater genetic diversity and thence adaptability. It helps not to be dependent on a specific type of food, especially one than can disappear, and to be ready and able to move on when things get difficult and go in search of better conditions. Animals that can burrow and hibernate may also have an advantage.

We now enter the Paleogene. The dust settles, the Sun shines once again and photosynthesis resumes. Now, our story of life on Earth has so far swept past rather frenetically in chapters spanning hundreds of millions of years, but as we get closer to *Homo sapiens* we will need to reach for shorter geological timescales (Figure 86). Each geologic period is divided into *epochs*, and the Paleogene opens with the Paleocene (with a 'c') epoch, spanning from 66 to 56 million years ago, the Eocene, from 56 to about 34 million years ago, and the Oligocene, from 34 to 23 million years ago.

The Paleocene begins cool and dry but the climate slowly warms through the next ten million years as carbon dioxide builds up in the atmosphere once more. The climate at the equator becomes tropical, and the land is hot and arid in both northern and southern hemispheres. The poles are cool and temperate. Forests and flowering plants rebuild themselves, and splashes of colour return. Palm trees grow in Greenland.

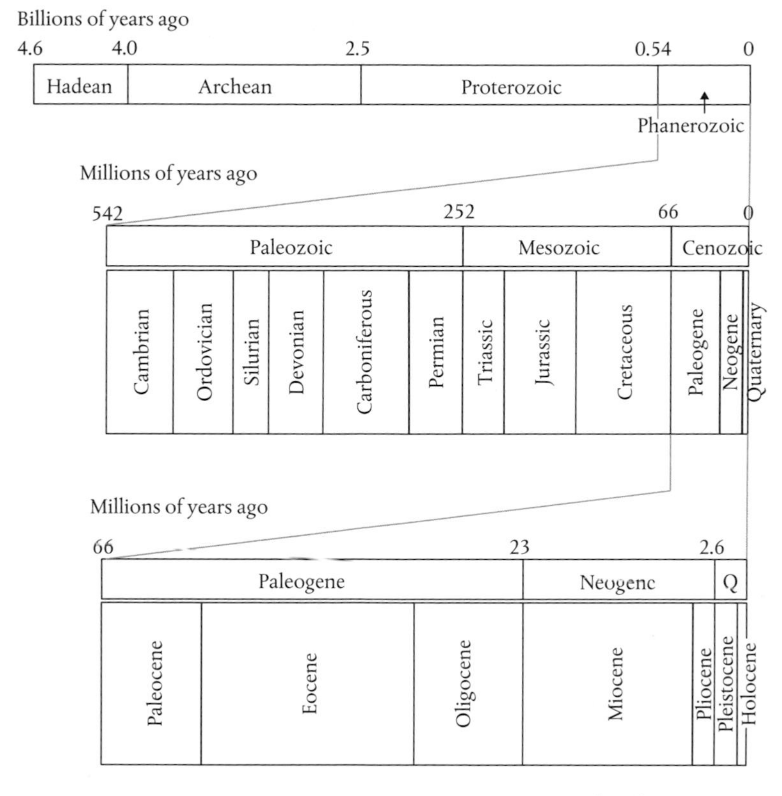

FIGURE 86 As we get closer to the time frame of human evolution, the geological timescale becomes more compressed. The Phanerozoic eon begins 542 million years ago and continues to the present day. This, in turn, is divided into eras, including the Paleozoic, Mesozoic, and Cenozoic. We have been following events thus far in relation to the successions of periods within these eras, from the Cambrian to the Paleogene, Neogene, and Quaternary. We now split each of these latter periods of the Cenozoic into epochs—the Paleogene into Paleocene, Eocene, and Oligocene; the Neogene into Miocene and Pliocene, and the Quaternary into Pleistocene and Holocene.

The warming continues, culminating in the Paleocene-Eocene Thermal Maximum, marked by another major excursion in the $^{13}C/^{12}C$ ratio in marine and terrestrial carbonate rock. Temperatures continue to rise in the early Eocene, before eventually falling back.

The disaster taxa have moved to occupy the newly permissive ecology of the Paleocene, and both birds and mammals now experience substantial adaptive radiations.

DARWIN'S FINCHES AND THE RADIATIONS OF THE MAMMALS

During the voyage of HMS *Beagle*, Darwin famously noted small evolutionary adaptations among finches populating different volcanic islands of the Galapagos Archipelago. He presented his findings at a meeting of the Geological Society in London on his return in 1837, and experts identified 12 different species of finch, an astonishing number given the relative proximity of the islands (15 different species are known today).

Darwin reasoned that these finches had evolved from a common South American ancestor, their adaptations (particularly of the shapes of their beaks) reflecting small differences in the ecosystems that they had come to inhabit. The finches had evolved in relation to their local environment, the shape of the beak reflecting the nature of their primary food—long thin beaks for teasing worms from the ground to short stubby beaks more suited to a diet of seeds. The lesson is that geographic isolation helps to foster the *diversification* of species.*

We now see this on a grand scale in the evolutionary radiations of the mammals in the Paleocene and Eocene. The last supercontinent is breaking up and the continents are on their way to their present-day configuration. But South America has yet to form a land bridge with North America. Africa is still separated from Eurasia by the Tethys Ocean. India has yet to collide with Asia. Australia is now separated from Antarctica and is making its way steadily northwards. The story of mammalian evolution is inextricably linked with the story of the Earth's continental movements.

Not much in the way of fossil evidence is available for mammals during the Paleocene, perhaps because mammals are still rather small creatures, but the record becomes richer in subsequent epochs.

From Eocene-epoch fossil beds discovered at Riversleigh in northwest Queensland we learn of Australia's rich marsupial heritage, present-day fauna barely doing justice to an extraordinary variety of ancient koalas, bandicoots, wombats, possums, kangaroos, rat-kangaroos, marsupial lions, and Tasmanian tigers. Among the fossils are many that can't be readily classified. These have been assigned to a colloquial order of extinct 'Thingodonts', of which one, now

---

* Darwin's broad conclusions have been confirmed by recent genetic analysis, particularly of one of the genes responsible for beak shape, called *ALX1* (ALX Homoebox Protein 1). See Lamichhaney *et al.*, *Nature*, **518** (2015), pp. 371–5.

named *Yalkaparidon*, appears variously to be some kind of bizarre mole or a primitive mammalian woodpecker.[1]

I don't propose to trace the evolutionary history of marsupials any further, but I can't resist mentioning the herbivorous *Diprotodon*. This appears in the fossil record about 1.6 million years ago and is believed to have become extinct about 48 000 years ago, shortly after the arrival on the continent of *Homo sapiens* (see later). It is the world's largest known marsupial, growing to the size of a hippopotamus. *Diprotodons* are thought to be depicted in Aboriginal rock art and may have been hunted to extinction.

The Messel Pit, an abandoned quarry near the village of Messel about 35 kilometres southeast of Frankfurt in Germany, is an abundant source of fossils and was declared a World Heritage site in 1995. This was once an Eocene-epoch lake. The fossils are exceptionally well preserved, showing fur, feathers, and 'skin shadows'.

The fauna here in Europe couldn't be more different from the marsupial-dominated Australian continent. There are some primitive marsupials, but the majority are placental mammals. They include bats, anteaters, rodents, hedgehogs, small primitive horses (*Eurohippus*), the carnivorous *Creodonts*, which share a common ancestor with the *Carnivora* (and which will eventually evolve to give us cats, tigers, bears, hyenas, wolves, and giant pandas), predatory cat-like mammals thought to be the ancestors of *Carnivora*, and some small primates.

Of course, the mammals are not alone. They share their environment with birds, reptiles (snakes, lizards, and several different kinds of crocodile), insects (beetles and giant ants), and fish.

The mammals of South America are no less extraordinary. They include primitive marsupials and placental mammals such as anteaters, armadillos, and sloths and a variety of hoofed or nearly hoofed animals, the forerunners to horses, giraffes, camels, deer, cows, and hippopotamuses. The Miocene witnesses the arrival of *Glyptodonts*, relatives of the armadillos but growing to three metres in length. The elephant-sized ground sloth *Megatherium* appears much later, towards the end of the Pliocene, six metres long and weighing in at four tonnes.

South America experiences a series of invasions, starting in the middle Eocene with the arrival of rodents and then small primates from Africa. The two continents are much closer to each other at this time and it is thought that many animals accidentally 'raft' across the oceans on driftwood. The formation of the Panama isthmus about three million years ago, joining North and South America, sparks the *Great American Interchange*. Many of the mammals of North America go south, and those of South America go north. The glyptodonts (and their much smaller relatives) and megatheria make it to North America, passing rodents, bears, cougars, jaguars, American lions, sabre-toothed tigers, and wolves going in the other direction.

On balance, the mammals from the cooler climate of North America have the edge; many species from the south more adapted to a tropical climate don't make it any further north than Central America.

It might seem that, after spending millions of years evolving from lobe-finned fishes to land-living tetrapods, figuring out how to adapt all those water-based biochemical and physiological systems, there would be no going back. But this is not how evolution works. Life adapts wherever it can find a suitable niche, and if a freshwater or ocean ecosystem appears to offer more favourable conditions than those that prevail on dry land, then land-living animals may come to eye it rather jealously. In this context, 'more favourable' may simply mean that the watery environment offers more food resources, but it can also mean that it provides shelter from land predators.

Pakicetus is a genus of extinct mammal found in Pakistan, in rock dated to the Eocene. It looks like almost any other kind of mammal from this period—four legs, a long snout, and a long tail, about the size of a dog. But, just as lobe-finned fish have fin bones that foreshadow adaptation to walking on land, so *Pakicetus* has limb bones and an inner ear structure more suited to a life underwater. It occupies a semi-aquatic habitat along the Tethys coastline. It stands, fully immersed in water but within sight of land, keeping a watchful eye on predators and stalking prey in both environments.

Now *Pakicetus* is certainly not a whale, but it is regarded as an archaic *cetacean*, an order which includes marine mammals such as whales, dolphins, and porpoises. And, just as we can follow the evolutionary adaptations from *Eusthenopteron* to *Casineria* (Figure 82), so we can track the adaptations that take us back to the sea, from *Pakicetus* to *Ambulocetus* (which could swim as well as walk), to remingtonocetids, to protocetids (in which the limbs are starting to look more like fins), and finally to late-Eocene basilosaurids, which look recognizably like whales and which grow up to 18 metres in length.[2]

Evolution does not reinvent gills. These are still mammals—they have lungs and need to breathe in oxygen from the air and exhale carbon dioxide. But they are nevertheless adapted to long periods underwater, literally holding their breath (up to two hours for some species). They return to the surface to take a gulp of air and to exhale, through a blowhole that was once the nasal opening at the end of the snout.

MEET THE PRIMATES

Seen against the background of this rather exuberant period of mammalian evolution, the primates appear really rather inconspicuous. In the warm climate of the middle Eocene, tropical forests cover much of the land. The primates are perfectly adapted to life in the trees. They are small, variously active during the day (diurnal) or the night (nocturnal), feeding on leaves, fruits, and insects.

Scientists still argue today about the characteristics that constitute the primate order. About a dozen or so features are recognized. These include eyes close together and pointing forwards to provide stereoscopic vision, a grasping foot with a 'divergent' big toe, grasping hands with an opposable thumb, and increased brain size. There are always exceptions, primates that don't share all these characteristics (for example, humans are primates but don't have divergent big toes). But to a large extent these are the kinds of features we might expect for an animal adapted to tree-living: nimble and quick-witted, scurrying along delicate tree branches, and jumping from branch to branch.

We recognize two primate sub-orders. The *strepsirrhine* ('wet-nosed') sub-order includes the diurnal adapoids which thrive in the Eocene, with most becoming extinct towards the end of the epoch. Modern-day strepsirrhines include lemurs and lorises. The *haplorrhine* ('dry-nosed') sub-order includes the nocturnal omomyids, an extinct family related to modern tarsier monkeys, and the anthropoids or so-called 'higher primates'.

In May 2009, the announcement of the discovery of a fossil primate from the Messel Pit caused quite a stir and some considerable controversy. Norwegian palaeontologist Jørn Hurum, based at the University of Oslo's Natural History Museum, and his colleagues claimed that this wonderfully preserved fossil primate (named Ida, after Hurum's daughter) is an adapoid with some key features shared with haplorrhines (Figure 87). They argued that it therefore constitutes a kind of 'missing link' in early primate evolution on the journey that leads eventually to us, concluding: 'Defining characters of [Ida] ally it with early haplorrhines rather than strepsirrhines'.[3] This was the year of Darwin's bicentennial

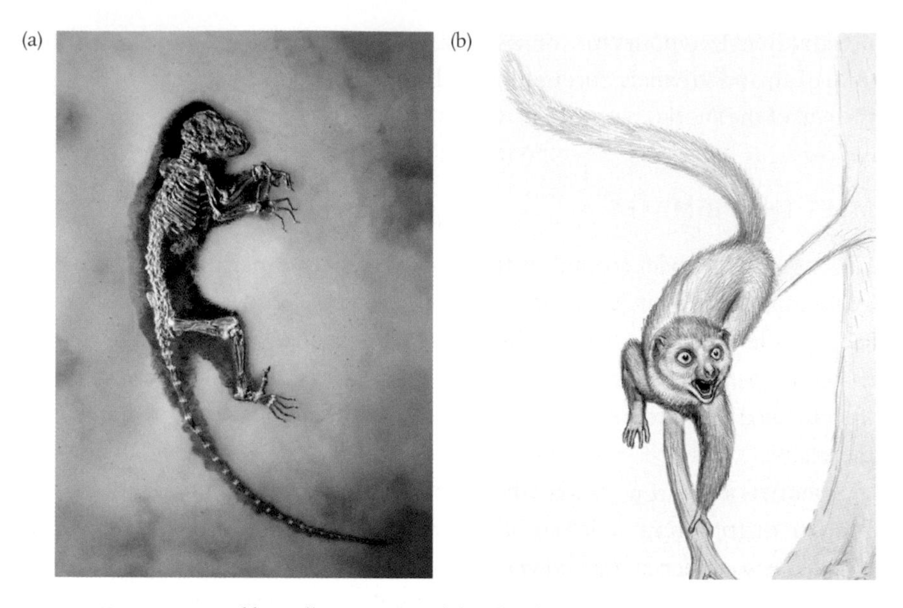

FIGURE 87 Hurum and his colleagues argued that fossil Ida (*Darwinius masillae*), image, (a) is a strepsirrhine primate with some of the characteristics of early haplorrhines, making it a kind of 'missing link' between the two primate sub-orders. The scientific consensus is that Ida is a strepsirrhine related to modern lemurs with nothing to tell us about the evolution of haplorrhines (and monkeys and apes).

(he was born in February 1809), and the fossil primate was named *Darwinius masillae*.

That many disagreed with this conclusion wasn't the principal cause of the controversy. Hurum is also widely known in Norway as a science popularizer. He has published several popular books on evolution and appears regularly on Norwegian radio and television. What angered the community most was the carefully orchestrated media blitz that accompanied the announcement. This was broadcast live by the BBC and was covered by another 40 international television channels. Google changed its logo for the day.

It was less a scientific announcement, more a product launch. It was accompanied by a website,* a television documentary (called *The Link*), and a popular book (also called *The Link*, written by Colin Tudge). The documentary was narrated by David Attenborough and was broadcast in Britain by the BBC a week later.

* http://www.revealingthelink.com

John Fleagle, an American anthropologist at New York's Stony Brook University and an internationally recognized expert on primate evolution, initially rejected the claim outright, declaring that Ida is an ancestral strepsirrhine related more to lemurs than monkeys and apes, and therefore sheds no light on the story of human evolution. He was an anonymous reviewer of the scientific paper on which the announcement had been based, and had requested that the authors 'tone down' their claims prior to publication. He wrote a more conciliatory essay on the subject the following year, but did not change his position, and the balance of scientific opinion remains firmly behind the identification of Ida as an ancestral lemur.[4]

Both adapoid and omomyid fossils dating from the Eocene and early Oligocene are relatively common. So-called 'molecular clock' estimates based on genetic analysis suggest that the primates may have originated 70–80 million years ago, before the end of the Cretaceous. If this is correct, it leaves us with a 14–24 million year gap in the primate fossil record. We can only assume that sometime during this period the strepsirrhines and haplorrhines went their separate evolutionary ways.

Figure 88 attempts to describe what happens next. The haplorrhine primates experience a series of divergences in the evolutionary tree that leads eventually to humans (but please once again recall all the caveats about representations based on tree structures). The first to separate into a distinct branch are the tarsiiformes, roughly around the middle of the Eocene. At this time these animals range over much of North Africa, Europe, Asia, and North America, but modern tarsier monkeys are found today only on the islands of South East Asia.

On the other branch are the anthropoids. The Fayum Depression, which lies west of the Nile and south of Cairo in Egypt, is a rich source of primitive anthropoid fossils. Among the best-known is *Propliopithecus zeuxis*. These are ape-like creatures about the size of a dog with long, projecting faces and pronounced canines. They likely live high up in the forest canopy, eating mostly soft fruits.

Towards the end of the Eocene the anthropoid line in turn diverges. The next to branch away are the platyrrhines (meaning 'flat-nosed'), also known as New World monkeys. These are small to mid-size creatures, one family of which possesses prehensile tails (meaning that the tail has adapted to grasp branches and objects).

The Earth's climate to this point has been fairly generous, but global temperatures are now falling. There are a number of possible reasons, but atmospheric

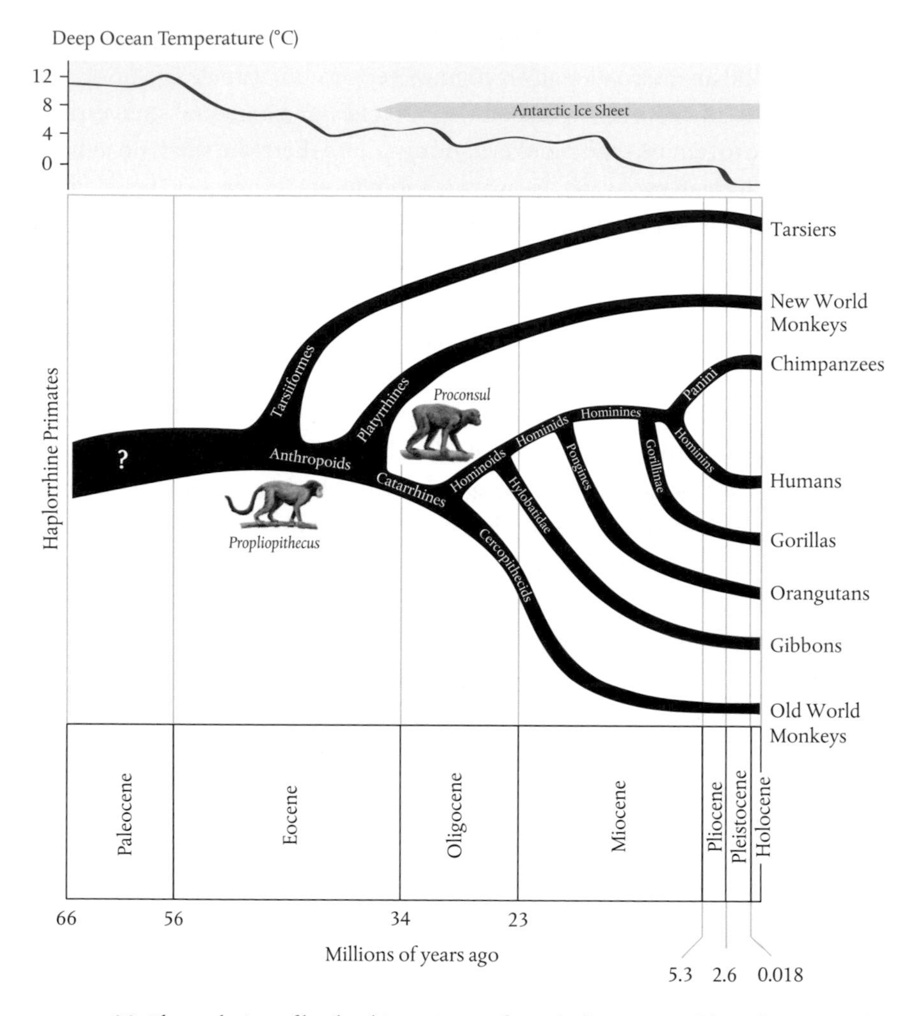

FIGURE 88 The evolution of haplorrhine primates from the beginning of the Paleocene to the present day. As there is little fossil evidence from the Paleocene, this section of the trunk of the evolutionary tree is labelled with a question mark (one estimate dates the divergence of haplorrhines and strepsirrhines to about 63 million years ago). The nature, location, and timing of these evolutionary divergences is strongly dependent on the global climate and, as temperatures decline from the end of the Paleocene (represented here in terms of deep ocean temperatures based on studies of oxygen isotope ratios), the tropical forests (and their primate inhabitants) retreat to Africa and parts of Asia.

carbon dioxide levels are declining, perhaps exacerbated by blooms of freshwater ferns in the Arctic Ocean becoming buried in sediment. Glaciers reappear at both poles and the Antarctic ice sheet rapidly grows. The shifting continents open up the Drake Passage, as South America becomes fully detached from

Antarctica, and the Antarctic Circumpolar Current thus created dredges cold water from the depths of the ocean to the surface.

The forests recede, threatening the existence of the primates that have come to depend on them. The fossil evidence suggests that further primate evolution now takes place largely in Africa and Asia.

The now much cooler mid-Oligocene witnesses a branching of the catarrhine ('down-nose') line into cercopithecids (Old World monkeys) and hominoids. Several early hominoid fossils have been discovered at sites in Kenya, perhaps the most notable being the skull of *Proconsul*, found in 1948 by Mary Leakey, palaeoanthropologist wife of Louis Leakey. This is a monkey-like creature likely to be found moving on all four legs along tree branches. However, its lack of a tail is very ape-like, and debates continue about precisely what kind of hominoid it is. The name means 'before Consul', which refers to a chimpanzee in London Zoo named Consul that had gained some notoriety.

Its hands are not unlike human hands in proportion, and it would appear to be able to grasp small objects with some precision. It lives in tropical forest, but some of the sites in which *Proconsul* fossils have been found are suggestive of a more seasonal environment, and it is possible that it may occasionally come down to the ground.

The next group to diverge is the hylobatidae, or 'lesser apes', at or around the end of the Oligocene. These are the gibbons, found today across parts of Asia, from northeast India to Indonesia and the Philippines.

And so, at or around the beginning of the Miocene, we arrive at the hominids or 'Great Apes'.

PLANET OF THE APES

Continental Africa now starts to make contact with Europe and Asia, and there is evidence to suggest that the formation of a land bridge between the two continents about 17 million years ago facilitates the migration of hominoids from Africa to Europe. Hominoid fossils dating back to this period have been found in Germany and in Turkey.

In 2004, Spanish anthropologist Salvador Moyà-Solà and his colleagues from the Miquel Crusafont Institute of Palaeontology discovered a partial skeleton dated 12–13 million years old at a site near Barcelona. This was a male, they surmised, about the size of a chimpanzee and probably feeding on fruit. Analysis of the creature's anatomy indicated many of the characteristics of the great apes, as well as some monkey-like features. They named it *Pierolapithecus catalaunicus* ('ape from Pierola in Catalonia'), and suggested that it represents a close relative of the last common ancestor of the great apes, inevitably provoking the usual 'missing link' stories in the popular media.

If this assignment is correct (and, as always, there are scientists who disagree), then *Pierolapithecus* sits on the hominid line of the evolutionary tree before it branches off to give pongines (orangutans) and hominines, the subfamily that will go on to evolve to give gorillas, chimpanzees, and humans.

This 'planet of the apes' persisted for several million years. We can get some sense of what these creatures might have been like from fossils dated to the late Miocene, about 9 or 10 million years ago. The evidence suggests two distinctly different groups. One group has teeth and jaws that are little different from those of *Proconsul*, suggesting a diet of fruit. Their limbs are more suited to a life spent largely in the trees, their movement based on using their powerful arms to swing between branches rather than walking along them, much like modern-day orangutans. Their habitats in southern Europe are subtropical and much wetter, with rainy (monsoon) seasons.

The second group is very different from *Proconsul*. These apes have stronger teeth with a thick enamel and strong jaws, suggesting a diet of hard fruits, nuts, and seeds and habitats likely to be much more seasonal, perhaps with long dry periods. They remain quite comfortable in the trees, but are just as much at home on the ground. They are not particularly built for suspending themselves from branches, and it seems likely that they cover large distances by walking (on all fours) across the ground rather than swinging through the trees.

These apes leave behind no evidence other than their fossilized bones and teeth. But we can make some educated guesses about how they lived.

Like humans, the living apes are social creatures. Gorillas form polygamous groups (called troops) centred on a dominant male, which is larger and distinguished by white—or 'silver'—hair on its back. Substantial size differences between similarly aged males and females is called *sexual dimorphism*. The dominant male carries many social responsibilities, such as mediating conflicts, protecting the troop, and leading it to food sources. The core of social life is derived from the bonds between the dominant male and his females (and, in turn, the bonds between females and their young infants). The younger males in the group are accepted for as long as their back hair remains black: as it begins to whiten they may be expelled and will move away to found their own troop.

In chimpanzee society the dominant male is not necessarily the largest but may be the most cunning or politically astute, adept at influencing the group (called a community) which includes many males, backed up by displays indicative of power or strength. The community is more complex and hierarchical than the gorilla troop, with females forming their own matriarchal structures which can give them priority access to food. Groups of powerful females can sometimes call the shots, ousting a dominant male that has lost their support and replacing him with another.

Both these examples of ape society are the result of many further millions of years of evolutionary and social development, and on this kind of timescale much can change, as we will soon see. But it doesn't take a great leap of imagination to picture everyday scenes from the lives of the late-Miocene hominines. Size differences that persist between the sexes suggest social structures that are more troop than community.

Indeed, many popular books and museum exhibits on the subject of human evolution try to convey impressions of ancestral ape life through drawings and dioramas. These certainly aid our comprehension, but there is nothing quite like being able to look into the eyes of these creatures as they look back at you.

The gorillas now diverge to follow their own evolutionary path, one that will eventually lead them to memorable encounters with Dian Fossey and David Attenborough, among others. Between about five and seven million years ago they are followed by the chimpanzees, the line of hominines separating into hominins and panini. All the evidence suggests that this final evolutionary parting of the ways happens in Africa.

At last, with about five million years left on the clock, we're on the line that leads us to humans.

Humans and chimpanzees possess similar numbers of genes (22,763 and 20,947, respectively). A recent study by Matthew Hahn at Indiana University in Bloomington, Illinois, and his colleagues determined that about 93.6% of these genes are common to both humans and chimps. This is a little less than original estimates made in the early 1970s, but more than enough to demonstrate our recent common ancestry. The differences are the combined result of a net gain of 689 new genes in humans and the net loss of 729 genes in chimpanzees since they diverged.[5]

I don't have much more to say about chimpanzees, other than to note that their line split further a couple of million years ago to give us modern chimps and bonobos (or 'pygmy chimps'). Such is our genetic proximity to our primate cousins that American scientist Jared Diamond titled his 1991 popular book *The Third Chimpanzee*.*

FOOTPRINTS ON THE SANDS OF TIME

Most (if not all) cartoon depictions of human evolution show a linear succession from a knuckle-walking ape on the left to an upright walking (bipedal) human on the right. Gould railed against these images in *Wonderful Life* and indeed, as we will see, the story of human evolution as we currently understand it is rather more complicated than this. But by focusing on bipedalism the cartoons do capture an important truth. Bipedalism is a defining feature of human origins.

Of course, evolution has invented bipedalism already many times and many mammalian species today are bipeds (think kangaroos). Nevertheless, at first sight, walking upright looks like a really dangerous thing for an ape to want to do. Gorillas and chimpanzees are both capable of doing this for short distances when carrying objects (such as food, or infant gorillas or chimps), but they do not make a habit of it. It requires a large-scale restructuring of the skeleton and musculature and actually slows down the walker, making them more vulnerable to a fast-moving predator. Anyone who has suffered back pain knows the price to be paid for their bipedalism.

So why do it? There are many hypotheses and not a great deal of hard evidence. There might have been a number of reasons. Although early hominins are still quite at home in the trees, it seems likely that they spend more and more time on the ground. Parts of the world are continuing to get drier, and perhaps the retreating forests have opened up wide savannahs that must now be crossed. Perhaps an erupting African volcano has smothered any vegetation close by with lava and ash, forcing the inhabitants to seek safer havens. Despite its risks, walking is an efficient way of covering large distances while minimizing the amount of energy that must be expended.

The fossil record for the period between about four and seven million years ago is relatively patchy, and a number of discoveries have been declared to be on the hominin evolutionary line leading to humans. These include *Sahelanthropus tchadensis* (6–7 million years old), *Orrorin tugenensis* (5.8–6 million

* This was published in the UK with the title *The Rise and Fall of the Third Chimpanzee*.

years old), *Ardipithecus kadabba* (5.2–5.8 million years old) and *Ardipithecus ramidus* (4.2–4.4 million years old). Interpretation of these bones (limited to a small handful in some cases) is tricky. Many hint at ape-like characteristics, some hint at human-like characteristics. *Ardipithecus ramidus* possesses a divergent big toe, more suited to tree-living and perhaps less indicative of the kind of habitual bipedalism that serves as a marker for hominin evolution. The debates continue.

The evidence becomes a lot less ambiguous just a few hundred thousand years later. Fossils of *Australopithecus anamensis* ('Southern ape from the lake') have been discovered in Kenya and Ethiopia and determined to be 4.1–4.2 million years old. The limb bones strongly suggest bipedal motion. Is it here that we find the first hard evidence for the beginnings of the hominin line?

Of course, from fossil bones that are often fragmentary and incomplete we can only make well-educated guesses about anatomy and physiology and infer how *Australopithecus anamensis* actually moved about. If only there was some other way of obtaining evidence for bipedalism . . .

So picture the scene. Three hominins are making their way—on foot—across an expanse of ground covered in ash recently deposited by a volcano that lies about 20 kilometres distant. The volcano still rumbles. It is raining, but our friends do not appear to be in any hurry. As they walk they leave footprints in the wet ash, in a pattern spanning some 24 metres in length (Figure 89). The footprints are preserved as the ash dries and hardens. Sometime after they have passed by, the volcano covers the area with further layers of ash.

When the ash layer containing the footprints is uncovered many years later, isotope and stratigraphic analysis indicate that they are about 3.6 million years old.

These are the *Laetoli footprints*, discovered by a team led by Mary Leakey in 1978. The site is some 45 kilometres south of Olduvai Gorge in Tanzania, close to the western rim of the Great Rift Valley, a divergent plate boundary where the African plate is in the process of rifting apart, and which has a string of volcanoes dotted along its length (including the now dormant Mount Kilimanjaro).

The Laetoli footprints have been attributed to *Australopithecus afarensis* ('Southern ape from Afar'), another *Australopithecus* species whose fossil remains have been found in Tanzania and Ethiopia. Perhaps the most famous representative of *Australopithecus afarensis* is Lucy, discovered in November 1974 at a site in Hadar in Ethiopia's Afar Depression, by American palaeoanthropologists Donald Johanson and Tom Gray. As much as 40% of Lucy's fossil skeleton was recovered and she was determined to be about 3.2 million

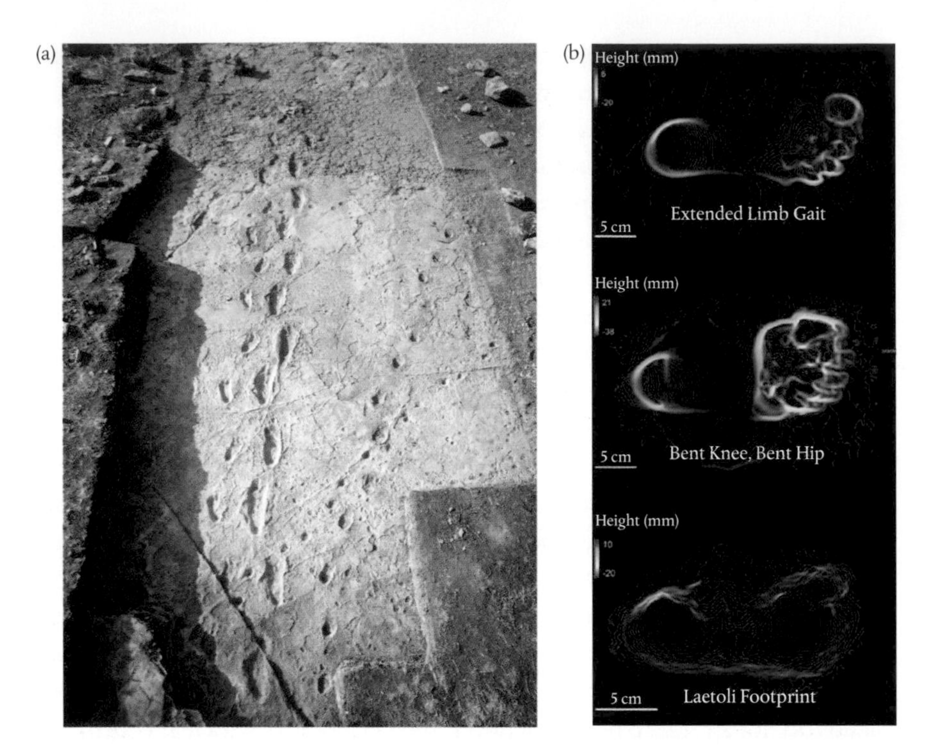

FIGURE 89 The Laetoli footprints (a) are approximately 3.6 million years old, caught in wet volcanic ash that hardened before being covered and preserved by fresh layers of ash. They are believed to have been made by three hominins of the species *Australopithecus afarensis*. (b) shows three-dimensional scans of experimental footprints made by a modern human walking with a regular, extended limb gait and with a more ape-like bent knee, bent hip gait (think chimpanzee). Note how the ape-like gait makes much deeper depressions particularly around the toes. These are compared with a similar three-dimensional scan of a Laetoli footprint. The gait appears to be much more human-like than ape-like. Adapted from David A. Raichlen, Adam D. Gordon, William E. H. Harcourt-Smith, Adam D. Foster, and Wm. Randall Haas, Jr., *PLoS ONE*, **5** (2010) e9769.

years old. Although she is formally known as AL 288-1, Johanson named her 'Lucy' because at the time of her discovery the Beatles' song *Lucy in the Sky with Diamonds* was being replayed constantly on a tape machine in their camp.

Analysis of the movements likely to have created the Laetoli footprints combined with the evidence from Lucy's skeletal structure suggest that bones and muscles had by this time adapted to a way of walking that is distinctly human-like rather than ape-like.[6] A similarly aged foot discovered in South Africa appears to show that the big toe remains divergent, though much less so than in *Ardipithecus*. But if the feet, legs, and hips are human-like, the arms have retained certain ape-like characteristics. This suggests that *Australopithecus* still needs to climb trees, presumably to escape from predators.

The discovery of a further, younger, species, called *Australopithecus africanus* ('Southern ape from Africa') actually predates all the others. The fossilized skull of a small child was found in 1924 at a limestone quarry in Taung, South Africa. The 'Taung child' was discovered by workers who passed it on to a director of the Northern Lime Company (which managed the quarry), who in turn passed it on to his son who proudly displayed it on his mantelpiece at home. Josephine Salmons was both a friend of the son's family and a student of a recently appointed Professor of Anatomy at the University of Witwatersrand, Raymond Dart. She spotted the fossil skull on the mantelpiece and brought it to Dart's attention.

Dart wrote a paper just 40 days later which was published in the journal *Nature* in February 1925. He named the fossil *Australopithecus africanus* and claimed that it may represent an important human ancestor, a claim that was (for all sorts of reasons) roundly rejected by the scientific establishment at the time. Ten years later a South African limestone cave called Sterkfontein yielded many more *Australopithecus africanus* fossils, including an almost complete skeleton determined to be more than two million years old. Twenty-five years after the initial discovery the establishment capitulated and accepted *Australopithecus* as a potential human ancestor.

Dart was proved right, but he had also argued that australopithecines were hunters, dragging their prey into the limestone caves. But there is no evidence that *Australopithecus* made any kinds of tools, let alone weapons, and they did not possess the capability to make fire.* It rather seems that they were largely vegetarians. Perhaps the reason so many examples have been found in Sterkfontein is because it was they who were dragged into the caves by predators.

The Sterkfontein Caves now form part of a World Heritage Site, named the 'Cradle of Humankind'. About 15 kilometres to the northeast lies the Malapa cave, part of the same limestone cave complex. In August 2008 Matthew Berger, the then nine-year-old son of palaeoanthropologist Lee Berger at the University of Witwatersrand, joined his father on a walk through the Malapa Nature Reserve. He spotted what looked to be an interesting fossil, which he brought to his father. Berger could barely believe what his son had found.

---

* But then science is always full of surprises. Recent discoveries of 3.3 million-year-old stone tools in West Turkana, Kenya, pre-date the oldest known tools by 700 000 years. Their makers could have been *Australopithecus afarensis*. See Harmand et al., *Nature*, **521** (2015), pp. 310–15.

Here, it seems, is another species, *Australopithecus sediba* ('Southern ape from the well'), determined to be about 1.95 million years old. This species looks to be the most human-like of all the australopithecines, with a pelvis more geared to efficient walking and running, and very human-like hands. The hands in particular hint at the possibility that *Australopithecus sediba* might have manipulated stone tools.

Berger and his colleagues claimed that *Australopithecus sediba* is transitional between *Australopithecus africanus* and early species of the genus *Homo*. But there's one slight problem. As we will see, fossil evidence for *Homo* predates *Australopithecus sediba*. This does not necessarily preclude the latter as an ancestor, as the origin of this species may predate the fossils that were discovered in 2008. Indeed, fossils of some *Homo* species also predate some of the examples of *Australopithecus africanus*. What this means is that several species of both *Australopithecus* and *Homo* may have overlapped for a time.

The story of *Australopithecus* doesn't quite end here. The species I've described here are considered slightly built or 'gracile', meaning slender. There are more heavily built 'robust' species that were initially assigned to the genus *Australopithecus* but which are now called *Paranthropus robustus* and *Paranthropus boisei*. These come later, 2–1.3 million years ago, perhaps reflecting a much more open, drier environment in which food sources such as nuts, seeds, and roots are more common. These species don't get much further, however, and it has been suggested that as early *Homo* species became more adept at tool-making and hunting, *Paranthropus* may have been driven to extinction.

So, who were the direct ancestors to the *Homo* lineage? Alas, there is no definitive answer. As I think you will have gathered by now, tracing the evolution of humans to some point of origin can be a rather contentious activity. Careers can be made. Or careers can be damaged (if sometimes only temporarily) by what the scientific community regards as overreaching and unjustified claims based on insufficient evidence. Furious disagreement is about the only thing that can be guaranteed.

As species diversify it becomes increasingly difficult to draw on the fossil evidence to discriminate between those species that continue and evolve further, those that evolve to different but obviously related forms, and those that simply go nowhere. Logic might dictate the drawing of a direct line from *Australopithecus africanus* or *Australopithecus sediba* to early species of *Homo*. But don't get too wedded to this notion. The fossil-hunting continues and new discoveries could transform the picture completely.

ECCE HOMO

The australopithecines represent a fundamentally important evolutionary step on the way to humankind but, although they walk upright, they retain many ape-like characteristics and remain distinctly 'ape-human'. Sexual dimorphism is still quite marked.

It's not difficult to work out what kinds of clues palaeontologists need to look for to tell the next part of the story. With two limbs now devoted to walking upright, the other two limbs and particularly the hands are free for other things. It's no coincidence that the oldest stone tools appear around 2.5 million years ago (but see the footnote on p. 323). We imagine that the greater hand–eye coordination demanded by tool-making accelerates the evolution of a larger brain (and a larger cranium to contain it). A larger brain and the development of vocal cords facilitates the articulation of sounds, and will eventually give us speech and language.*

The first stone tools were discovered at Olduvai Gorge during an expedition led by Louis Leakey in 1931. The Gorge is about 100 metres deep and 25 miles long. It sits in the eastern Serengeti Plain in northern Tanzania, close to the Rift Escarpment. About half a million years ago a stream diverted by seismic activity cut into the Gorge, exposing steep cliffs showing an alternating sequence of lake, river, and volcanic sediments spanning two million years.

Leakey called them 'Oldowan tools'. They are made by flaking pieces off handy-sized lumps of volcanic rock by striking them with another stone. The flakes themselves can be quite sharp, and are likely more important as tools than the 'core' left behind. Although we can't be certain precisely what the finished tools are used for, we can guess that principal uses are cutting, chopping, scraping, and pounding. These tools leave their marks on animal bones discovered nearby, indicating that the tool users are butchering carcasses. Their diet now clearly includes some meat.

* Scientists define one final set of prehistoric time periods based on the introduction and subsequent development of stone tools, although there is no international consensus of how these should be divided. In Europe, the 'Palaeolithic', or 'Old Stone Age' spans from 2.5 million years ago to the beginning of the Holocene about 12 000 years ago. The Lower Palaeolithic runs from 2.5 million to about 300 000 years ago, the Middle Palaeolithic from 300 000 to about 40 000 years ago and the Upper Palaeolithic from 40 000 to 12 000 years ago. The Neolithic ('New Stone Age') spans from about 12 000 to 6500–4000 years ago.

But who is using the tools found at Olduvai Gorge? The answer was a long time coming. Mary Leakey discovered *Paranthropus boisei* at Olduvai Gorge in 1959. But bone fragments discovered at the site a year later appeared more promising. The teeth of this species are smaller compared to *Australopithecus* or *Paranthropus*, and there is evidence of a larger cranial capacity. Here, surely, is the toolmaker. Louis Leakey felt that there was sufficient evidence to declare the discovery of a new species of early *Homo*, which he called *Homo habilis* ('Handy Man'), in a paper published in *Nature* in 1964.

Not everyone agreed. Things got more complicated in the early 1970s with the discovery of other fossil remains in Koobi Fora, on the eastern side of Lake Turkana in northern Kenya. Richard Leakey (son of Louis and Mary) found more stone tools, more examples of *Paranthropus*, and a skull that indicated an even larger brain. Roughly speaking, australopithecine skulls suggest a brain volume of the order of 400–550 cubic centimetres (cm^3). *Homo habilis* is a little larger, ranging from 550 to about 690 cm^3. The skull found at Koobi Fora suggests a brain volume of about 750 cm^3.

Was this yet another species? Some have argued that this is just another example of *Homo habilis*, putting the differences down to intra-species variability. Others have argued that the differences are so great that these warrant assignment to another species, named *Homo rudolfensis* (Lake Turkana was previously known as Lake Rudolf). More examples of *Homo rudolfensis*, dated between 1.78 and 1.95 million years ago, were described by Maeve Leakey (wife of Richard) and her colleagues in a paper published in *Nature* in August 2012.

We're so used to living our lives as the sole surviving species of the genus *Homo* that it's difficult to imagine sharing the world with another. And, given the problems we have just getting along with members of our own species, it's hard to imagine that another species, especially one that we might deem inferior, could be expected to fare well. But, if we're prepared to accept the arguments in favour of both *Homo habilis* and *rudolfensis*, then it seems that at least two species of humans co-existed in close proximity in this part of eastern Africa about 1.7–2 million years ago.

Well, make that three.

Fossils attributable to *Homo ergaster* ('Working Man', but read on) were discovered in 1975 (a mandible discovered in southern Africa in 1949 is also thought to belong to *Homo ergaster*). A virtually complete skeleton of a young male, thought to be around 9 years old at the time of his death, was discovered at Lake Turkana in 1984 by palaeoanthropologists Kamoya Kimeu and Alan Walker. This specimen is known as 'Turkana boy' or 'Nariokotome boy'. He was already tall (about 1.5 metres or 5 feet), with a slender physique and a brain volume of about 900 cm^3. He is about 1.5 million years old.

Homo ergaster is now thought to be the African ancestor to *Homo erectus* ('Upright Man'), and some palaeoanthropologists choose not to distinguish these as distinct species. Although not everyone agrees, I propose to drop *ergaster* in favour of *erectus*. Examples of *Homo erectus* have been found in Indonesia ('Java Man'), China ('Chinese Man of Peking'), and Georgia. Although there are differences between African and Asian examples, it is thought that the species originates in East Africa and migrates to Asia, probably as a result of simply extending the range over which members of the species will forage. They then adapt to the environments where they settle.

Much of this scenario remains hotly disputed, but if it is broadly right then *three* species of *Homo* overlap with each other in eastern Africa: *Homo habilis, rudolfensis,* and *erectus*. This overlap is thought to have lasted several hundred thousand years (Figure 90). Nobody can be sure if one or other descended from the third, or if they all evolved from a common ancestor. It's also not possible to say if members of these different species are ever in exactly the same place at the same time and so interact with each other. Given that these early humans likely range over large areas in search of food and water, I find it difficult to imagine that they are not at least aware of each other.

What is clear is that both *Homo habilis* and *rudolfensis* disappear from the fossil record by 1.4 million years ago. We don't really know why, but we can speculate. The continued cooling of the planet through the Miocene and Pliocene has resulted in the onset of a series of ice ages at the beginning of the Pleistocene. Ice sheets spread downwards from the poles into the northern and southern hemispheres, establishing a cycle of glacial advance and retreat which will repeat on timescales of 40 000 and 100 000 years right to the present day. As forests retreat and open savannah expands, our East African inhabitants are forced to scratch out an ever more meagre existence.

Homo erectus possesses a brain volume now approaching 1000 cm^3. The species has developed more advanced tool-making techniques that are used to

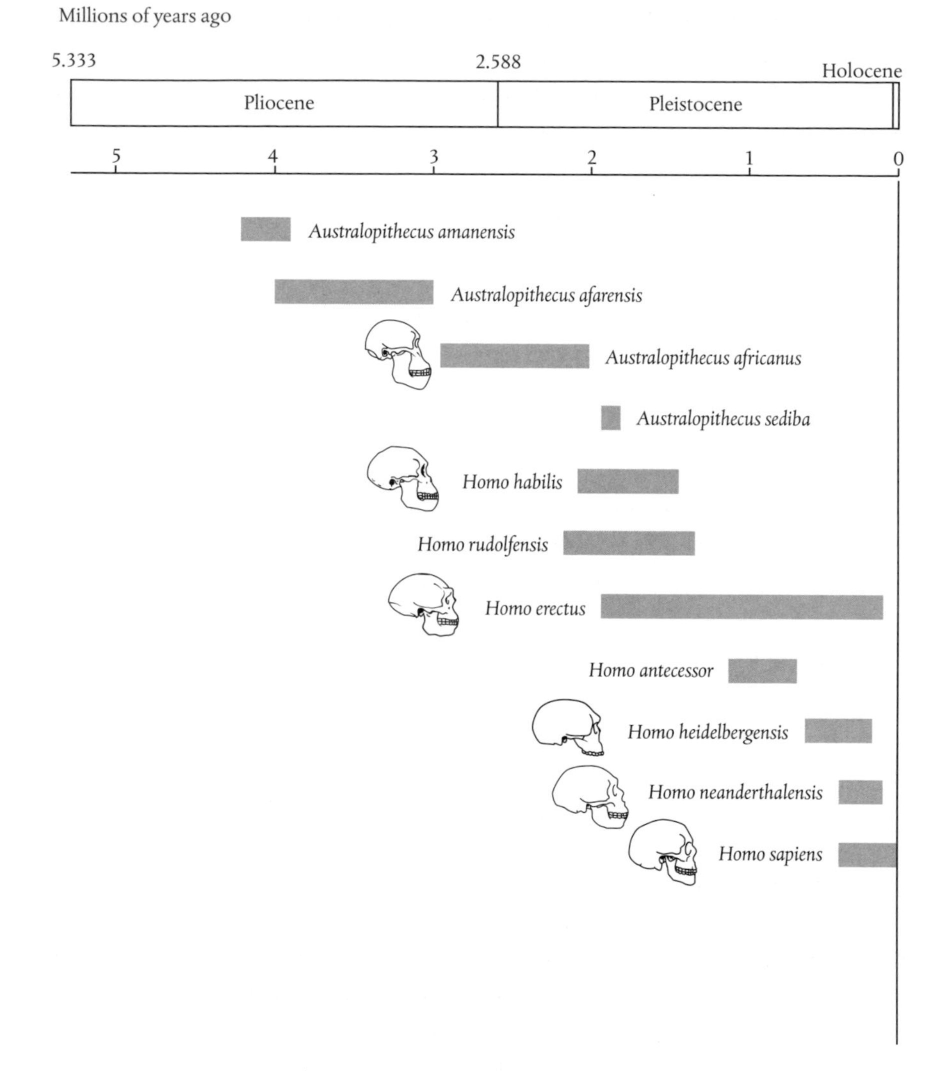

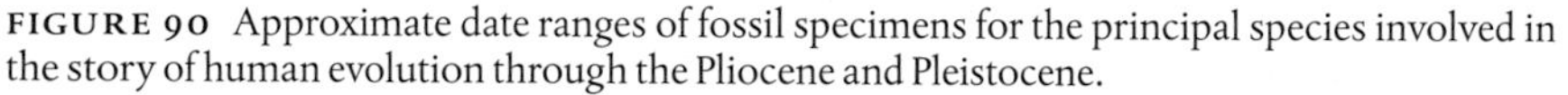

FIGURE 90 Approximate date ranges of fossil specimens for the principal species involved in the story of human evolution through the Pliocene and Pleistocene.

produce hand axes (called 'Acheulian tools', named for Saint-Acheul, a suburb of Amiens in northern France, where they were first discovered in 1859). If the ranges of these early humans do overlap, then there can be little doubting which species is best adapted.

Homo erectus carries the flag for humankind for another 1.3 *million* years, and some fossil remains are determined to be only about 200 000 years old. If the success of a species is measured simply in terms of its longevity, then *erectus* is the most successful of all human species to date.*

ON THE TRAIL OF THE WISE MAN

You may be wondering what, precisely, warrants the assignment of a jumbled collection of fossilized bones and associated manufactured objects, such as stone tools, to a distinct human species. For sure, there can be little doubting the value of a discovery of a new human species on a palaeontologist's CV, which explains why discoveries are sometimes rushed into print. Until the 1950s, there was a tendency to assign all newly discovered human fossils to distinct species, leading to a great proliferation of names, such as *Homo chapellensis, grimaldi, mediterraneus, spyensis,* and many more. Eventually, sense prevailed. But the fact remains that differences between species can sometimes be very slight, and very subtle.

This is why it has proved so difficult to achieve a consensus on some of the finer details of human evolution. It is also why assignments tend to change over time. What are seen as examples of a distinct human species today may be recognized as examples of another species tomorrow. Indeed, some palaeoanthropologists have swung the pendulum in the opposite direction, arguing that there are no real differences, just a lot of variability on some basic common themes within a single human species.

The saying goes that 'bones never lie' and, of course, there is a sense in which the bones are straightforward, incontrovertible facts from the story of human evolution. But, as in all aspects of scientific endeavour, facts are open to interpretation and are rarely—if ever—completely unambiguous.

Why is it so difficult? Why not simply invoke the *biological species concept,* which says that species are differentiated by their reproductive compatibility. Defined this way, distinct species cannot interbreed. This doesn't make inter-species sex impossible, but what it does mean is that the offspring from such a union are likely to be sterile. For example, horse-donkey hybrids (mules)

* Of course, if *Homo sapiens* survives for another million years or so then the records will have to be rewritten. Stay tuned.

tend to be sterile because of a mismatch in the numbers of chromosomes carried by their parents.*

Alas, this is not a helpful distinction in this case. There are many examples of inter-species breeding in mammals and, as we will see very shortly, there is evidence to indicate past interbreeding between currently accepted and recognizably different human species, including our own. It's probably best to remember that biological taxonomy, including the use of the species category, is a human invention designed to make sense of the otherwise bewildering complexity of life on Earth. Such categorizations are bound to be a little fuzzy around the edges.

Perhaps we can think of it this way. Let's just pretend for a minute that we could bring *Homo erectus* back to life and clothe him or her in some undistinguished but modern garb. Let's further suppose that, following their remarkable resurrection, they behave in an unremarkable way, not overawed or bewildered by their modern surroundings. We pass them by in the street. We notice the way they walk and we look briefly into their face and—most importantly—into their eyes. If we pass by without pulling up short, thinking that there is something distinctly different (if not downright odd) about this person, I would suggest that the differences are insufficient to consider them as belonging to a separate species.

I'm confident that we would find *Homo erectus* rather peculiar by our modern *sapiens* standards. The shape of the head, the elongated, protruding face and the distinct brow ridge (called the supraorbital torus) would make this individual rather ape-like and aggressive in appearance.

This 'passer-by' test sounds pretty unscientific, but in his book *The Origin of Our Species*, British palaeoanthropologist Chris Stringer relates the story of eccentric anthropologist Grover Krantz, who strapped a brow ridge modelled on *Homo erectus* to his own head for six months in order to explore its possible anatomical benefits. He found that it helped to shade his eyes from the Sun and kept his hair out of his eyes when running. But what resonated most with Stringer was that on a dark night it would also scare people out of their wits. 'I think [the brow ridge] may even have had a signaling effect in ancient humans,' Stringer wrote, 'accentuating aggressive stares, especially in men'.[7]

* Our nearest primate relatives, chimpanzees and gorillas, have 24 pairs of chromosomes compared to humans who have 23 pairs. It seems that after diverging and going our separate ways, two of our smaller chromosomes became fused together to form what we now call chromosome 2.

Our story is now complicated by the fact that *Homo erectus* has migrated from Africa, passing first into Asia and thence into Europe. So far, all the evolutionary 'action' has taken place in Africa, but the success of *Homo erectus* means that human evolution is now no longer restricted to this single geographical location.

Consequently, we find remains of *Homo antecessor* ('Pioneer Man'), thought to be European descendants of *Homo erectus*, in a limestone cave system in northern Spain. *Antecessor* is dated to about 850 000 years ago, with a brain volume of the order of 1000 cm^3, similar to *erectus*. The fossils are of several individuals, some of them children, and bear signs of cannibalism.

Evidence from other locations is pretty meagre. Stone tools attributed to *antecessor* have been found in France and in both Suffolk and Norfolk in England. In May 2013 coastal erosion of cliffs near the village of Happisburgh (pronounced 'Hayesborough') in Norfolk exposed fossil footprints dated between 1 million and 780 000 years ago. They are believed to have been made by five *antecessor* adults and children, with adult heights estimated to range from 1.6 to 1.73 metres. In several footprints the toes are clearly visible.[8]

It's not quite clear what happens to *antecessor*, and the species disappears from the fossil record about 700 000 years ago. It seems likely that they are edged into extinction by the appearance of an African descendant of *erectus*, called *Homo heidelbergensis*.

One of the best-preserved of all human fossils was found in 1921 at the Broken Hill mine in what was then northern Rhodesia and is now Zambia. The fossil skull was so striking that it scared away the workers who had found it, buried in sediment. Their supervisor, Tom Zwigelaar, was less superstitious and later posed for a photograph holding the skull in his left hand. It was later acquired by the Natural History Museum in London, where it is currently on display. A plaster cast of the skull then on display in the Museum so deeply impressed a young Chris Stringer, visiting with his parents, that he would go on to study it as part of his PhD and would spend a large part of a highly successful academic career fitting it into the story of human evolution. He has spent 15 years trying to date the skull accurately (recent estimates suggest that it is something of the order of 300 000 years old).

The skull was named '*Homo rhodesiensis*' ('Rhodesian Man'), but was subsequently recognized to be an example of another human species, *Homo heidelbergensis* ('Heidelberg Man'), which had been discovered in 1907 in a sandpit at Mauer, near Heidelberg in Germany. The species naming convention is such that the earlier find takes precedence, so *rhodesiensis* became *heidelbergensis*, leaving us in a rather awkward situation. Heidelberg Man is now thought to be descended from *Homo erectus*. The species name tells where it was first discovered, not from where it originates.

Heidelbergensis retains the strong brow ridge but has a less projecting face, a prominent nose, and a brain volume of 1100–1400 cm^3, overlapping the average 1350 cm^3 of contemporary *Homo sapiens*. Stone tools associated with *heidelbergensis* are more sophisticated and there is some evidence that, 500 000 years ago, some could have been making stone-tipped spears for hunting. They may have communicated using very early forms of vocalization, perhaps humming or mimicking animal sounds.

It is not certain precisely from where *heidelbergensis* originates but, like *erectus*, it migrates across Africa, Asia, and Europe. Fossils attributed to *heidelbergensis* have been found in Germany, France, Britain, Greece, Italy, Israel, and China. An extensive collection of fossil bones discovered at La Sima de los Huesos (the Pit of the Bones) at Atapuerca in northern Spain were originally thought to be those of *heidelbergensis* but are now believed to belong to a descendant, *Homo neanderthalensis*.

Once again, the patterns of migration of a species provide opportunities for evolutionary development in different places, and we can identify two lines of descent from *heidelbergensis*. One of these is *Homo neanderthalensis* ('Man from the Neander Valley'), built to cope with the colder European and western Asian climate. The other originates in Africa, and is called *Homo sapiens* ('Wise Man').

The Neanderthals have had something of a bad press. They have often been depicted as hairy, cave-dwelling savages lacking in intelligence and with no social graces (in 1866, the German naturalist Ernst Haeckel had proposed the name *Homo stupidus*). They are in some ways more robust than their African cousins, and they do live in caves. But it seems they also honour their dead and bury them in these same caves, fortunately for us increasing the chances of fossilization and preservation. Bone fragments belonging to more than 500 Neanderthals have been discovered so far, spread across Europe and western Asia, with about 20 near-complete skeletons.

The earliest Neanderthals appear about 400 000 years ago and they disappear from the fossil record about 35 000 years ago. The Neanderthal head is a little larger than that of modern humans, and they have long faces with a prominent nose and a double-arched brow ridge. I doubt that a Neanderthal would pass the 'passer-by' test.

With volumes typically 1400 cm^3, their brains are on average larger than those of living humans, but according to some analyses they have a smaller frontal lobe and a larger occipital lobe (for processing visual inputs) at the back, perhaps adaptations that aid the kind of visual acuity needed for hunting or for moving around in low light levels. Brain volume alone is now no longer the best indicator of 'intelligence' and, though still very crude, we should instead compare the ratio of brain volume to body mass, known as the encephalization quotient (EQ). *Homo heidelbergensis* has an EQ of the order of 3.4–3.8, the Neanderthals 4.3–4.8 and early *Homo sapiens* 5.3–5.4 (Figure 91).

The Neanderthals are short and stocky, with thickened bones and a powerful, muscular physique, well suited to the European Ice Age climate. Some are fair-skinned. They make a variation on stone tools called 'Mousterian', named

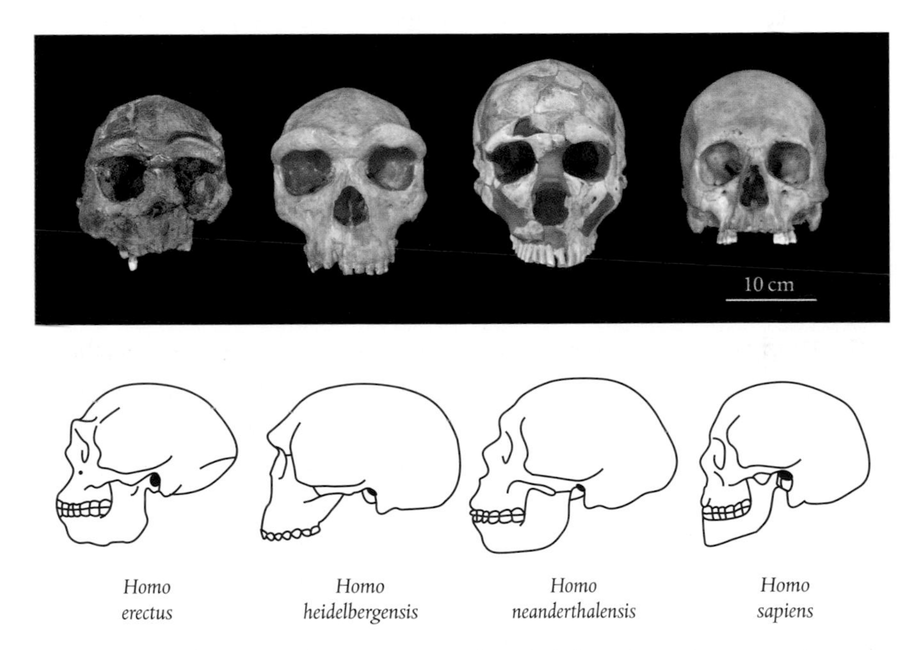

FIGURE 91 The bones never lie. Comparison of fossil skulls of Asian *Homo erectus* (specimen Sangiran 17 from Java, determined to be about 700 000 years old), African *Homo heidelbergensis* (Broken Hill 1, Zambia, about 300 000 years old), European *Homo neanderthalensis* (La Ferrassie 1, France, perhaps 70 000 years old), and Asian *Homo sapiens* (a modern human skull from Polynesia).

for Le Moustier, the cave in France where these were first recognized. These are scrapers, points, and knives, some of which are affixed to wooden handles to make spears. It's likely that they use animal skins as rudimentary clothing. It also seems likely that they know how to make fire. They probably speak a very primitive language.

A further early human group, called Denisovans, was discovered in 2008 in Denisova Cave in the Altai Mountains of southwestern Siberia. Palaeoanthropologists dated the remains (just a finger bone and a molar tooth) to about 50 000 years ago. The Denisovans are thought to be descended from *Homo heidelbergensis* and represent a kind of Asian sister-group to the Neanderthals rather than a distinct species.

The Neanderthals appear well-adapted to their environment, which begs some questions concerning why they (and the Denisovans) are now extinct. I'll go on to consider the reasons below, but I didn't want to turn my attention fully to *Homo sapiens* without first mentioning one further human species.

Flores is one of the Lesser Sunda Islands that lie east of Java, Indonesia. In 2003, a team of Australian and Indonesian palaeoanthropologists discovered some remarkable fossil remains of another human species at Liang Bua Cave, in western Flores. It was named *Homo floresiensis*. The fossils are remarkable because they suggest that *Homo floresiensis* grew no more than about 1 metre (less than three and a half feet) tall. Given the success of Peter Jackson's film trilogy based on J. R. R. Tolkien's *Lord of the Rings*, the new species inevitably became known informally as 'Hobbits'.

Rather like the Hobbits of Tolkien's invention, *Homo floresiensis* have small brains (about 400 cm^3) but they are adept tool-makers. Their fossils were found with accompanying evidence of suitably scaled stone tools and animal butchery. They may be descendants of *Homo erectus*, the first human species to migrate from Africa, who evolved smaller sizes in response to the limited island environment (a process called insular dwarfism). Analysis of their bone structures suggest that they could even be descendants of an earlier hominin species that migrated from Africa 2 million years ago. Nobody knows quite what to make of them.

OUT OF AFRICA

Let's take stock. About 200 000 years ago the planet is home to several human species. There's the venerable *Homo erectus* which originated in Africa about 2 million years ago and has spread across Europe and Asia. From *erectus* comes *antecessor* (in Europe), *heidelbergensis* (in Africa, Europe, and Asia), and—possibly—*floresiensis* (in Asia). From *heidelbergensis* comes *neanderthalensis* (in Europe), possibly the Denisovans (in Asia) and *sapiens* (in Africa). If we count the Denisovans as a sister group to the Neanderthals, that's at least six different human species, with five overlapping in time (*antecessor* disappeared about half a million years earlier).

Fossil evidence for one of the earliest *Homo sapiens* was found in Florisbad, about 45 kilometres northwest of Bloemfontein in South Africa. The age of the Florisbad skull was re-determined in 1996 to be about 260 000 years old. Two adult skulls and a child's lower jaw found at Jebel Irhoud in Morocco have been determined to be about 200 000 years old. A re-dating of fossils first discovered by Richard Leakey in 1967 at Omo Kibish, on the Omo River in southwestern Ethiopia, suggests 195 000 years. Adult and child skulls found at Herto Bouri in Ethiopia's Afar Depression have been dated at 160 000 years old.

The cranium is somewhat smaller, the face is much flatter, the nose less prominent, the chin more prominent, and the aggressive brow ridge is now largely gone. These are recognizably modern humans, with features unlikely to raise an eyebrow in the 'passer-by' test.

Although there are alternative theories of the origins of *Homo sapiens*, there is now something of a consensus building among palaeoanthropologists for what's known as the Recent African Origins model, also referred to as 'Out of Africa'. This model has been championed for more than 30 years by Chris Stringer and Gunter Brauer at the University of Hamburg. Aside from the patterns in the fossil evidence, there is also the evidence of our own genes.

I mentioned in Chapter 9 that when complex cells first evolved, some of the bacterial endosymbionts became mitochondria, retaining some local autonomy by holding onto a small circular chromosome, containing about 16 000 base pairs, and the apparatus to make a selection of proteins. In a fertilized human egg the mitochondrial DNA (mtDNA) contained in this chromosome is simply cloned from the female, with little or no genetic contribution from the male. These genes are therefore passed only from mother to daughter: males cannot pass them on.* The mtDNA mutates much more quickly than nuclear DNA, allowing recent ancestry to be traced through genetic

* Next time you see your mother remember to thank her for her mitochondrial DNA.

analysis. A 1987 study suggested that *Homo sapiens* does indeed have a last common human ancestor, mutation rates suggesting that she existed about 200 000 years ago. She became known as 'Mitochondrial Eve'. And she was from Africa.

This study was heavily criticized and, while some of the procedures applied were questionable, further research has supported the initial conclusions. Recent estimates date Mitochondrial Eve to between 100 000 to 200 000 years ago.* Similar studies of the Y-chromosome, which traces ancestry exclusively through the male line, places 'Y-chromosomal Adam' most likely in Africa between ages that vary from one study to the next but average out at around 200 000 years, though it is extremely unlikely that Adam and Eve were contemporaries. And, of course, unlike the biblical Adam and Eve, they were not alone.

Stringer has pulled all the evidence together to produce a reasonably definitive picture of our recent evolutionary history, a picture that will doubtless change in response to new discoveries as and when they are made. This picture (Figure 92) tells the story more or less as I have told it here.

It shows that the modern form of *Homo sapiens* appears in Africa around 200 000 years ago and spreads across Africa and, subsequently, Eurasia. We are witness to their success. The African migrants are likely to be a relatively small group, perhaps numbering a few thousand. This may not be a single exodus, but possibly a number of migrations closely spaced in time.

They reach North Africa and cross into western Asia about 100 000 years ago, passing into China about 60 000 years ago. They head south and east into Australia about 50 000 years ago, and advance into Europe about 45 000 years ago. It seems likely that they cross the Bering Strait into Alaska about 15 000 years ago, moving south into North America about 13 000 years ago, and then into South America about 12 000 years ago.

In Europe and Asia they encounter the Neanderthals and the Denisovans. Analysis of both Neanderthal and Denisovan DNA extracted from fossil remains, and comparisons with our modern human genome spill some interesting secrets. About 2% of the DNA of modern, non-African humans can be

* A March 2013 study by a team from the Max Planck Institute for Evolutionary Anthropology dates Mitochondrial Eve to 160 000 years old. A further study reported in August 2013 suggested an age range between 99 000 and 148 000 years.

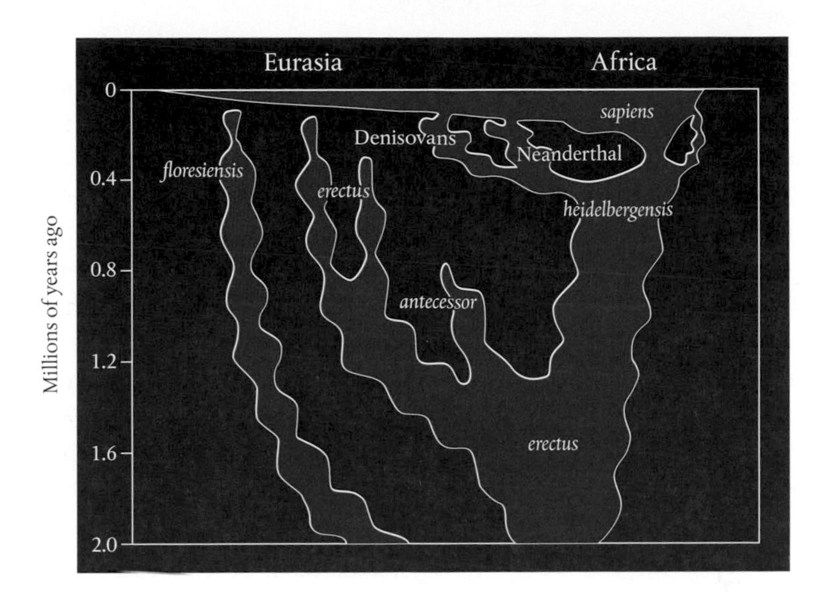

FIGURE 92 A tree diagram describing the history of recent human evolution over the past 2 million years based on both fossil evidence and DNA research, and the distribution of humans across Africa and Eurasia. The origin of *Homo floresiensis* is uncertain, but the species may have descended from *Homo erectus* or an earlier hominin species.

traced to the Neanderthals.* A small proportion of Neanderthal DNA is found in Denisovans and 4–6% of Denisovan DNA can be found in the DNA of modern Melanesians (natives of Vanuatu, Solomon Islands, Fiji, and Papua New Guinea) and Australian aborigines. It seems that Tibetans inherited genes specifically adapted for life at high altitude from Denisovan ancestors.

This rather striking evidence of interbreeding suggests that *Homo sapiens* does not simply emerge from Africa and 'take over', driving all other human species to extinction. A recent re-evaluation of the fossil evidence suggests that the Neanderthals and *Homo sapiens* co-existed in Europe for 5000 years, so the displacement of the Neanderthals certainly did not happen overnight.

A likely scenario is that, although the Neanderthals and Denisovans are well suited to cope with a climate that can be cold, harsh, and brutal at times, the appearance of a more innovative and adaptable species simply pushes them over the edge. Europe and Asia are still in the grip of a glacial period that

---

* Genome studies suggest that contemporary East Asians possess 1.7–2.1% Neanderthal DNA, and Europeans 1.6–1.8%. See Fu *et al.*, *Nature* **514** (2014), p. 448.

began 110 000 years ago and will continue to the beginning of the Holocene, 12 000 years ago. As food sources become scarce, it is *Homo sapiens* that is better able to survive. We'll take a look at some of the specifics in the next chapter.

Homo erectus, antecessor, heidelbergensis, and *floresiensis* are gone. Now the Neanderthals and Denisovans disappear although, to a very modest extent, these ancient human lineages live on in us, the lone survivors.

12

COGITO ERGO SUM

The Origin of Human Consciousness

We're not quite done yet. The story of human evolution does not end with the first appearance of *Homo sapiens* in Africa 200 000 years ago, or even with the subsequent migrations from Africa into Europe, Asia, and the rest of the world. Our curiosity about our origins demands that we try to explain why it is *this* particular species that goes on to dominate the planet.

In other words, what is it that makes us 'special' in evolutionary terms? The answer, of course, is our seemingly unique form of *consciousness*. We have developed an ability to represent the external world to ourselves in our minds in ways that offer much, much more than survival advantages. Consciousness gives us the capability of abstract, symbolic thought, and to create complex mental lives. It allows us to communicate with others and, for some of us at least, to live out our lives in extraordinarily complex social structures. It enables us to create uniquely human cultures. No other creatures on Earth can do this.

This is a simple and straightforward answer, but it begs a lot of further difficult questions. What is consciousness and how does it work? How and when in our evolutionary story does it arise? Answering these questions will go some way to aiding our understanding not only of our origins but also of our own humanity.

Unfortunately, the study of consciousness is the only discipline I've come across that is structured principally in terms of its *problems*. So, we have the 'hard problem' of consciousness, the 'mind–body' problem, the problem of 'other minds', and many more. These problems have sponsored much discussion and many words but—at present—there appear to be no solutions.

But, hey, let's think positively. Unsolved problems are in many ways more fascinating. There has been much scientific progress which we will review below, after first dealing with some of the philosophical baggage.

Let's explore what we think we might know and see what it tells us.

THE HARD PROBLEM

Consciousness is rather like life (see the opening paragraphs of Chapter 8). We recognize it when we experience it ourselves and we have learned to recognize it in others (or so we think, anyway), but it is difficult to write down a definition that everyone will agree to.

Actually, I'm guessing that you don't really need a definition. You know what it feels like to have conscious experiences of the external world and you have what I might call an inner mental life. You have thoughts, and you think about those thoughts. You know what your own consciousness *is* or at least what it feels like. This is not very scientific but it's probably good enough to be going on with.

The American philosopher Ned Block has argued that consciousness can be divided into two distinct types: *Phenomenal consciousness* (or P-consciousness) is the processing of external stimuli to produce mental experiences (what some philosophers call *qualia*). In our conscious minds we experience visual images, sounds, smells, taste, and touch, including any emotions that such inputs might engender, such as instinctive joy, anxiety, or fear.

Access consciousness (or A-consciousness) is the 'higher-level' accessing, processing, and integration of mental information (including information recalled from memory) for the purposes of focusing attention, reasoning, reporting, decision-making, and controlling subsequent behaviour. P-consciousness is about functional and emotional experiences derived from raw sensory inputs. A-consciousness is about thinking what to do about them.

Not all philosophers agree with this distinction and to save words in what follows (and to preserve your patience) it's probably best if you automatically assume this for any and all such proposals. Philosophy tends to be rather more about arguing, less about agreeing.

In an influential paper published in 1995 in the *Journal of Consciousness Studies*, Australian philosopher and cognitive scientist David Chalmers argued that all the different aspects of A-consciousness represent 'easy' problems. He wrote:[1]

> The easy problems are easy precisely because they concern the explanation of cognitive *abilities* and *functions*. To explain a cognitive function, we need only specify a mechanism that can perform the function. The methods of cognitive science are well-suited for this sort of explanation, and so are well-suited to the easy problems of consciousness.

Frankly, I'm not so sure this is all that easy. But Chalmers' argument is about the *relative* intractability of these problems, and for him P-consciousness is where it gets really hard.

Let's take a specific example. Imagine a red rose, lying on an expanse of pure white silk. We might regard the rose as a thing of beauty, its redness stark against the silk sheen of brilliant nothingness. What, then, creates this vision, this evocative image, this tantalizing reality? More specifically, what in the external world creates this wonderful experience of the colour red?

That's easy. We google 'red rose pigment' and discover that roses are red because their petals contain a subtle mixture of chemicals called anthocyanins, their redness enhanced if grown in soil of modest acidity. So, anthocyanins in the rose petals interact with sunlight, absorbing certain wavelengths of the light and reflecting predominantly red light into our eyes. We look at the petals and we see red. This is surely quite straightforward.

But, hang on. What, precisely, is 'red light'? Our instinct might be to give a scientific answer. Red light is electromagnetic radiation with wavelengths between about 620 and 750 nanometres. It sits at the long-wavelength end of the visible spectrum, sandwiched between invisible infrared and orange.

But light is, well, light. As we saw in Chapter 1, it consists of tiny wave–particles of energy which we call photons. And, no matter how hard we look, we will not find an inherent property of 'redness' in photons with this range of wavelengths. Aside from differences in wavelength, there is nothing in the physical properties of photons to distinguish red from green or any other colour.

We can keep going. We can trace the chemical and physical changes that result from the interactions of photons with cone cells in your retina all the way to the stimulation of your visual cortex at the back of your brain. Look all you like, but you will not find the *experience* of the colour red in any of this chemistry and physics. It is obviously only when you somehow synthesize the information being processed by your visual cortex in your conscious mind that you experience the sensation of a beautiful red rose.

We could invent some equivalent scenarios for all our other human senses—taste, smell, touch, and hearing. But we would come to much the same conclusion. Just *how* is all this supposed to work? This is the hard problem. Chalmers again:

> The really hard problem of consciousness is the problem of *experience*... When we see, for example, we *experience* visual sensations: the felt quality of redness, the experience of dark and light, the quality of depth in a visual field. Other experiences go along with perception in different modalities: the sound of a clarinet, the smell of mothballs. Then there are bodily sensations, from pains to orgasms; mental images that are conjured up internally; the felt quality of emotion, and the experience of a stream of conscious thought. What unites all of these states is that there is something it is like to be in them. All of them are states of experience.

The problem is hard because we not only lack a physical explanation for how this is supposed to happen, we don't even really know how to state the problem properly.

Okay. If the 'how' problem is too hard, could we at least generate some clues by pondering on *where* these experiences might happen?

THE MIND–BODY PROBLEM

The French philosopher René Descartes is rightly regarded as the father of modern philosophy. In his *Discourse on Method*, first published in 1637, he set out to build a whole new philosophical tradition in which there could be no doubt about the absolute truth of its conclusions. From absolute truth, he argued, we obtain certain knowledge. However, to get at absolute truth, he felt he had no choice but to reject as being absolutely false everything in which he could have the slightest reason for doubt. This meant rejecting all the information about the world that he received through his senses.

Why? Well, first, he could not completely rule out the possibility that his senses would deceive him from time to time. Just as an optical illusion creates the impression of a square or a triangle that does not really exist (Figure 93(a) and (b)),

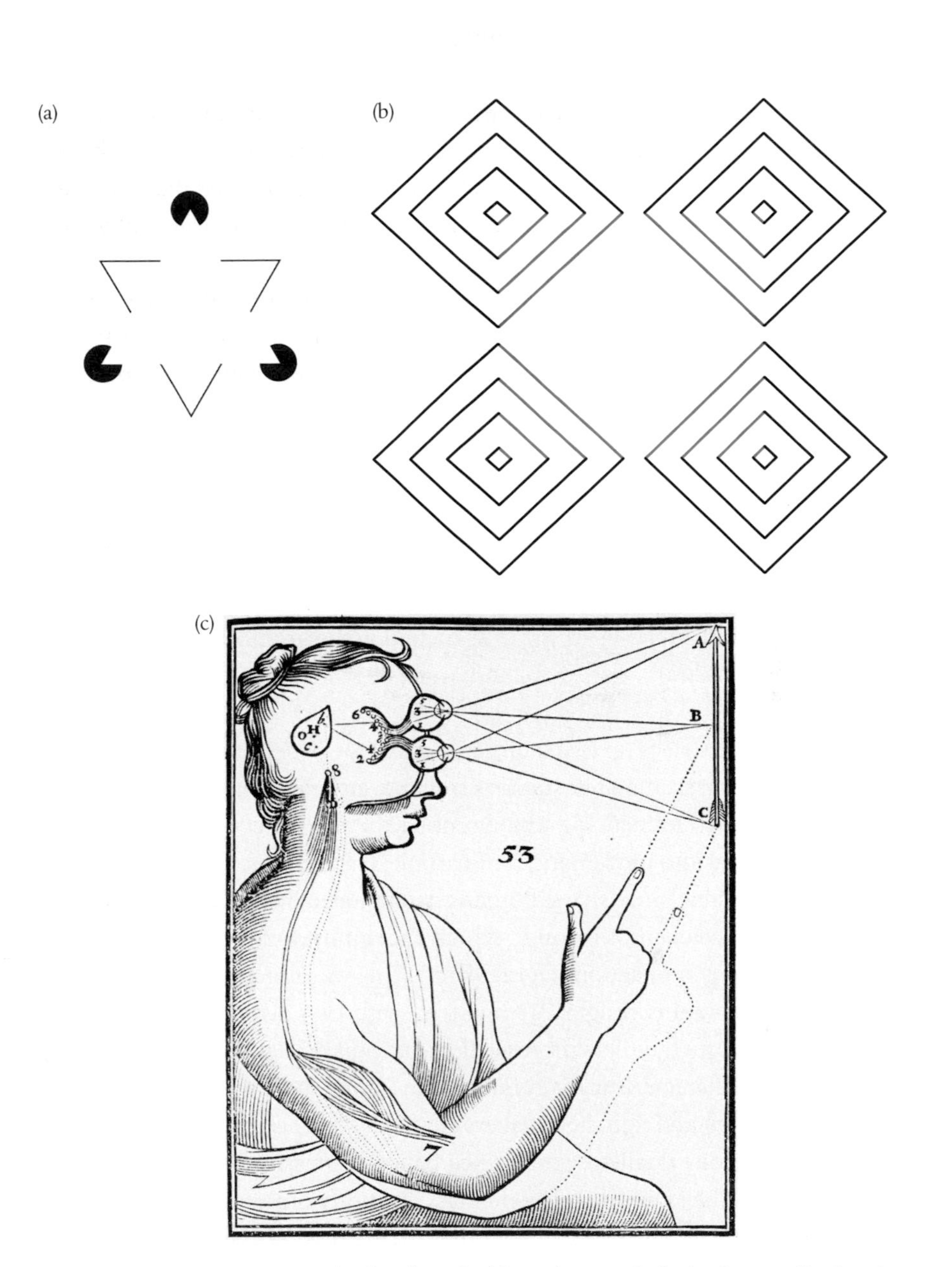

FIGURE 93 The Kanizsa triangle, first described by Italian psychologist Gaetano Kanizsa in 1955 (a) and the square neon colour spread (b) are examples of cognitive fiction illusions—our P-consciousness is tricked into perceiving geometrical objects that aren't there. Descartes argued that his senses cannot be trusted to deliver a faithful representation of the external physical world. He went on to argue that mind and body are distinct and separate substances, capable of independent existence. He identified the pineal gland at the centre of the brain as the 'seat' of consciousness, the place where the unphysical mind interacts with the physical brain (c).

so all his senses might play him false. Second, he could not be certain that his perceptions and experiences were not part of some elaborate dream. Finally, he could not be certain that he was not the victim of a wicked demon or evil genius with the ability to manipulate his sensory inputs to create an entirely false impression of the world around him.

Descartes felt that there was at least one thing of which he could be certain. He could be certain that he was a being with a conscious mind that has thoughts. He argued that it would seem contradictory to hold the view that, as a thinking being, he does not exist. Therefore his own existence was also something about which he could be certain. *Cogito ergo sum*, he concluded. I think therefore I am.

The external physical world is vague and uncertain, and may not appear as it really is. But the conscious mind seems very different. Descartes went on to reason that this must mean that the conscious mind is separate and distinct from the physical world and everything in it, including the unthinking machinery of his body, and his brain. Consciousness must be something 'other', something unphysical.

Descartes had to face up to the difficult challenge of determining how something with no physical manifestation could nevertheless make itself felt in the physical world, by influencing and directing the machinery—how a thought can be translated into movement, for example. His solution was to identify the pineal gland, a small pear-shaped organ that lies deep in centre of the brain, as the 'seat' of consciousness through which the unphysical mind gently nudges the physical body into action (Figure 93(c)). This was not an entirely arbitrary selection. The brain is roughly symmetric, with two distinct cerebral hemispheres joined by a bundle of fibres called the *corpus callosum*. The pineal gland at least has the characteristic of uniqueness in a brain that otherwise looks like a duplication of left and right hemispheres.

This mind–body dualism (sometimes called Cartesian dualism) is entirely consistent with belief in the soul or spirit. The body is merely a shell, or host, or mechanical device used for giving outward expression and extension to the unphysical thinking substance. My mind defines who I am whereas my body is just something I use (perhaps temporarily). Descartes believed that although mind and body are joined together, connected through the pineal gland, they are quite capable of separate, independent existence.

Now I can't really fault Descartes' logic concerning the relationship between the external physical world and my conscious representation of it. This is what philosophers call the 'brain in a vat' scenario. For all you know, some time ago an evil scientist removed your brain and is keeping it alive, feeding it with signals that give you *impressions* of an external world, such that you are '... sitting and reading these very words about the amusing but quite absurd supposition that there is an evil scientist who removes people's brains from their bodies and places them in a vat of nutrients which keep the brains alive'.[2] This uncertainty is the basis for many science fiction novels and movies, such as the 1999 Hollywood blockbuster *The Matrix*.

But to conclude from this that the conscious mind must therefore be unphysical involves a rather bold leap of logic, one that I believe is indefensible. There are few scientists and philosophers who embrace this kind of dualism today (although, as I've discovered, it's possible to find some in some unusual and unexpected places). The trouble is that by disconnecting the mind from the brain and making it unphysical we push it beyond the boundaries of science and make it completely inaccessible. Science simply can't deal with it.

In *The Concept of Mind*, first published in 1949, the British philosopher Gilbert Ryle wrote disparagingly of Cartesian-style mind–body dualism, referring to it as the 'ghost in the machine'. In his 1991 book *Consciousness Explained*,* the American philosopher Daniel Dennett said:[3]

> This fundamentally anti-scientific stance of dualism is, to my mind, its most disqualifying feature, and is the reason why... I adopt the apparently dogmatic rule that dualism is to be avoided at all costs. It is not that I think I can give a knock-down proof that dualism, in all its forms, is false or incoherent, but that, given the way dualism wallows in mystery, *accepting dualism is giving up*.

If we reject dualism, and we have no solution to the 'hard problem', where does this leave us?

Now, we've been here before, and I think you know the strategy for finding a way forward. We proceed by acknowledging our ignorance and making some reasoned *assumptions*. We assume that, however it works, consciousness arises as a direct result of the neural chemical and physical processes that take place in the brain. Our experience of a red rose has a *neural correlate*—it

* Readers unfamiliar with Dennett's book might wonder why, if Dennett has already explained consciousness, I haven't simply given his explanation here. I should perhaps point out that detractors of the arguments presented in Dennett's book have tended to dismiss it as *Consciousness Ignored* or *Consciousness Explained Away*. Please read it and make up your own mind.

corresponds to the creation of a specific set of chemical and physical states involving a discrete set of neurons located in various parts of the brain.

In philosophical terms, this is known as 'materialism'.

NEURAL CORRELATES

Now I wouldn't suggest making this materialist assumption unless I had some good grounds for believing in its validity. The average brain weighs about 1.4 kilogrammes and has the consistency of cold porridge. It is a rather delicate organ and, tragically for many people, rather a lot can go wrong with it. But looking at the effects of brain damage on the conscious responses of those affected can tell us a lot about the connections between the brain and the conscious mind.

For example, stroke victims suffering damage to the primary visual cortex, located in the occipital lobe at the back of the brain, may experience *blindsight*. They can no longer see objects or movement but nevertheless seem strangely aware of them and can even point them out with a high probability of success, even though they believe they are just making random guesses. The patients in such cases appear to have no experiences of these objects (through P-consciousness), yet when they are asked to access information about them and report back (through A-consciousness), the information is nevertheless there.

Whereas those suffering blindsight think they are blind but can actually 'see', those suffering *visual anosognosia* are actually blind but think they can see. They report back fabricated experiences or memories not to deceive, but because they are convinced their experiences or memories are real.

In *prosopagnosia*, sufferers have a conscious awareness of objects (and especially people) but without the associations provided by long- or short-term memory. Although the objects or people are entirely familiar, the sufferer no longer recognizes them, an affliction memorably described in the case of Dr P in Oliver Sacks' *The Man Who Mistook His Wife For A Hat*. The title says it all.

These tragic cases provide important clues that the origins of conscious experiences arise in specific parts of the brain and that these are changed or become impaired when these parts are damaged.

In the 1950s the Canadian neurosurgeon Wilder Penfield carried out operations on over one thousand patients suffering some form of brain damage.* In

* Including his own sister.

seeking out the damaged area, Penfield and his associates would expose the surface of a patient's brain and stimulate it with electrodes. As there are no pain receptors in the brain (why would you need them there?) this whole procedure could be carried out under local anaesthetic. The patients would remain conscious and simply tell the surgeons what they felt.

Using this approach, Penfield was able to identify which parts of the brain are responsible for processing sensory inputs and which are responsible for motor responses required to trigger bodily movement. These maps can now be found in any textbook on neuroscience, and are often represented in terms of the projection of a homunculus ('little person') on the surface area of both halves of the cerebral cortex, suitably distorted in shape to reflect the strength of the sensation or response caused. Such diagrams show the relative importance in both sensation and movement of the hands (especially fingers and thumb), face (especially lips), and tongue.*

On stimulating different areas, the patients would sometimes report sensations and would sometimes make involuntary movements. And when parts of their right and left temporal lobes were stimulated, some patients reported seeing and hearing things (sometimes separately, sometimes together). Some reported thoughts, recalled memories and experienced visual flashbacks. Some reported dream-like experiences.

An effective remedy for epileptics suffering abnormal electrical activity on one side of the brain is to cut the corpus callosum, the bundle of fibres connecting the brain's two hemispheres. The result is a 'split brain'. In truth, the two halves are still connected by central regions of the brain and the patients therefore retain a strong sense of a single self and, in fact, suffer no loss of faculties that is otherwise characteristic of brain damage.

However, research on split-brain patients conducted in the 1960s by American neurobiologists Roger Sperry and Michael Gazzaniga demonstrated that the left and right hemispheres have very different functions. The left hemisphere tends to dominate in verbal, analytical, and sequential processing tasks and the right hemisphere tends to specialize in spatial and creative synthetic tasks.

---

* The maps are almost mirror images, but the division between sensing and motor response is not completely black and white. The motor cortex also receives some sensory input and stimulation of the somatic sensory cortex can sometimes produce movement.

Visual input from the left side of the visual field—the right side of the retina—is processed by the right visual cortex, located towards the back of the right hemisphere and vice versa for the right side of the visual field. Tactile input from the left hand is processed in the right hemisphere and vice versa for the right hand. In a right-handed patient, speech is handled in the left hemisphere. A word flashed into a split-brain patient's right visual field, and hence to the left hemisphere, is immediately recognized: the patient can say the word and write it down. However, a word flashed into the same patient's left visual field, and hence to the right hemisphere, is recognized but as the hemispheres are now disconnected the patient can now no longer say it or write it down.

Fortunately, neuroscientists no longer have to rely on patients that have had to undergo such drastic surgery. They have access to a battery of technologies, such as functional magnetic resonance imaging (fMRI) and positron emission tomography (PET), which can probe the workings of the brain in exquisite detail in non-invasive ways. Experiencing something or thinking about something stimulates one or more parts of the brain. As these parts get to work, they draw glucose and oxygen from the bloodstream. An fMRI scan shows where the oxygen is being concentrated, and so which parts of the brain are 'lighting up' as a result of some sensory stimulus, thought process, or memory. A PET scan makes use of a radioactive marker in the bloodstream but otherwise does much the same thing, though with lower resolution.

I had cause to have an MRI scan in the summer of 2013 (Figure 94(a)). This picture does not show any part of my brain 'lighting up' (it wasn't an fMRI scan), but I can assure you that I was conscious throughout (I was struck by how noisy it was in there). I was just pleased to see that my brain does actually seem to fill my head.*

From such studies we learn that the human brain has been rather cobbled together, the result of a long history of evolutionary adaptations of the 'make do' variety. The parts of the brain responsible for processing sensory inputs of the kind that shape P-consciousness are liberally distributed over much of the cortex (Figure 94(b)). The parts of the brain involved in the more executive functions involved in A-consciousness are located in the frontal lobes.

* I'd had my doubts in the past.

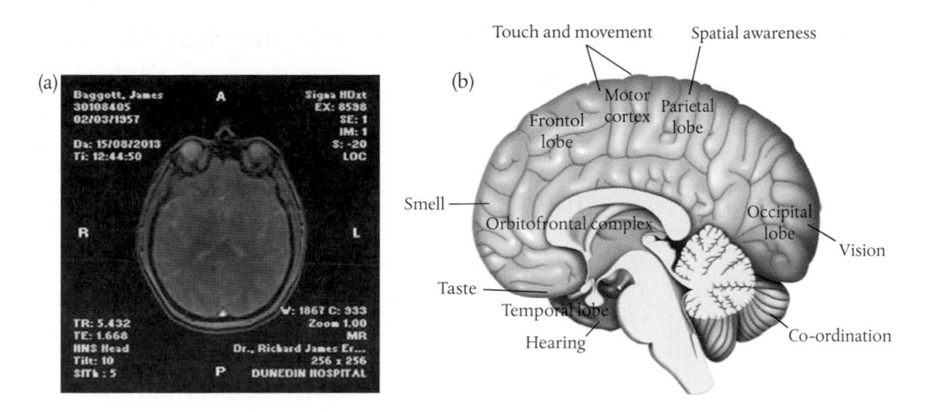

FIGURE 94 I had cause to have an MRI scan in August 2013, (a). This image shows the bilateral symmetry of my brain, with near-equal left and right hemispheres, and some hints of structure (called the limbic system) in the centre. Neuroscientists have mapped those parts of the brain cortex that are involved in sensory perception of the kind involved in P-consciousness (b). Much of the cortex is deployed in the tasks of sensing the external physical world, with the frontal lobe involved in higher-level processing of the kind required by A-consciousness.

The brain is obliged to work in a distributed manner. For a time it was thought that we would find memories stored in a specific location, as 'engrams' in the brain's equivalent of random access memory, or RAM. But we now know that this is not how memories work. The sight and smell of a red rose are stored in those parts of the brain that handle vision and smell and which were involved in the original P-consciousness experience. Recalling its memory involves triggering the same parts of the brain and so 're-experiencing' it. Penfield and his colleagues would not have been aware that other parts of the brain were involved as they stimulated patients' memories by applying their electrodes on selected surface areas.

In fact, memories seem not much different from thoughts. British psychologist Alan Baddeley has proposed a model of short-term 'working memory' involving separate, distributed subsystems which hold visuo-spatial and verbal information and a kind of 'central executive' which coordinates and integrates this information and retrieves memories.

It works well enough, but it's hardly optimal. It's more Heath Robinson (or Rube Goldberg for American readers) than Frank Lloyd Wright, with function following form rather than the other way around. I doubt that the brain would win any design or systems architecture awards.

BUT WHERE DO I FIND 'ME'?

Studies of patients suffering some form of brain damage, and the use of modern technology have allowed neuroscientists to build up a detailed picture of the ways in which different parts of the brain work together to produce conscious experiences, though they still can't tell us *how* these experiences happen—they don't solve the 'hard problem'.

But what about the 'grand synthesis' that seems to take place in my mind, the synthesis that gives me a continuous sense of self, of who I am? Are there parts of my brain where I feel joy and sorrow, where I hold my sense of morality, where (I think) I exercise free will, where I make sense of the world and plan for the future? Are there parts of my brain where I might expect to find *me*?

The answer is broadly yes. It is the dorsolateral (meaning upper side) prefrontal cortex.

How do we know? Again, from studies of patients whose prefrontal cortex has suffered damage of some kind. Consider the tragic tale of Phineas Gage who, in 1848, was working as a foreman for the Rutland and Burlington Railroad, blasting rock to clear the way prior to laying down new track. One afternoon during the summer, he suffered a terrible accident. He had drilled a hole in a rocky outcrop and filled it with gunpowder, fuse and sand. As he gently 'tamped' the sand with a specially fashioned iron rod—something he had done many times before—the powder exploded prematurely, driving the rod up through his left cheek, through his brain's frontal lobe and out the top of his head.

Surprisingly, this did not kill him. Even more surprisingly, he remained conscious, talking coherently and even making jokes as his colleagues rushed him to a doctor in the nearby town of Cavendish. Computer simulations some 120 years later showed that the rod had missed those regions of his brain that specialize in language and motor function, but it had damaged areas of his prefrontal cortex (Figure 95).

Gage recovered from his injuries within two months, but he emerged from the experience a different person. The damage to his brain:[4]

> ... compromised his ability to plan for the future, to conduct himself according to the social rules he previously had learned, and to decide on the course of action that ultimately would be most advantageous to his survival.

Gage became temperamental, abusive, and shiftless, in marked contrast to his behaviour and personality before the accident. He lost his job at the railroad,

FIGURE 95 Phineas Gage, pictured here (a) in a daguerreotype, suffered a terrible injury in 1848 while working for the Rutland and Burlington Railroad. An iron rod was blasted up through his left cheek, through his brain's frontal lobe and out the top of his head. Computer simulations performed some 120 years later (b) allowed Antonio Damasio and his colleagues to determine which parts of his brain had been damaged. Gage survived the accident, but damage to his frontal lobe changed his personality.

drifted into farm labour, and spent time as a circus exhibit. He died twelve years after his accident.

The Portuguese-American neurobiologist Antonio Damasio suggests that Gage had not suffered any injuries that had reduced his powers of reason—his brain could still function as an analytical machine. But he was now 'unbalanced'. Stripped of the emotional checks and balances that had previously guided his behaviour, Gage could no longer make the 'right' choices. He was no longer himself.

As I mentioned above, the right hemisphere typically cannot communicate in speech or writing, but a split-brain patient Paul S, who had previously suffered damage to his left hemisphere, had over time developed significant speech function in his right hemisphere and had learned to write with his left hand. When his brain was split, it was therefore possible for him to communicate separately from both hemispheres.

Gazzaniga described what this was like:[5]

> Instead of wondering whether or not [the patient's] right hemisphere was sufficiently powerful to be dubbed conscious we were now in a position to ask [the patient's] right side about its view on matters of friendship, love, hate, and aspirations... When [the patient] first demonstrated this skill, my student, Joseph LeDoux, and I... just stared at each other for what seemed an eternity. A half-brain had told us about its own feelings and opinions, and the other half-brain, the talkative left, temporarily put aside its dominant ways and watched its silent partner express its views.

The feelings and opinions of Paul's right hemisphere did not always coincide with those of his left. When Paul was asked what he wanted to be, his right hemisphere wrote 'automobile racer'. When asked the same question, his left hemisphere wrote (with his right hand) 'draughtsman'.

I personally find all this pretty compelling. Neuroscience in its modern form was established only in the second half of the last century, and our understanding has come an awfully long way in that relatively short time. We may never solve the 'hard problem', but I think we may go a long way towards demonstrating that consciousness is not a 'thing', but rather a *process* that occurs in a developed brain.

One final point before we move on. By grounding consciousness and mind in neuronal activity occurring in the brain, we have to acknowledge what this implies, which is that consciousness is not the preserve of humankind. On 7 July 2012, a prominent international group of cognitive neuroscientists, neuropharmacologists, neurophysiologists, neuroanatomists, and computational neuroscientists met together at the University of Cambridge, UK. After some deliberations, they agreed the *Cambridge Declaration on Consciousness*, which states:[6]

> ... the weight of evidence indicates that humans are not unique in possessing the neurological substrates that generate consciousness. Non-human animals, including all mammals and birds, and many other creatures, including octopuses, also possess these neurological substrates.

ON THE PATH TO THE GREAT LEAP FORWARD

If you buy these arguments then you might be willing to accept that consciousness and mind are some kind of 'higher-level' processes or functions of the material brain. Let's just log this for now so that we can get back to our story.

Recall that at the beginning of this chapter we asked when in our evolutionary history these capabilities might have arisen. Actually, we don't yet have a good answer for this, either. A brain leaves impressions on the inside of the brain case which we can decipher to give us clues about the proportions of the main lobes and hence brain development, but thoughts don't fossilize. However, we can use the information available to us to develop scenarios that I believe contain more than a grain of truth in them.

The logic runs something like this. The human species that emerges in Africa some 200 000 years ago is not blessed with a massively large brain. But the brain of *Homo sapiens is* different in that it has a larger frontal lobe which, as we have just learned, is where we will eventually find the 'higher-level' mental capabilities that define us individually as human beings. Genetic mutations arising earlier in the *Homo* lineage have already created the anatomical possibility for speech and have enhanced the brain's potential for language.

These pieces now come together. Humans develop language which considerably accelerates the growth of neural pathways in the brain, specifically those involved in abstract or symbolic thinking, on which our intrinsically human consciousness is built. Thinking is founded on language and its introduction now gives a massive boost to the development of the conscious mind.

In parallel with this, an ability to communicate allows the creation of larger and more complex social groups and structures. This creates new and important feedback loops and results in the development of a *social brain* capable of anticipating and dealing with the behaviours of others. More neural pathways are forged and consciousness is raised higher.

This is not a linear sequence (first this, then that), and it certainly doesn't happen overnight. But I strongly suspect that the origin and early evolution of human consciousness is driven by a complex interlinking of genetics, language, and society.

The seeds are sown. Things happen slowly at first, *Homo sapiens* behaving little differently from other archaic human species, making use of a limited set of primitive tools, wearing rudimentary clothing, and living in caves in small groups of hunter-gatherers. But the signs are there. A number of artefacts dating to between 70 000 to 100 000 years old have been discovered at Blombos Cave in Blombosfontein Nature Reserve on the South African coast about 300 kilometres east of Cape Town. These include engravings in red ochre (such as the Blombos Stone, Figure 96(a)), engraved bone and beads made from marine shells. These may represent the earliest forms of abstract representation by humans.

It seems that we eventually reach a tipping point, about 40 000 to 50 000 years ago, although some scientists dispute this.* This is the *Great Leap Forward*, or the 'human revolution', a flowering of human innovation and creativity involving a transition to what is known as *behavioural modernity*.

After the transition we see much more sophisticated tools ('later Stone Age' in Africa, and 'Upper Palaeolithic' in Europe and western Asia) including eyed needles of bone used to make seamed clothing. Instead of being content with tools made from whatever materials are locally available, these modern humans *import* tools made from materials sometimes hundreds of miles distant, presumably through elaborate trading and exchange networks.

We see the beginnings of a definitively human culture in finely worked carvings and sculptures, decorative jewellery, musical instruments, and in cave art, such as the paintings found in Chauvet Cave in the Ardèche near Avignon in southern France, which are determined to be 35 000 years old (Figure 96(b)). It has been suggested that these caves serve a purpose other than providing a gallery in which to exhibit works of art, and may have been used in initiation rituals.

---

* The concern is that the evidence may be far from complete, and we may not yet have the full story.

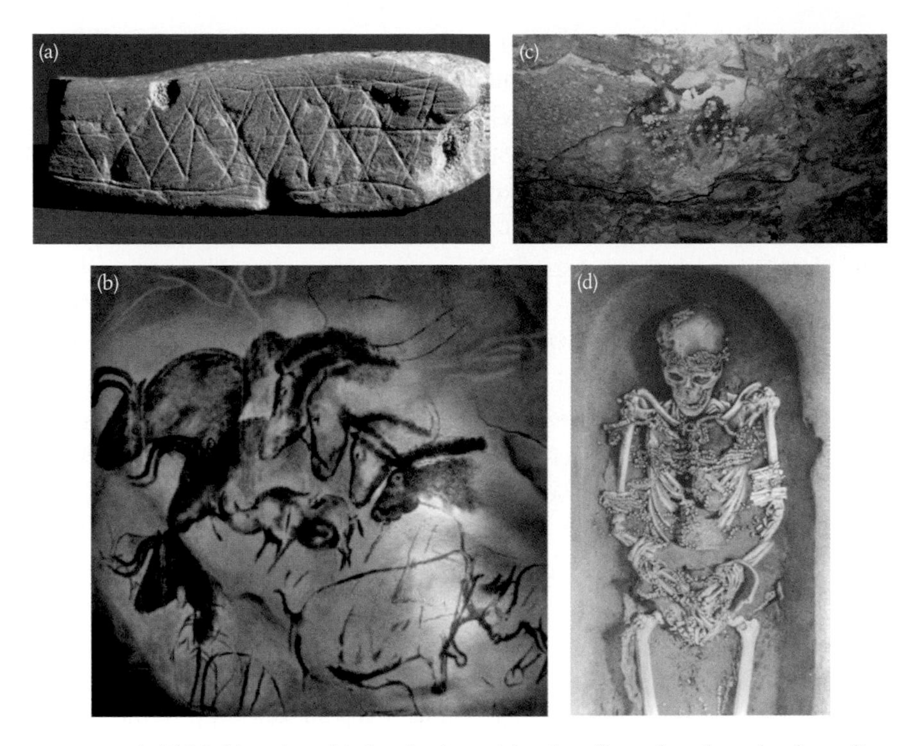

FIGURE 96 (a) Markings in red ochre (an iron-rich mineral) are thought to be the earliest examples of abstract expression by humans. Examples were discovered in Blombos Cave in South Africa and are between 70 000 and 100 000 years old. After the Great Leap Forward, thought to have occurred between 40 000 and 50 000 years ago, such expression reaches new heights, for example in paintings found in Chauvet Cave in southern France (b), determined to be 35 000 years old. Although most of the examples of this flowering of human creativity are found in Europe, examples have recently been discovered in Asia, such as these hand stencils found in a cave system in Sulawesi, Indonesia (c). Ritual burials at Sungir in Russia, dated at 28 000 to 30 000 years old, show rapidly developing craftwork (d). This adult male was buried in clothing decorated with 3000 beads.

For a long time the Great Leap Forward appeared to be a specifically European development, reflecting what some paleoanthropologists perceived as a rather Euro-centric bias in the reading of the archaeological record. American anthropologists Sally McBrearty and Alison Brooks, at the University of Cincinnati and George Washington University, respectively, published an influential paper in 2000 which attempted to set the record straight.[7] But the simple truth is that for a long time the prime examples of a flourishing human creativity were all European.

That changed in October 2014 with the publication in *Nature* of evidence of paintings discovered in a cave system on the Indonesian island of Sulawesi.[8]

These include twelve human hand stencils, each produced by spraying wet pigment over a hand pressed against the cave wall (Figure 96(c)), and two paintings of animals. The oldest hand stencil was determined to be at least 39 900 years old, using uranium isotopes sampled from associated rock deposits. European examples of such hand stencils are known, but this is so far the oldest such stencil in the world. Likewise, one of the animal paintings, determined to be at least 35 400 years old, is the oldest example of its type.

We find evidence for very specifically human rituals at Sungir, about 200 kilometres east of Moscow in Russia, where a team of archaeologists working in the 1950s uncovered two graves. These are ritual burials, the adult male—no doubt a local chieftain of some kind—is adorned with around 3000 pierced beads made of ivory, thought to have been stitched to his clothing (Figure 96(d)).

A second grave bears the remains of two children, believed to be a boy and a girl, presumably sacrificed. The children are similarly adorned, each with around 5000 beads. It is estimated that these beads would have taken about 10 000 hours to produce. The boy has what looks to have been a belt of pierced arctic fox teeth, and both have an ivory pin at the throat that may have been used to secure a cloak. The site is dated at 28 000 to 30 000 years old.

The Great Leap Forward could not have happened without genetic adaptations enabling the development of speech and language and complex social structures. So let's now look at each of these in turn.

FOXP2 AND THE LANGUAGE INSTINCT

If humans are to speak and use language, then there are a couple of anatomical prerequisites that evolution must first satisfy. The articulation of sounds requires anatomical changes in which the larynx (or 'voice box') descends and the tongue changes shape and descends into the pharynx at the back of the throat. This creates an airway called the 'supralaryngeal vocal tract'. The neck lengthens to accommodate the lowered larynx and to make it possible to swallow. This all comes at a cost. Opening up the airway risks blocking it with food, causing potentially fatal choking.

Anatomy is one thing, but language is not possible without the neural structures in the brain to support it. Using language requires an ability to anticipate what is to be vocalized, plan and then to move lips and tongue accordingly:[9]

> Human speakers plan ahead. As one begins to say the word 'too,' one's lips 'round' (protrude and narrow), anticipating the rounded [u] vowel. One's lips are not rounded at the start of the word 'tea' because the following vowel is not rounded.

Important clues to the genetic basis of language in humans were uncovered in 1998 by Oxford geneticists Simon Fisher and Anthony Monaco and their colleagues. They studied an extended London-based English family of 30 members spanning three generations (known only as the 'KE family'), first brought to the attention of the scientific community in 1990 by Myrna Gopnik, a linguistics professor at McGill University. About half the family suffer a severe speech and language disorder called developmental verbal dyspraxia, which renders them unable to plan and then perform the movements necessary for grammatically correct language.

The geneticists traced the problem to a mutation of a gene known as FOXP2. This sits on chromosome 7 and codes for a protein called forkhead box protein P2, consisting of more than 700 amino acids. Position 553 in this sequence is normally occupied by the amino acid arginine. Among those suffering speech and language impairment in the KE family, position 553 is occupied instead by histidine. It's incredible the damage that one misplaced amino acid in more than 700 can do.

Now FOXP2 is not (as sometimes described in the popular press) the 'language gene' or the 'grammar gene'. It seems that FOXP2 may actually be another command-and-control gene which, like the HOX genes, affect the expression of hundreds of others, particularly in the developing embryo. One of the reasons that a single amino acid variation causes such damage is that position 553 is close to that part of the protein which binds to other chromosomes and is fundamentally important in regulating other genes.

FOXP2 is not unique to humans. Variations of the gene are widespread in mammals. The FOXP2 gene in chimpanzees produces a protein that differs from the human version only in *two* amino acids. In mice the difference is three amino acids and in zebra finches seven amino acids. As far as we can tell, the Neanderthals and Denisovans possessed FOXP2 genes *identical* to modern humans.

Clearly, possession of FOXP2 doesn't automatically result in the development of language skills, suggesting that while it is necessary, it is not sufficient.

(As Steven Pinker put it in his 1994 book *The Language Instinct*, removing the distributor from a car engine stops the car moving, but this doesn't mean the car is controlled by the distributor.) It would seem that FOXP2 has a broader role in literally wiring the brain for the *capability* of language (among other things). It encourages the growth of interconnections between neurons which, when combined with a larger frontal lobe and the anatomical changes necessary for vocalizing speech, make the human instinct for language inevitable.

Could Neanderthals talk? Opinions vary. In *The Singing Neanderthals*, published in 2005, British archaeologist Steve Mithen argues that, while the Neanderthals would appear to have had the necessary anatomical structures for speech, they nevertheless lacked the 'cognitive fluidity' of the kind needed for symbolic thought and language.* It seems they had some, but not all, of the ingredients. Instead, Mithen believes that the Neanderthals developed a form of 'sing-song' communication which he calls 'Hmmmmm'.

OTHER MINDS, INTENTIONALITY, AND THE SOCIAL BRAIN

Descartes was pretty confident in his conclusion that he is a being with a conscious mind that has thoughts. But what about everybody else? Now, professional 'mind-readers' can no more read minds than professional magicians can do real magic. Our minds are shaped by our own personal, subjective conscious experiences and remain firmly locked inside our heads. We have no access to other minds, so how can we know whether they work in the same way that my mind and, presumably, your mind works?

It gets worse. You look human, you behave in all respects like a human, but for all I know you could be a 'philosophical zombie', lacking in any kind of conscious experience and mind.† This is what philosophers call the 'problem of other minds'.

Most people (who are not philosophers) don't lose too much sleep over this. We make use of a multitude of both verbal and non-verbal signals to deduce that others do indeed have minds and that these work in many ways like our own. We learn at a relatively young age (and largely unconsciously) to 'mentalize' by forming our own *theory of mind*.

* As I write, a 'drawing', discovered at Gorham's Cave in Gibraltar, is being claimed as an example of Neanderthal cave art. It looks like an unfinished game of noughts and crosses scratched into the cave wall but palaeontologist Clive Finlayson suspects that it might be a crude map.

† The basis for a new blockbuster movie or TV series, perhaps? 'Dawn of the Non-sentient Beings'? 'World War P'? 'The Walking P-zombies'? Maybe not.

If you do stop to think about it, then you may quickly come to the conclusion that thinking is always thinking *about* something. This aspect of mental content, this 'aboutness', is referred to by philosophers as *intentionality*. This means more than just intending things, such as I intend to buy a house or you intend to steal a car. Intentionality is the ability of a conscious mind to represent to itself physical objects, people, and states of affairs (real or imaginary).

Intentionality is organized in a hierarchical structure. First-order intentionality is self-awareness. Second-order intentionality is the representation of another mind—'I know what you're thinking'. Higher orders are increasingly more convoluted and complex, with a typical fifth-order statement being something like: 'I believe that you suppose that A wants B to believe that C intends to steal a car.' It is thought that children acquire a full theory of mind by about five years of age, and adults can typically operate comfortably at about fifth- or sixth-order.*

How does this work? There are a couple of alternatives, but the most likely mechanism involves simulating other minds within our own. We put ourselves in the position of the person whose behaviour we are monitoring, we 'see the world through their eyes' and simulate what is likely to be going on in their minds based on what would likely go on in our own under these same circumstances.

This is not an instinct that we are born with, but it develops through childhood and into young adulthood. It is not necessary for us to have had a long acquaintance with another person to imagine what they might be experiencing and thinking. We just need to believe (or assume) that their mind works in many ways like ours. We can apply this same approach to someone we've never met. In fact, we can apply it to fictional characters (including non-human characters) that don't even exist. We do this every time we read a novel.

I'm sure you won't be surprised to learn that there's a neuronal basis for this, too. In the late 1990s, a team of Italian neuroscientists led by Giacomo Rizzolatti obtained evidence for the existence of what have become known as *mirror neurons* in the brains of macaque monkeys. It was subsequently established that humans also have mirror neurons, widely distributed over many different parts

---

* Not all philosophers accept the need for this kind of convoluted structure. For example, in his 1995 book *The Construction of Social Reality*, American philosopher John Searle argues for a form of *collective intentionality*. This is a (presumably genetically based) predisposition towards sociality. We have thoughts that are about 'I', but we also have thoughts about 'we'.

of the brain. These are active when we observe someone doing something, responding to something, or expressing emotions, and they literally mirror the neuronal activity in the observed subject's brain.

With the aid of mirror neurons we mimic an observed activity subconsciously in our brains without actually performing the activity itself. A mechanism has been identified which inhibits the neurons responsible for bodily movement, so we don't end up comically aping our observed subject (although I defy anyone to watch the dental torture scene in John Schlesinger's *Marathon Man* without physically wincing). Our ability to put ourselves in an observed subject's position—to have *empathy*—is hard-wired in our brains.

British evolutionary psychologist Robin Dunbar pulls these, and many other, threads together in the *social brain hypothesis*:[10]

> There is a general consensus that the prime mover in primate brain evolution (and perhaps even that of all mammals and birds) is the evolution of more complex forms of sociality. Although other aspects of behaviour (notably ecological inventiveness) also correlate with brain size, these turn out to be consequences of having a large brain rather than its evolutionary cause.

This is an elaborate theory, constructed from many observations from the fossil record and the results of contemporary experimental neuroscience. Neural imaging in particular shows that when we draw on our theory of mind when dealing with others, we engage parts of the prefrontal cortex and two areas in the temporal lobe. The higher the level of intentionality involved, the greater the demand for neuronal capacity, specifically in the orbitofrontal complex. Chimps and orangutans can cope at the level of second-order intentionality but can go no higher; they don't have the neuronal capacity.

So, how does this work? It's tempting to imagine that this must have been all about *hunting*. After all, for these early modern humans, surely success in hunting is going to make all the difference in terms of survivability. Anything that aids both effectiveness (larger prey means more calories) and efficiency (expending fewer calories to bring down the larger prey without getting injured or killed) is going to be advantageous. Language and sociality, which allow a group of hunters to develop trust, anticipate, plan, and communicate, are surely key.

I wouldn't want to underestimate the importance of language and sociality to hunting, but it's worth noting that there are many animal species (such as hyenas) which are adept pack hunters, able to coordinate their actions without language and bring down substantially larger prey. We have grounds for believing that the Neanderthals were also very successful hunters, and we shouldn't be blinded by their ultimate fate and forget that they survived on Earth for 300 000 years.

Dunbar and his colleagues suggest that the driving forces for language and sociality among anatomically modern humans are actually rather different. Language allows us to communicate and exchange information. About what? Well, we might assume this was information about aspects of the physical environment that are valuable to know: the location of fresh water or a dangerous predator or the best way to bring down a woolly mammoth.

But language also allows us to 'socialize', to build relationships with each other through 'gossip'. It allows us to make declarations, probably very important in an early hunter-gatherer society: 'I'm off hunting for a couple of days. She's with me, so nobody mess with her while I'm gone.' Another argument suggests that language is a form of male 'display' and therefore important in sexual selection. Perhaps, as Stephen Fry once observed in a 1990s TV commercial for Heineken Export lager, females have always been attracted to a smooth-talking bar steward.

There's plenty of evidence to suggest that the exchange of physical-world information tends to take place only when the necessary social bonds have first been established. The personal often takes precedence over the practical.

In monkeys and apes, bonding within a community is achieved through social grooming. This takes time, and Dunbar and his colleagues have built some relatively simple time-budget models to estimate the amount of time likely to be consumed by foraging, feeding, moving, and enforced resting (digesting, sleeping, and avoiding overheating). The balance of time is then available for grooming. When combined with the average time required for a one-on-one grooming 'session', this can be used to establish a natural size for the community that can be successfully bonded in this way. For a chimpanzee community the average size is around 55.

But language enables bonding through forms of socializing that are one-to-many, such as telling stories and jokes. These early stories are likely to be about the origins of the physical world and of the people in the community, invoking various gods and mystical creatures in an attempt to explain some of the vicissitudes of life or, at least, enabling members of the community to be reconciled with them. Shamanistic rituals, involving trance-state-induced journeys to the spirit world or the 'other side', stir the emotions and forge a set of shared beliefs, the 'glue' which binds the community together, establishing its collective identity and providing a sense of life purpose.

Through the use of language and the engagement of the neural circuitry involved in the social brain, early modern humans are able to build larger communities. The limit of community size is now determined only by the neural capacities of its members. Based on extrapolations from relationships between social group size and the ratio of frontal lobe to subcortical brain size among apes, Dunbar estimates that the natural maximum group size for modern humans is about 150. This is sometimes referred to as 'Dunbar's number'.

Of course, today we live in substantially larger communities or even in cities numbering millions of inhabitants. But this is not how Dunbar is using the concept of 'community'. We might live and work among a large number of people, but how many of these do we *genuinely* know (and care about)? How many do we count as family and friends (even Facebook friends)? And how many of these are merely acquaintances, with whom we have no real relationship to speak of? Researchers have ranged widely in their studies over many different kinds of community—ancient and modern—and the number 150 keeps popping up.

Here are just a few examples. One study of the lengths of Christmas card distribution lists produced an average number of card recipients of 154. The Domesday Book lists villages with an average population size of 150. The recommended ideal size for an Anglican Church congregation is 200. The American company W. L. Gore and Associates, Inc. (known most widely for their Gore-Tex fabric) employs about 10 000 people, but builds factories designed to accommodate only 150, based on the observation that operating with a larger number starts to 'get clumsy'. 'People used to ask me, how do you do your long-term planning,' Gore associate Bob Hen told author Malcolm Gladwell, 'and I'd say, that's easy, we put 150 parking spaces in the lot, and when people start parking on the grass, we know it's time to build a new plant.'[11]

The creation of strong, socially bonded communities is extraordinarily powerful. As the glaciers advance towards the end of the Pleistocene, modern humans not only have the advantages of their inventions—tools and seamed clothing—they also have a support infrastructure based on social communities; their own and neighbouring communities with which they trade. They are much better able to cope with the exigencies of life on a planet that can sometimes be rather unfriendly and unforgiving.

ARE WE STILL EVOLVING?

I guess the answer is: why wouldn't we be? Evolution works through random mutations which create new genes. These are likely to be slightly different takes on existing genes, but as we saw in the case of FOXP2 a single changed nucleotide placing a single different amino acid in the resulting protein can have a big biological impact. The new genes may be advantageous, neutral, or harmful. Breeding within a population and between populations may cause these new genes to 'flow'.

These genes' survivability in the human genome depends on a number of factors. Obviously, if the genes are so harmful that they inhibit or even prevent their carriers from passing them on to successive generations then they are likely to be soon eliminated. Random sampling of genes in the processes of sexual reproduction may also serve to eliminate new genes, or accentuate them—a phenomenon known as 'genetic drift'.

One famous example of recent human evolution concerns lactose intolerance. Lactose is a sugar found in milk, and human milk contains a high concentration (about 9%). This is tricky to digest, requiring an enzyme—called lactase—which switches on in early infancy. However, once a child is weaned, the gene expressing the enzyme switches off as it is no longer needed. But, while many populations in sub-Saharan Africa, South East Asia and particularly China are indeed lactose-intolerant in adulthood, most Europeans and Americans of European descent are not.

Why not? The answer is startlingly simple. Hunter-gatherers spread across Europe and the rest of the world in communities of about 150 or so, as we've seen. This is the dominant mode of living throughout human prehistory. Then, around the beginning of the Holocene about 12 000 years ago, communities in various parts of the world stop wandering and begin to settle down. They discover the possibility of *agriculture*.

This might seem like a 'no-brainer', but the consequences are far from all positive. Geographically fixed communities are more vulnerable, particularly to passing raiders. The notion of 'safety in numbers' means rapid growth in the size of settlements beyond the comfortable number of 150, creating new social stresses that must now be managed.

For sure, food production is now somewhat more predictable (though dependent on sometimes unpredictable weather patterns) and it becomes possible to build food stores that can be used when harvests are poor. But the transition to agriculture changes the diet quite drastically. The intake of carbohydrates increases at the expense of proteins. Vitamin intake declines dramatically. As Dunbar writes:[12]

> People from Neolithic settlements were [physically] smaller than contemporary hunter-gatherers (always a sign of an inadequate diet) and had more signs of dietary stress in their bones. Indeed, estimates of the nutritional returns from historical and contemporary hand-tool farming suggest that the energetic returns from farming are significantly lower than from foraging.

The domestication of cattle—dairy farming—follows in Europe between 5000 and 10000 years ago. This brings access to a range of nutritional dairy products which can do much to improve the agricultural diet (despite the domestication of animals, meat is still a relative luxury). But dairy products contain lactose, to which humans are intolerant. A mutation to the gene which controls the lactase enzyme, keeping it switched on in adulthood, is thought to have appeared about 8000 years ago. These dietary pressures now encourage the selection of this mutated gene, and it quickly spreads through the European population.

Something to think about when next you add cow's milk to your tea.*

Let's consider one further example. As I mentioned in Chapter 9, a mutation in the human HERC2 gene affects the expression of the adjacent OCA2 gene, turning brown eyes blue (see the footnote at the beginning of Chapter 9). This mutation is thought to have arisen between 6000 and 10 000 years ago.

If we look at the incidence of blue eyes in present-day populations across Europe, we find that they are most common in Northern Europe, and specifically the Baltic region. In their book *The 10,000 Year Explosion*, American anthropologists Gregory Cochran and Henry Harpending speculate that the mutation originates in a village in Lithuania about 6000 years ago (a 'best guess' based on the available evidence). They explain its transmission through European and North African populations through an elaborate history involving the exploits

---

* A comment obviously aimed at European readers and American readers of European descent.

of the Vandals, who roam around and torment Europe for several centuries and who famously sack Rome in 455. Further transmission of the mutated gene is then continued through the 16th to 19th centuries by the activities of pirate corsairs.

Perhaps the most obvious signs of continuing human evolution are those involved in small adaptations to local climatic conditions. Exposure to tropical sunshine creates selection pressures for genes that support the production of melanin, a pigment which darkens the skin and affords some protection against the Sun's ultraviolet radiation. We can assume the early modern humans migrating out of Africa would have been dark-skinned.

Unfortunately, melanin protects the skin but it also suppresses the body's production of vitamin D. This is not an issue in places where sunlight is plentiful, but as modern humans occupy colder parts of Europe and East Asia, where the Sun shines rather less, they risk problems of vitamin D deficiency, such as childhood rickets and various other bone disorders. There are few dietary sources (vitamin D_2 is available in mushrooms and alfalfa, vitamin D_3 in fatty fish such as salmon and tuna).

Once again, single-nucleotide mutations to a selection of genes involved in melanin production (such as the gene SLC24A5, which sits on the long arm of chromosome 15) serve to reduce pigmentation, lighten the skin and restore vitamin D production. It is thought that these mutations appear in Europe about 11 000 to 19 000 years ago.[13] It this is correct, it means that the ancestors of modern Europeans (and Americans of European descent) are dark-skinned for about 30 000 years. About twice as long as they have been light-skinned, in fact.

Genetics shapes our fate. We use language and the social brain to build strong communities founded on a common set of beliefs, which over time become religious beliefs, and shared cultural values. Language underpins culture, and the emergence of local dialects or language variations helps to foster an even stronger sense of community identity.

Other communities may come to share these same beliefs and values, but there will inevitably be some who are in some way different. An 'us' and 'them' mentality is not only impossible to avoid, it is possibly welcomed, as 'not like us' can in itself be a defining attribute.

Small, single-nucleotide mutations in a handful of genes are amplified by selection pressures in favour of adaptations to the local climate. Differences in skin colour, eye shape, and eye-and-hair colour combinations are utterly trivial in the story of human evolution, but they all happen to be on the surface and they are *very* visible. These racial differences feed a growing 'us' and 'them' polarization.

Culture, religion, and race is a highly volatile mix. Let's throw in some avarice, lust for power, and delusions of grandeur. The rest, as they say, is history.

EPILOGUE

I will end the story here. Now, in the Preface I warned you about the nature of the scientific enterprise. By applying the Copernican Principle, taking 'us' out of the reckoning and looking dispassionately at what we find in the world around us, we approach a truth about creation that is rather more free from our specifically human prejudices. When we do this we discover that we appear to be really rather incidental, no more than a footnote in an appendix to the cosmos.

The creation story according to modern science leaves us under no illusions. In the unfolding history of our universe, it seems there is nothing inevitable about *Homo sapiens*.

Given the set of initial conditions, the specific mix of dark energy, dark matter, and baryonic matter and the imprint of quantum fluctuations amplified by cosmic inflation, the evolution of space, time, and energy to give stars and galaxies and eventually third-generation stars with accompanying planetary systems appears entirely predictable and inevitable, based on our established theories of physics and chemistry. We can go further. There is also something inevitable about the existence of planets orbiting in their star's 'habitable zone', providing liquid water and abundant raw materials suitable for the establishment of life.

If we are prepared to accept life as a 'cosmic imperative', then we accept that at least simple, single-celled prokaryotic life is an inevitable consequence of chemistry on a sunlit, warm, wet, habitable planet. Through the blind mechanisms of evolution, life inevitably spreads and fills the available ecological space. We might be encouraged to accept a certain inevitability in relation to endosymbiosis and the evolution of multicellular organisms, as life is enabled by circumstance to develop greater and greater complexity.

But then we see that there's nothing at all inevitable in the 500-million-year journey that takes us from *Charnia* and *Dickinsonia* or the Cambrian fauna to *Homo sapiens*. For this is a journey punctuated by the cruelties of unpredictable chance, the whips and scorns of time. The evolutionary radiations of the mammals (and then of primates, then humans) would not have been possible were it not for the preceding pattern of mass extinctions. These rested on an intimate but profoundly complex dance between the planet's moving continents, its oceans and atmosphere and, in one case at least, that ultimate blunt instrument of chance, an asteroid that happened to be in this place, at that time.

For us to be here today, the highly successful dinosaurs first had to die. It's really not that hard to conceive alternative scenarios in which the asteroid narrowly missed, or the volcanoes didn't erupt and in which, in consequence, we're not here to bear witness. And even with the ecological landscape cleared for us to evolve into, it is apparent that there are many paths that evolution could have taken, only a few of which lead to the possibility of modern humans.

It would seem that this was *never* about us, and it's hard not to feel really rather humbled by this knowledge.

Some will reject what this creation story tells us about ourselves, believing that we must be more important than this, that there must be more to our lives; that we're here for a purpose. By its very nature, science can't offer certainty, and there are (and always will be) gaps in our scientific knowledge and understanding in which those seeking purpose and meaning may find some solace.

We are some way off having a scientific theory that tells us unambiguously why our universe 'began' with the specific set of initial conditions that it so clearly possessed.* We may never have such a theory. Likewise there is currently no 'standard model' for the origin of life, and life, distinct from the collection

* Much is being made by some contemporary theoretical physicists of superstring theory's inability to specify a unique way of 'compactifying' the six extra spatial dimensions it requires, rolling them up and tucking them away so that we can't detect them. With a landscape of an estimated 10^{500} different possibilities to choose from, and no basis for making a choice, these theorists have speculated that nature actually didn't choose. They argue that all these different possibilities exist in a 'multiverse', of which only one universe—the one we happen to find ourselves in—has the initial conditions and properties compatible with life. I argued in my book *Farewell to Reality* that this is wrongheaded, lazy thinking, and is not scientific.

of biophysical and biochemical mechanisms that underpin it, appears to be a 'force' that remains imperturbably mysterious. Consciousness too remains frustratingly inscrutable and presently beyond the reach of understanding. But I'm more optimistic that life and consciousness will one day yield at least some of their secrets.

I believe that the message from the scientific story of creation is that we find ourselves here through a set of extraordinary historic circumstances and outrageous good fortune. Exploit the gaps if you must, but it's more likely that there is no purpose to be found in the structure and nature of the physical world: we are its accidental by-products, not its reason for being.

Purpose and meaning are human inventions and we must look not outwards but *inwards* to find them, by engaging our social brains. We are not privileged in the sense that this has all been created especially for our benefit. But being here—now—is nevertheless still an extraordinary privilege. Our task is, as it has always been, to work out what we do about that.

If the rise of intelligent life on Earth was accident rather than design, what does this tell us about the possibility of intelligent life elsewhere in the universe? I suspect that some scientists have long harboured the delusion that, over time, evolution is somehow an irresistible driving force for more complex forms of life and, eventually, human-style intelligence.* If my reading of our creation story is correct, then it's rather hard to see anything particularly inevitable about the development of human intelligence.

We certainly have no evidence for intelligent, extra-terrestrial life. I wouldn't rule it out (absence of evidence isn't evidence of absence) but I suspect it will continue to be hard to find, the search for it based more on an act of faith than a conclusion of science.

I've ended this story with the origin of human consciousness but of course this is not the end. Most readers will be familiar with the broad brush strokes of human history, from the first Neolithic settlements which appear around 11 000 years ago, the rise (and fall) of the first cities, city-states, and empires,

* For example, the 'Drake equation' developed in 1961 by American astronomer Frank Drake, was designed to provide a focus for debate at the first meeting of a group of scientists involved in the search for extra-terrestrial intelligence (SETI). It is not a true mathematical equation, but rather a collection of factors that, when multiplied together, give some sense of the probability of the existence of civilisations in our Milky Way galaxy that might be capable of radio communications. In their original estimates, Drake and his colleagues assumed that the probability of life emerging on a planet in its star's habitable zone is 100%. They also assumed that 100% of these planets would then go on to develop intelligent life.

the foundation of world religions, to the invention of the scientific method, the industrial revolution, and the birth and early childhood of the modern world.[1]

As a human population we've grown increasingly accustomed to the idea of rapid and sometimes overwhelming change. We've grown used to the coming and going of our specifically human creations, such as buildings and institutions, as the notion of 'progress' has shaped aspects of our physical and social landscape. But our lifespans are such that in older age we inevitably develop a strong sense of permanence for much of our natural landscape, of mountains and plains and rivers, of oceans and sky, and of the universe of visible stars. We know that these are not unchanging, but this change is barely noticeable, except occasionally when disaster strikes.

The scientific story of creation takes us well beyond the boundaries of our unaided perception and simple reckoning. It opens our eyes to greater truths. Aside from a few gaps in our understanding, it allows us to answer most of our 'big questions'. We now know the nature of the material world and we know how it works. We know what the universe is and how it was formed, although we're obliged to speculate for times earlier than a trillionth of a second after its creation. We know what life is, or at least what it does, though we might struggle to define it. We know where we come from and broadly how we have evolved. We know how and why we think, and what it means to be human. And we know how we know.

There's still plenty of mystery, and much yet to be learned. Now fixing on a definitive date for the origin of science is a little tricky, but in our 24-hour 'day of creation' it seems reasonable to conclude that we have been applying a scientific methodology to explore the origins of the universe and of ourselves for just three thousandths of a second.

We've barely started.

APPENDIX

Powers of Ten

The scientific story of creation inevitably involves some mind-boggling numbers, large and small. To avoid a wearisome repetition of numbers in the millions of billions of trillions or thousandths of millionths of billionths, scientists express these numbers as powers of ten. If this way of expressing large and small numbers is unfamiliar, this appendix provides a handy reference.

TABLE 3: **Positive powers of ten**

Number	Name	Power of Ten	Prefix	Symbol
1	One	10^0		
10	Ten	10^1	deca	da
100	Hundred	10^2	hecto	h
1000	Thousand	10^3	kilo	k
1 000 000	Million	10^6	mega	M
1 000 000 000	Billion	10^9	giga	G
1 000 000 000 000	Trillion	10^{12}	terra	T

The speed of light, c, is measured to be 299 792 458 metres per second, which we can write as 299.792458 million metres per second or 2.99792458×10^8 metres per second.

TABLE 4: **Negative powers of ten**

Number	Name	Power of Ten	Prefix	Symbol
0.1	Tenth	10^{-1}	deci	d
0.01	Hundredth	10^{-2}	centi	c
0.001	Thousandth	10^{-3}	milli	m
0.000001	Millionth	10^{-6}	micro	μ
0.000000001	Billionth	10^{-9}	nano	n
0.000000000001	Trillionth	10^{-12}	pico	p

Excited hydrogen atoms emit red light with a wavelength of 656.28 nanometres (this is the so-called Hα line). We can re-express this number as 656.28 billionths of a metre, 656.28×10^{-9} metres or $6.5628 \text{ x } 10^{-11}$ metres.

NOTES

CHAPTER 1: IN THE 'BEGINNING'

1 Ludwig Wittgenstein, *Tractatus Logico-Philosophicus*, translated by C. K. Ogden, Kegan Paul, Trench, Trubner & Co. Ltd., London, 1922, p. 90.
2 The kinetic energy of an object of mass m and speed v is given by $\frac{1}{2}mv^2$.
3 The precise relationship is $F = Gm_1m_2/r^2$, where F is the force of gravity, m_1 and m_2 are the masses of object 1 and object 2, respectively, and r is the distance between them. G is Newton's gravitational constant, an empirical constant of proportionality measured to be of the order of 6.673×10^{-11} m^{3}kg^{-1}s^{-2}.
4 Albert Einstein, 'How I Created the Theory of Relativity', lecture delivered at Kyoto University, 14 December 1922, translated by Yoshimasa A. Ono, *Physics Today*, August 1982, p. 47.
5 Wheeler, p. 235.
6 Albert Einstein, *Proceedings of the Prussian Academy of Sciences*, **142** (1917). Quoted in Isaacson, p. 255.
7 In his autobiography, Ukrainian-born theoretical physicist George Gamow wrote: 'When I was discussing cosmological problems with Einstein he remarked that the introduction of the cosmological term was the biggest blunder he ever made in his life.' George Gamow, *My World Line: An Informal Autobiography*, Viking Press, New York, 1970, p. 149. Quoted in Isaacson, pp. 355–6.
8 Lemaître wrote: 'Everything happens as though the energy in vacuo would be different from zero.' G. Lemaître, *Annales de la Société Scientifique de Bruxelles*, Serie A, **53**, pp. 51–85. Quoted in Nussbaumer and Bieri, p. 171.
9 The Planck scale is a mass-energy scale with a magnitude around 10^{19} billion electronvolts, where the quantum effects of gravity are presumed to be strong. It is characterized by measures of mass, length, and time that are calculated from three fundamental constants of nature: the gravitational constant, G, Planck's constant h divided by 2π, written $\hbar$ (pronounced 'h-bar') and the speed of light, c. The Planck mass is given by $\sqrt{\hbar c/G}$ and has a value around 1.2×10^{19} billion electronvolts, or 1.2×10^{28} electronvolts. The Planck length is given by $\sqrt{G\hbar/c^3}$ and has a value around 1.6×10^{-35} metres. The Planck time is given by $\sqrt{G\hbar/c^5}$ (the Planck length divided by c), and has a value around 5×10^{-44} seconds.
10 See Guth, p. 86.

CHAPTER 2: BREAKING THE SYMMETRY

1 Louis de Broglie, from the 1963 re-edited version of his PhD thesis. Quoted in Pais, *Subtle is the Lord*, p. 436.
2 Tom Stoppard, *Hapgood*, Faber and Faber, London, 1988.

3 In 1905, Einstein suggested that the energy (E) carried by an individual light-quantum is proportional to its frequency (given by the Greek symbol nu, ν), according to $E = h\nu$, where h is Planck's constant, which has a value of 6.626×10^{-34} joule-seconds. In many ways, this result rivals his most famous equation, $E = mc^2$.
4 CERN press release, 14 March 2013.
5 Lederman, p. 22.
6 The masses of the elementary particles are reported by the Particle Data Group and can be found online at pdg.lbl.gov. The mass of the electron is reported to be about 0.510998928 MeV/c^2 (million electron volts divided by c^2). The mass of the proton is 938.272046 MeV/c^2, giving a ratio of electron-to-proton masses of 0.000544617.

CHAPTER 3: THE LAST SCATTERING SURFACE

1 The characteristic decay time of a free neutron is reported by the Particle Data Group as a 'lifetime' of 880.0±0.9 seconds. This is the time it takes for neutrons to decay to $1/e$ of their initial number, where $e = 2.71828$ is the base of natural logarithms. The 'half-life' is the time taken for neutrons to decay to half their initial number, which can be calculated from the lifetime by multiplying it by ln(2), or 0.69315. This gives a half-life of about 610 seconds.
2 If at the start of primordial nucleosynthesis there are 88 protons and 12 neutrons in every 100 nucleons (a roughly seven-to-one ratio), then the 12 neutrons become incorporated into six ^{4}He nuclei. These six ^{4}He nuclei also account for 12 protons, leaving 76 free. But the ^{4}He nuclei are four times heavier than an individual proton (assuming that the masses of the proton and neutron are the same, which they nearly are). So, the percentage by mass of protons is 76% and the percentage of ^{4}He nuclei is $6 \times 4 = 24\%$.
3 Hans Bethe, quoted by Ralph Alpher and Robert Herman, *Physics Today*, August 1988, p. 28.
4 Steven Weinberg, *The First Three Minutes*, p. 131. This quote is often cited by theorists arguing in favour of the variety of unproven and largely untestable theories that litter contemporary theoretical physics (see my book *Farewell to Reality*). Weinberg was not saying that we should invest belief in every wild theoretical speculation. I believe he was cautioning us against the natural tendency to be rather conservative when extrapolating even well tried and tested theories to the extraordinary conditions that likely prevailed at the beginning of the universe.
5 Robert Dicke, quoted by Dennis Overbye, p. 130.
6 Robert Dicke, quoted by David Wilkinson, 'Measuring the Cosmic Microwave Background Radiation', in P. James E. Peebles, Lyman A. Page, Jr., and R. Bruce Partridge (eds.), *Finding the Big Bang*, Cambridge University Press, 2009, p. 204.
7 The de Broglie relationship is written $\lambda = h/p$, where λ is the wavelength, h is Planck's constant (6.626×10^{-34} joule-seconds) and p is the momentum. This means that a wavepacket state containing a broad range of wavelengths could be expected to exhibit a broad range of momenta.
8 Heisenberg established that the uncertainty in position multiplied by the uncertainty in momentum must be greater than, or at least equal to, Planck's constant h.
9 George Smoot, quoted in Singh, p. 462.

CHAPTER 4: SETTING THE FIRMAMENT ALIGHT

1 See S. D. M. White and M. J. Rees, *Monthly Notices of the Royal Astronomical Society*, **183** (1978), pp. 341–58.

2 The end of the Dark Ages is signalled by something called re-ionization, the moment in the history of the universe when there's sufficient intensity of ultraviolet light from newly formed stars to strip electrons from the clouds of neutral hydrogen atoms that have prevailed as the principal form of baryonic matter since the moment of recombination. The age of 550 million years is derived from analysis of the polarization of the cosmic background radiation reported by scientists working on the Planck satellite, see: http://www.esa.int/Our_Activities/Space_Science/Planck/Planck_reveals_first_stars_were_born_late

3 Shingo Hirano, Takashi Hosokawa, Naoki Yoshida, Hideyuki Umeda, Kazuyuki Omukai, Gen Chiaki, and Harold W. Yorke, astro-ph/1308.4456v1, 21 August 2013.

4 Howard E. Bond, Edmund P. Nelan, Don A. VandenBerg, Gail H. Schaefer, and Dianne Harmer actually estimate an age for HD 140283 of 14.5±0.8 billion years. The large uncertainty in this measurement (much of which is attributable to uncertainty in the star's initial oxygen abundance) means that it does not necessarily conflict with the measured age of the universe, 13.8 billion years. See astro-ph/1302.3180v1, 13 February 2013.

5 For stellar materials at temperatures greater than or equal to 30 000 kelvin, the opacity κ is proportional to $T^{-3.5}$, an approximate relationship first established by the Dutch physicist Henrik Kramers in 1923. Doubling the temperature therefore reduces the protostar's opacity by a factor $2^{3.5}$, or 11.3.

6 Albert Einstein, letter to Paul Langevin, 16 December 1924. Quoted in Moore, p. 187.

7 Hubble's law can be expressed as $v = H_0D$, where v is the velocity of the galaxy, H_0 is Hubble's constant for a particular moment in time and D is the so-called 'proper distance' of the galaxy measured from the Earth, such that the velocity is then given simply as the rate of change of this distance. Although it is often referred to as a 'constant', H_0 varies with changes in the rate of expansion of the universe. Nevertheless, the age of the universe can be roughly estimated as $1/H_0$. A value of H_0 of 67.8 kilometres per second per megaparsec (about 2.2×10^{-18} per second) gives an age for the universe of 45×10^{16} seconds, or 14.3 billion years.

8 How did I create Figure 30? I started with a square 4 × 4 array of barred spiral galaxies on a Microsoft PowerPoint slide. I grouped these, duplicated the resulting object and increased its scale height and width, having first taken care to lock the aspect ratio. This increase in scale represents the expansion of the universe. But this expansion doesn't apply to the individual galaxies themselves (the galaxies don't get noticeably bigger due to the expansion), so I ungrouped the larger object and replaced the galaxies with ones of the original size before re-grouping. I changed the colour of the galaxies in the original object to grey, and created Figure 30(a) by lining up the galaxies in row 2, column 2 and Figure 30(b) by lining up the galaxies in row 3, column 4.

9 The redshift parameter z is related to the speed of the object v according to the equation $z=\sqrt{[1+(v/c)]/[1-(v/c)]}-1$, where c is the speed of light. In instances where v is much less than c, this can be approximated as $z \cong v/c$. You can check this for yourself by

putting $v = 0.001c$. The full equation for z then reduces to $z = \sqrt{\frac{(1+0.001)}{(1-0.001)}} - 1 \cong 0.001$, which is just v/c.

10 The equations are $z = v/c$ and $v = H_0 D$. Measurement of the redshift allows us to estimate the speed of the object from $v = zc$. We can then estimate the proper distance from $D = v/H_0$, or $D = zc/H_0$. So, a redshift z of 0.00333 implies a speed v of 0.00333 × 299,792 kilometres per second = 998 kilometres per second and a distance D of 998 kilometres per second/67.8 kilometres per second per megaparsec, or 14.7 megaparsecs (47.9 million light-years).

11 J. P. Ostriker and P. J. E. Peebles, *Astrophysical Journal*, **186** (1973), p. 467. In fact, their arguments concerning the stability of galaxies turned out to be incorrect, although they still managed to draw the correct conclusion (Simon White, personal communication, 30 October 2014).

12 For example, in Modified Gravity (MOG) theory, developed by John W. Moffat at Toronto University, Einstein's gravitational field equations are modified in a way that strengthens gravity at large distances in the presence of large masses. The theory can account for the galaxy rotation curves without invoking the existence of dark matter.

13 Douglas Adams, *The Hitch Hiker's Guide to the Galaxy*, Pan Books, London, 1979, p. 62.

CHAPTER 5: SYNTHESIS

1 It's worth noting that the molecular energy level diagram for the hypothetical molecule He_2, formed from two helium atoms, doesn't look much different from the diagram shown in Figure 37. However, we now have four electrons to dispose of, two from each atom. Two of these go into the bonding orbital and the other two go into the anti-bonding orbital, giving a bond order of two minus two divided by two, or zero. No bond is formed. He_2 does not exist as a stable molecule.

2 Note that the bonding in sodium chloride is different from that depicted in Figure 37, which is referred to as *covalent* bonding. Rather, sodium chloride provides an example of a molecule with an *ionic* bond. Instead of sharing electrons in one or more molecular orbitals, an electron is transferred completely from the sodium atom to the chlorine atom, forming Na^+ and Cl^- ions, which then bind together through their mutual electrostatic attraction.

3 See M. Nielbock, R. Launhardt, J. Steinacker, A. M. Stutz, Z. Balog, H. Beuther, *et al.*, astro-ph/1208.4512v3, 8 September 2012.

4 The lyrics to Joni Mitchell's 'Woodstock' can be found online at a number of sites, such as: http://www.songlyrics.com/joni-mitchell/woodstock-lyrics/.

5 The cosmological 'look-back' time is related to the cosmological redshift, z, the Hubble constant, and the density parameters Ω_M and Ω_Λ. Astrophysicists often refer to distant times or ages not in terms of thousands, millions, or billions of years ago but rather in terms of z. I guess the reason is relatively simple. Cosmological parameters have tended to change slightly with every satellite mission or data release, thereby changing the estimated ages of everything. However, redshifts based on the measured frequencies of light tend to be somewhat more permanent. I found a useful calculator on Ned Wright's cosmology site: http://www.astro.ucla.edu/~wright/CosmoCalc.html. This

allows you to input the latest values of the cosmological parameters and calculate a variety of things, including the age for any specific value of z.

6 The cosmic scale factor, usually given the symbol $a(t)$, is a measure of the relative size of the universe at time t compared to the present, for which $a(\text{now}) = 1$. It can be determined from the redshift z according to the equation $a(t) = 1/(1 + z)$. So, a Type Ia supernova with a redshift $z = 1$ corresponds to a scale factor of 0.5, or 50% of the current size of the universe.

7 Adam G. Reiss, Alexei V. Filippenko, Peter Challis, Alejandro Clocchiatti, Alan Diercks, Peter M. Garnavich, *et al.*, *The Astronomical Journal*, **116**, September 1998, p. 1009. The italics are mine.

8 Brian Schmidt, quoted in Panek, p. 240.

CHAPTER 6: SOL

1 Ernst Zinner, *Science*, **300** (2003), pp. 265–7.

2 S. Mostefaoui, G. W. Lugmair, P. Hoppe, and A. El Goresy, *Lunar and Planetary Science*, **XXXIV** (2003), Abstract 1585. Note that since this work, a new measurement of the half-life of ^{60}Fe has been reported, see: G. Rugel, T. Faestermann, K. Knie, G. Korschinek, M. Poutivtsev, D. Schumann, *et al.*, *Physical Review Letters*, **103** (2009), 072502.

3 M. Gritschneder, D. N. C. Lin, S. D. Murray, Q.-Z. Yin, and M.-N. Gong, astro-phSR / 1111.0012v1, 31 October 2011.

4 Emanuel Swedenborg, *The Principia*, Volume II, translated by J. R. Rendell and I. Tansley, The Swedenborg Society, London, 1912, p. 183, quoted in Gregory L. Baker, *The Physics Teacher*, October 1983, p. 444.

5 The angular momentum of a rotating object (usually given the symbol L) can be calculated by multiplying the moment of inertia (I) with the angular speed of rotation (ω), or $L = I\omega$. We can obtain ω from the period of rotation (T, the time taken for one complete rotation), since $\omega = 2\pi/T$, in units of radians per second. The formula for I differs for different physical systems, but these can be looked up in a textbook or online (e.g. see http://en.wikipedia.org/wiki/List_of_moments_of_inertia). For a point mass rotating about a central point, I is equal to the mass of the object, m, multiplied by the square of the distance between the object and the centre of rotation, r, or $I = mr^2$. So, for the orbital motions of the planets around the Sun we have $L = I\omega = 2\pi mr^2/T$. We can now look up values of the masses, orbital radii, and orbital periods of the planets and use these to calculate L. We get the following:

	Mass	Orbital Radius	Orbital Period	L	% of Total L
	10^{24}kg	10^6km	Days	10^{43}kgm^2/s	
Mercury	0.33	57.91	87.97	0.0001	0.00%
Venus	4.87	108.21	224.70	0.0018	0.06%
Earth	5.97	149.51	365.26	0.0027	0.08%
Mars	0.64	227.94	686.97	0.0004	0.01%

(continued)

(continued)

	Mass	Orbital Radius	Orbital Period	L	% of Total L
	10^{24}kg	10^{6}km	Days	10^{43}kgm²/s	
Jupiter	1868.60	778.57	4332.59	1.9012	58.44%
Saturn	586.46	1,433.45	10,579.22	0.8145	25.04%
Uranus	86.81	2876.68	30,799.10	0.1696	5.21%
Neptune	102.43	4503.44	60,193.03	0.2510	7.71%

Note that in performing these calculations, we need to turn kilometres into metres and days into seconds. For a uniform solid sphere, $I=\left(\frac{2}{5}\right)mr^2$ and so for the spin motion of the Sun around its axis we have $L=\left(\frac{4\pi}{5}\right)mr^2/T$. If we look up the values of the mass, radius, and sidereal rotation period of the Sun we get:

	Mass	Radius	Rotation Period	L	% of Total L
	10^{24}kg	10^{6}km	Days	10^{43}kgm²/s	
Sun	1.99×10^{6}	0.70	25.05	0.1120	3.44%

The total angular momentum of the solar system is the sum of these values of L and is equal to 3.25×10^{43} kgm²/s. These calculations show that although the Sun accounts for 99.9% of the total mass of the solar system, it has only 3.4 per cent of the total angular momentum.

We can ask what the rotation period of the Sun would need to be in order for it to account for 99.9% of the total angular momentum. In this case, the total angular momentum of the eight planets (3.14×10^{43} kgm²/s) accounts for just 0.1% of the total, requiring the Sun to have $L = 3.15 \times 10^{46}$ kgm²/s. Rearranging the equation for L gives $T=\left(\frac{4\pi}{5}\right)mr^2/L$, from which we deduce that T would need to be 77 seconds.

6 The relationship between luminosity and surface temperature is given by $L = 4\pi R^2 \sigma T_{eff}^4$, where L is the luminosity of the star, R is the radius, T_{eff} is the effective surface temperature and σ is the Stefan–Boltzmann constant, a constant of proportionality between the energy radiated by a black body per unit surface area and its temperature, with a value of about 5.67×10^{8} Wm⁻²T⁻⁴.

CHAPTER 7: TERRA FIRMA

1 See, for example, Anne M. Hofmeister, and Robert E. Criss, *Planetary and Space Science*, **62** (2012), p. 111.

2 See http://exoplanet.eu
3 By definition, Earth lies in the Sun's habitable zone. The Sun has a luminosity 1 $L_{\odot}$ and the Earth orbits at a distance of 150 million kilometres, or 1 astronomical unit (AU). The habitable zones of stars with different luminosities scale with the square root of the luminosity. A star with a luminosity of 0.25 $L_{\odot}$ has a habitable zone at an orbital distance of $\sqrt{0.25}$ = 0.5 AU. A star with 0.5 $L_{\odot}$ has a habitable zone at $\sqrt{0.5}$ = 0.71 AU. A star with 2 $L_{\odot}$ gives $\sqrt{2}$ = 1.4 AU, 4 $L_{\odot}$ gives $\sqrt{4}$ = 2 AU, and so on.
4 See Erik A. Petiguraa, Andrew W. Howard, and Geoffrey W. Marcya, *Proceedings of the National Academy of Sciences*, **111** (2013), p. 19273.
5 Robin M. Canup, *Icarus*, **168** (2004), p. 433.
6 Robin M. Canup, *Science*, **338** (2012), p. 1052.
7 C. Goldblatt, K. J. Zahnle, N. H. Sleep, and E. G. Nisbet, *Solid Earth* **1** (2010), p. 1.
8 Matija Ćuk and Sarah T. Stewart, *Science*, **338** (2012), p. 1047.
9 Simon A. Wilde, John W. Valley, William H. Peck and Colin M. Graham, *Nature*, **409** (2001), p. 175. Concerns that loss or migration of lead atoms from within the zircon crystals might have distorted the results and produced a misleading age were laid to rest by subsequent analysis, see John W. Valley, Aaron J. Cavosie, Takayuki Ushikubo, David A. Reinhardt, Daniel F. Lawrence, David J. Larson, *et al.*, *Nature Geoscience*, **7** (2014), p. 219.
10 For example, the mean radius of Earth's orbit is about 150 million kilometres and it obviously takes on average a little over 365 days to complete one round trip. For Mars, the equivalent figures are 229 million kilometres and 687 days. So, for Earth the ratio of the cube of the mean radius to the square of the period is 25.05, for Mars it is 25.09 billion billion cubic kilometres per square day. The equivalent figure for Mercury is 25.09, for Venus 25.09, Jupiter 25.14, Saturn 25.44, Uranus 25.10, and Neptune 25.21.

CHAPTER 8: THE COSMIC IMPERATIVE

1 See, for example, Langmuir and Broecker, p. 389.
2 Jacques Monod, *Chance and Necessity: An Essay on the Natural Philosophy of Modern Biology*, Alfred A. Knopf, New York, 1971, p. 180.
3 Christian de Duve, *Philosophical Transactions of the Royal Society A*, **369** (2011), p. 620.
4 Large organic molecules quickly become tedious to draw out by hand, and chemists have adopted a number of ways to reduce this tedium without losing the most important features of a chemical structure. As most of the atoms involved in organic substances are carbon and hydrogen, these atoms are often not drawn at all but are *implied* by the structure. For example, the amino acid phenylalanine has a side chain which consists of a phenyl group. This is a derivative of benzene (C_6H_6) which forms a hexagonal ring with alternating single C-C and double C=C bonds and with each carbon atom bonded to a single hydrogen atom. Such compounds form a class of substances called 'aromatic', as some exhibit distinctly aromatic smells. In the phenyl group, one of the hydrogen atoms in benzene has been replaced by a —CH_2 group. The benzene ring is then drawn as a simple hexagon, each apex representing a carbon atom with the single bonds represented by single lines and double bonds by double lines. The hydrogen atoms are not shown at all, but students of chemistry quickly learn that this

doesn't mean they're not there. If in doubt, count up the number of bonds connecting each carbon atom. There should always be four. If there are only two or three, then the third and fourth are C—H bonds. Atoms other than carbon can occupy apexes in such cyclic structures, and when they do they are drawn explicitly. Examples include oxygen and nitrogen.

5 Charles Darwin, letter to John D. Hooker, 1 February 1871, see http://www.darwinproject.ac.uk/editors-blog/2012/02/15/darwins-warm-little-pond/
6 J. Oró, A. Lazcano, and P. Ehrenfreund, in Thomas, Hicks, Chyba, and McKay (eds.), p. 13. However, see also C. F. Chyba and K. P. Hand in the same volume, p. 195. These latter estimates illustrate the sensitivity to assumptions concerning Earth's primitive atmosphere.
7 Stanley L. Miller, *Science*, **117** (1953), p. 528.
8 Nick Lane and William F. Martin, *Cell*, **151** (2012), pp. 1406–16.

CHAPTER 9: SYMBIOSIS

1 Darwin, p. 484.
2 Carl Woese, *Proceedings of the National Academy of Sciences*, **95** (1998), p. 6858.
3 See J. Thomas Beatty, Jörg Overmann, Michael T. Lince, Ann K. Manske, Andrew S. Lang, Robert E. Blankenship, *et al.*, *Proceedings of the National Academy of Sciences*, **102** (2005), pp. 9306–10. The green sulphur bacteria were found in the environment of a 'black smoker' hydrothermal vent along the East Pacific Rise.
4 Stanley Miller, quoted in Knoll, p. 88.
5 Martin Brasier, Nicola McLaughlin, Owen Green, and David Wacey, *Philosophical Transactions of the Royal Society B*, **361** (2006), p. 889.
6 J. William Schopf, *Philosophical Transactions of the Royal Society B*, **361** (2006), p. 882.
7 Abigail C. Allwood, John P. Grotzinger, Andrew H. Knoll, Ian W. Burch, Mark S. Anderson, Max L. Coleman, and Isik Kanik, *Proceedings of the National Academy of Sciences*, **106** (2009), pp. 9548–55. This quote appears on p. 9555.
8 David Wacey, Matt R. Kilburn, Martin Saunders, John Cliff, and Martin D. Brasier, *Nature Geoscience*, **4** (2011), pp. 698–702.
9 The authors wrote: '... we conclude that the ancient [microbially induced sedimentary structure]-forming microbial mats in the Dresser Formation were dominated by microbes mimicking the behavior of modern cyanobacteria.' See Nora Noffke, Daniel Christian, David Wacey, and Robert M. Hazen, *Astrobiology*, **13** (2013), pp. 1121–2.
10 S. J. Mojzsis, G. Arrhenius, K. D. McKeegan, T. M. Harrison, A. P. Nutman, and C. R. L. Friend, *Nature*, **384** (1996), pp. 55–9.
11 Yoko Ohtomo, Takeshi Kakegawa, Akizumi Ishida, Toshiro Nagase, and Minik T. Rosing, *Nature Geoscience*, **7** (2014), pp. 25–8.
12 Oleg Abramov and Stephen J. Mojzsis, *Nature*, **459** (2009), pp. 419–22.
13 The naming and dating of geologic time periods varies over time and can vary from one publication to the next. In this book, I've based all the references to geologic time periods on the International Chronostratigraphic Chart, v2013/01, which is produced by the International Commission on Stratigraphy. See http://www.stratigraphy.org.

14 See, for example, J. Cameron Thrash, Alex Boyd, Megan J. Huggett, Jana Grote, Paul Carini, Ryan J. Yoder, *et al.*, *Scientific Reports*, **1** (2011).
15 For example, Cox *et al.* present evidence for an ancestral role for a form of archaea called Crenarchaeota (Cymon J. Cox, Peter G. Foster, Robert P. Hirt, Simon R. Harris, and T. Martin Embley, *Proceedings of the National Academy of Sciences*, **105** (2008), pp. 20356–61). Subsequently, further detailed genetic analysis led Kelly *et al.* to identify the eukaryote ancestor as a member or sister group of a phylum of Crenarchaeota called Thaumarchaeota (S. Kelly, B. Wickstead, and K. Gull., *Proceedings of the Royal Society B*, 29 September 2010).
16 Lane, *Life Ascending*, p. 112.
17 For the original study see: Jochen J. Brocks, Graham A. Logan, Roger Buick, and Roger E. Summons, *Science*, **285** (1999), pp. 1033–6. The quote is taken from a recent review: Shuhai Xiao, *Evolution from the Galapagos: Social and Ecological Interactions in the Galapagos*, Volume 2 (2013), p. 110.
18 John P. Grotzinger, David A. Fike, and Woodward W. Fischer, *Nature Geoscience*, 4 (2011), pp. 285–92.
19 See Jonathan L. Payne, Alison G. Boyer, James H. Brown, Seth Finnegan, Michał Kowalewski, Richard A. Krause, Jr., *et al.*, *Proceedings of the National Academy of Sciences*, **106** (2009), pp. 24–7, especially Fig. 1, p. 25.
20 Knoll, p. 164.
21 S. Schröder and J. P. Grotzinger, *Journal of the Geological Society, London*, **164** (2007), pp. 175–87.

CHAPTER 10: A SONG OF ICE AND FIRE

1 But there were hints. English palaeontologist John W. Salter published papers in 1856 and 1857 reporting the discovery of fossils in Ediacaran-age rocks in Longmynd, Shropshire, and these were referenced in Darwin's *Origin of Species*. Salter thought they were caused by marine worms and trilobites, but it now seems that these structures are microbial. See Richard H. T. Callow, Duncan McIlroy, and Martin D. Brasier, *Ichnos*, **18** (2011), pp. 176–87.
2 This quote appears in Fortey, pp. 59–60 and is derived from the sixth British edition of Darwin's *Origin of Species*, published in 1872 (and in which the quote appears on p. 286). Darwin's original 1859 first edition contains a slightly different set of words: 'To the question why we do not find records of these vast primordial periods, I can give no satisfactory answer . . . But the difficulty of understanding the absence of vast piles of fossiliferous strata, which on my theory no doubt were somewhere accumulated before the Silurian epoch, is very great.' These passages appear on p. 307. Thanks to the Online Variorum of Darwin's *Origin of Species*, with which I was able to reconcile these differences: http://darwin-online.org.uk/Variorum.
3 This quote comes from the first episode of David Attenborough's documentary series *First Life*, first broadcast in Britain by the BBC on 5 November 2010. In *Attenborough's Journey*, a separate documentary chronicling the making of the series, he goes on to lament: 'I can't remember where I heard about the discovery of a *Charnia*, but I certainly kicked myself. I thought, I could have been part of history. I could have discovered that. Why didn't I bother to look?'

4 The system and conventions that biologists use to classify all the different forms of life that we know of is called biological taxonomy. It has its origins in the taxonomy developed in 1735 by the great Swedish botanist Carolus Linnaeus and its modern form is presented in a simplified representation shown in Figure 97, which shows a sequence of taxonomic *ranks* starting with domain (of which, as we know, there are three—Bacteria, Archaea, and Eukarya) and ending with species. The simple fact that we need six ranks between top and bottom tells you everything you need to know about the diversity and complexity of life on Earth.

Lying beneath domain is the second taxonomic rank of kingdom. Within the domain of Eukarya (the domain of all life based on complex eukaryote cells) there are four kingdoms—Protists (a very diverse group of eukaryote microorganisms), Fungi, Animalia (animals), and Plantae (plants).

Each of these has one or more *phyla*. Biologists would classify life forms to different phyla based on the notion of a 'body plan'. We can think of this as the large-scale structure of the organisms in terms of the nature and placement of their body parts, their morphology (size, shape, pattern, colour) and their physiology (for example, did they crawl on four legs or walk on two?). Today, genetic sequencing provides a much more reliable measure of the 'relatedness' of different phyla, sometimes throwing up distinct genetic differences between seemingly similar forms while tying together seemingly very different forms.

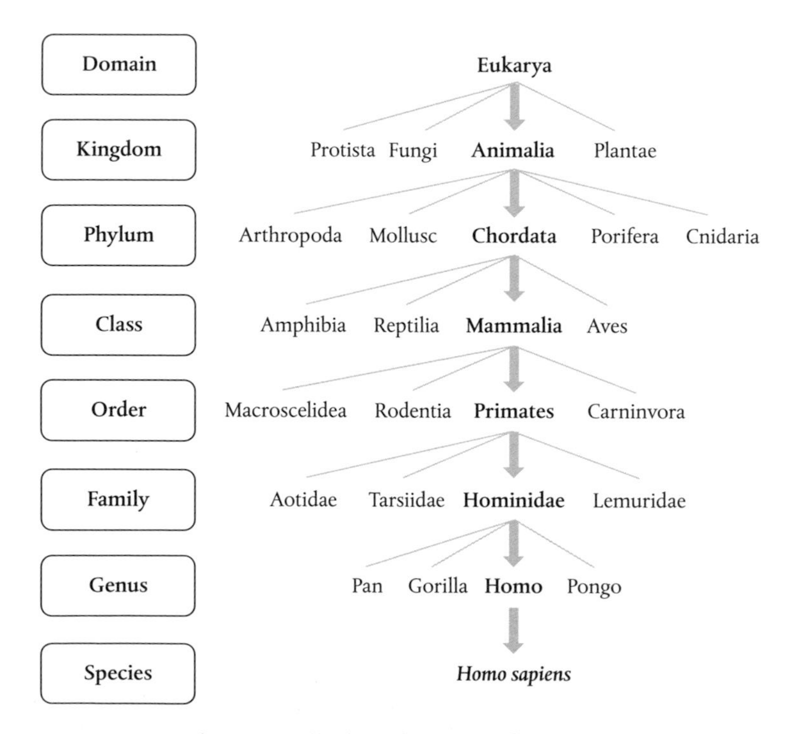

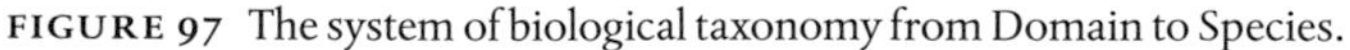
FIGURE 97 The system of biological taxonomy from Domain to Species.

So, the phyla of Animalia include Arthropoda (insects, spiders, and crustaceans), Mollusca (molluscs), Porifera (sponges), and Cnidaria (jellyfish, corals, and sea anemones). It also includes Chordata, which are essentially bilaterally symmetric animals with—among other things—a hollow dorsal nerve cord. A sub-phylum of the Chordata includes the vertebrates: animals with a backbone. Chordates include the classes of Amphibia (frogs, toads, and salamanders), Reptilia (all reptiles), Mammalia (all mammals, distinguished by females possessing mammary glands, among other things), and Aves (birds).

Within the class Mammalia are the orders Macroscelidea (elephant shrews), Rodentia (all rodents), Primates and Carnivora (a diverse group which includes carnivorous cats, tigers, bears, hyenas, wolves, and giant panda, which are in fact largely herbivorous, but never mind). Within the order Primates are Aotidae (night monkeys), Tarsiidae (tarsier monkeys), Hominidae (the 'great apes'), and Lemuridae (lemurs). The Hominidae include the genus Pan (chimpanzees), Gorilla, Homo (humans), and Pongo (orangutans).

Finally, the genus Homo today has only one surviving species, *Homo sapiens* (it is an accepted standard to italicize the names of species, leading with the genus, which is capitalized).

This is wickedly complex, and it probably doesn't help for you to know that the structure shown above is substantially simplified. When physicists began to be overwhelmed in the 1950s by an ever-expanding particle 'zoo' and names proliferated, Enrico Fermi responded to a question from one young physicist thus: 'Young man, if I could remember the names of these particles, I would have been a botanist.' (Enrico Fermi, quoted by Helge Kragh, *Quantum Generations* as 'physics folklore', p. 321.)

Not to worry. We won't need to remember all the names or the complex taxonomic ranks of animals. I will try to keep the discussion at the levels of phyla, genus, and species.

5 See Fortey, p. 99.

6 Knoll, p. 199.

7 This section owes much to the visualization of the Cambrian fauna at the foot of the Cathedral Escarpment provided by Conway Morris in his excellent book *The Crucible of Creation*, see chapter 4, pp. 63–115.

8 H. B. Whittington, *Philosophical Transactions of the Royal Society B*, **271** (1975), pp. 1–43. This quote appears on p. 2.

9 Darwin, p. 73.

10 J. J. Sepkoski's Online Genus Database is available at http://strata.geology.wisc.edu/jack/. This site also features the diagram on which Figure 80 is based. Interested readers might also wish to visit the FossilPlot website, which builds on Sepkoski's original database and provides a freeware program in Microsoft Excel which can be used to construct a variety of different plots. This site is maintained by Idaho State University supported by a grant from the US National Science Foundation: see http://geology.isu.edu/FossilPlot/Index.htm.

11 Film footage of the first Moon landing is available to view on YouTube. Armstrong actually said 'That's one small step for man; a giant leap for mankind,' whereas the script called for him to contrast the small step that he needed to take (as a man) to place his foot on the Moon's surface with the significance of this step for all mankind.

12 Jennifer A. Clack, *Evolution: Education & Outreach,* **2** (2009), p. 222.
13 Erwin, p. 58.
14 Seth D. Burgess, Samuel Bowring, and Shu-zhong Shen, *Proceedings of the National Academy of Sciences,* 10 February 2014.
15 Fortey, p. 235.
16 Luis W. Alvarez, *Alvarez: Adventures of a Physicist,* Basic Books, New York, 1987, p. 254.

CHAPTER 11: THE HUMAN STAIN

1 See, for example, Robin M. D. Beck, *Biological Journal of the Linnean Society,* 97 (2009), pp. 1–17.
2 There's a nice illustration of these creatures in J. G. M. Thiewissen, Lisa Noelle Cooper, John C. George, and Sunil Bajpai, *Evolution: Education & Outreach,* **2** (2009), pp. 272–88. See Figure 27, p. 286.
3 Jens L. Franzen, Philip D. Gingerich, Jörg Habersetzer, Jørn H. Hurum, Wighart von Koenigswald, and B. Holly Smith, *PLoS ONE,* **4** (2009), p. e5273. The quotation appears on p. 25.
4 For a retrospective on the fossil Ida controversy five years on, see http://sciencenordic.com/fossil-ida-%E2%80%93-five-years
5 Jeffrey P. Demuth, Tijl de Bie, Jason E. Stajich, Nello Cristianini, and Matthew W. Hahn, *PLoS One,* **1** (2006), p. e85
6 David A. Raichlen, Adam D. Gordon, William E. H. Harcourt-Smith, Adam D. Foster, and Wm. Randall Haas, Jr., *PLoS ONE,* **5** (2010), p. e9769.
7 Stringer, p. 32.
8 See Nick Ashton, Simon G. Lewis, Isabelle De Groote, Sarah M. Duffy, Martin Bates, Richard Bates, *et al., PLoS ONE,* 9 (2014), p. e88329.

CHAPTER 12: COGITO ERGO SUM

1 David J. Chalmers, *Journal of Consciousness Studies,* **2** (1995), pp. 200–19.
2 Hilary Putnam, *Reason, Truth and History,* Cambridge University Press, 1981, p. 6.
3 Dennett, p. 37.
4 Damasio, p. 33.
5 M. S. Gazzaniga, *Cornell University Medical College Alumni Quarterly,* **45** (1982), pp. 8–9. This quotation is reproduced in Thompson, p. 417.
6 Cambridge Declaration on Consciousness, 7 July 2012.
7 Sally McBrearty and Alison Brooks, *Journal of Human Evolution,* **39** (2000), pp. 453–563.
8 M. Aubert, A. Brumm, M. Ramli, T. Sutikna, E. W. Saptomo, B. Hakim, *et al., Nature,* **514** (2014), pp. 223–7.
9 Philip Liebermann, *Current Anthropology,* **48** (2007), p. 44.
10 Dunbar, p. 59.
11 Bob Hen, quoted by Malcolm Gladwell, *The Tipping Point,* Abacus, 2001, p. 185.
12 Dunbar, p. 310.
13 Sandra Beleza, António M. Santos, Brian McEvoy, Isabel Alves, Cláudia Martinho, Emily Cameron, *et al., Molecular Biology and Evolution,* **30** (2012), pp. 24–35.

EPILOGUE

1 If you're concerned that I'm ending just when the story is really starting to get interesting, then I'd like to draw your attention to a movement called 'big history', whose proponents seek to relate a naturalistic, scientifically informed human history from the big bang to modern civilization. This is an emerging discipline, driven by academic historians who have produced some highly readable accounts, such as David Christian's *Maps of Time* and Cynthia Stokes Brown's *Big History: From the Big Bang to the Present* (see the Bibliography, below).

These accounts are less concerned to tell the creation story according to modern science, with all its gaps, uncertainties, and speculations, which has been the purpose of this book. The balance is instead firmly biased towards human history from the beginning of the Holocene, 12 000 years ago, with non-human and human prehistory (everything from the big bang to Neolithic settlements) accounting for about a quarter of the story.

An online, interactive version of this kind of big history is available from the ChronoZoom Project (www.chronozoom.com) which was begun by Walter Alvarez and Ronald Saekow at the University of California, Berkeley, in 2009 and is now an open source community project supported by Microsoft Research and various educational institutions. Check it out.

BIBLIOGRAPHY

Archibald, John, *One Plus One Equals One: Symbiosis and the Evolution of Complex Life*, Oxford University Press, 2014.

Baggott, Jim, *The Quantum Story: A History in 40 Moments*, Oxford University Press, 2011.

Baggott, Jim, *Higgs: The Invention and Discovery of the 'God Particle'*, Oxford University Press, 2012.

Brasier, Martin, *Secret Chambers: The Inside Story of Cells and Complex Life*, Oxford University Press, 2012.

Brown, Cynthia Stokes, *Big History: From the Big Bang to the Present*, The New Press, New York, 2007.

Carroll, Sean, *The Particle at the End of the Universe: The Hunt for the Higgs and the Discovery of a New World*, Oneworld Publications, London, 2012.

Carter, Rita, *Mapping the Mind*, 2nd edition, Phoenix, 2010.

Chalmers, A. F., *What is This Thing Called Science?* (3rd edition), Hackett, Indianapolis, 1999.

Chalmers, David J. (ed.), *The Philosophy of Mind*, Oxford University Press, 2002.

Christian, David, *Maps of Time: An Introduction to Big History*, University of California Press, 2004.

Cochran, Gregory and Harpending, Henry, *The 10,000 Year Explosion: How Civilization Accelerated Human Evolution*, Basic Books, New York, 2009.

Conway Morris, Simon, *The Crucible of Creation: The Burgess Shale and the Rise of Animals*, Oxford University Press, 1998.

Cox, Brian and Forshaw, Jeff, *Why Does* $E = mc^2$? Da Capo Press, Cambridge, MA, 2009.

Crease, Robert P., *A Brief Guide to the Great Equations: The Hunt for Cosmic Beauty in Numbers*, Robinson, London, 2009.

Cushing, James T., *Philosophical Concepts in Physics*, Cambridge University Press, 1998.

Damasio, Antonio R., *Descartes' Error*, Macmillan, London, 1991.

Davies, Paul, *The Origin of Life*, Penguin Books, London, 1999.

Darwin, Charles, *On the Origin of Species by Means of Natural Selection or the Preservation of Favoured Races in the Struggle for Life*, John Murray, London, 1859.

Dawkins, Richard, *The Blind Watchmaker*, Penguin, London, 1988.

De Duve, Christian, *Vital Dust: Life as a Cosmic Imperative*, Basic Books, New York, 1995.

Dennett, Daniel, *Consciousness Explained*, Penguin, London, 1991.

DeSalle, Rob and Tattersall, Ian, *The Brain: Big Bangs, Behaviors, and Beliefs*, Yale University Press, 2012.

Dunbar, Robin, *Human Evolution*, Penguin, London, 2014.

Edelman, Gerald, *Bright Air, Brilliant Fire: On the Matter of the Mind*, Penguin Books, London, 1994.

Erwin, Douglas H., *Extinction: How Life on Earth Nearly Ended 250 Million Years Ago*, Princeton University Press, 2008.

Farmelo, Graham (ed.), *It Must be Beautiful: Great Equations of Modern Science*, Granta Books, London, 2002.

Feynman, Richard P., *QED: The Strange Theory of Light and Matter*, Penguin, London, 1985.

Finlayson, Clive, *The Improbable Primate: How Water Shaped Human Evolution*, Oxford University Press, 2014.

Fortey, Richard, *Life: An Unauthorised Biography*, Harper Collins Publishers, London, 1997.

Gamow, George, *Thirty Years that Shook Physics*, Dover Publications, New York, 1966.

Gell-Mann, Murray, *The Quark and the Jaguar*, Little, Brown & Co., London, 1994.

Goldsmith, Donald, *The Runaway Universe: The Race to Find the Future of the Cosmos*, Perseus Publishing, New York, 2000.

Gould, Stephen Jay, *Wonderful Life: The Burgess Shale and the Nature of History*, Penguin Books, London, 1989.

Gribbin, John, *Q is for Quantum: Particle Physics from A to Z*, Weidenfeld & Nicolson, London, 1998.

Guth, Alan H., *The Inflationary Universe: The Quest for a New Theory of Cosmic Origins*, Vintage, London, 1998.

Halpern, Paul, *Collider: The Search for the World's Smallest Particles*, John Wiley & Son, Inc., New Jersey, 2009.

Heil, John, *Philosophy of Mind*, Routledge, London, 1998.

Hoddeson, Lillian, Brown, Laurie, Riordan, Michael, and Dresden, Max, *The Rise of the Standard Model: Particle Physics in the 1960s and 1970s*, Cambridge University Press, 1997.

Isaacson, Walter, *Einstein: His Life and Universe*, Simon & Schuster, New York, 2007.

Kennedy, J. B., *Space, Time and Einstein: An Introduction*, Acumen, Chesham, 2003.

Kirshner, Robert P., *The Extravagant Universe: Exploding Stars, Dark Energy and the Accelerating Cosmos*, Princeton University Press, 2002.

Knoll, Andrew H., *Life on a Young Planet: The First Three Billion Years of Evolution on Earth*, Princeton University Press, 2003.

Kragh, Helge, *Quantum Generations: A History of Physics in the Twentieth Century*, Princeton University Press, 1999.

Krauss, Lawrence M., *A Universe from Nothing: Why There is Something Rather than Nothing*, Simon & Schuster, London, 2012.

Lane, Nick, *Oxygen: The Molecule that Made the World*, Oxford University Press, 2002.

Lane, Nick, *Life Ascending: The Ten Great Inventions of Evolution*, Profile Books, London, 2009.

Langmuir, Charles H. and Broecker, Wally, *How to Build a Habitable Planet: The Story of Earth from the Big Bang to Humankind*, Princeton University Press, 2012.

Lederman, Leon (with Dick Teresi), *The God Particle: If the Universe is the Answer, What is the Question?* Bantam Press, London, 1993.

Mayr, Ernst, *What Evolution Is*, Phoenix, London, 2002.

Meredith, Martin, *Born in Africa: The Quest for the Origins of Human Life*, Simon & Schuster, 2011.

Mo, Hojun, van den Bosch, Frank, and White, Simon, *Galaxy Formation and Evolution*, Cambridge University Press, 2010.

Moore, Walter, *Schrödinger: Life and Thought*, Cambridge University Press, 1989.

Morgan, Michael, *The Space Between Our Ears: How the Brain Represents Visual Space*, Weidenfeld & Nicolson, London, 2003.

Nambu, Yoichiro, *Quarks*, World Scientific Publishing, Singapore, 1981.

Nussabaumer, Harry and Bieri, Lydia, *Discovering the Expanding Universe*, Cambridge University Press, 2009.

Ostriker, Jeremiah P. and Mitton, Simon, *Heart of Darkness: Unravelling the Mysteries of the Invisible Universe*, Princeton University Press, 2013.

Overbye, Dennis, *Lonely Hearts of the Cosmos: The Quest for the Secret of the Universe*, Picador, London, 1993.

Pääbo, Svante, *Neanderthal Man: In Search of Lost Genomes*, Basic Books, New York, 2014.

Pais, Abraham, *Subtle is the Lord: The Science and the Life of Albert Einstein*, Oxford University Press, 1982.

Panek, Richard, *The 4 per cent Universe: Dark Matter, Dark Energy and the Race to Discover the Rest of Reality*, Oneworld, Oxford, 2011.

Pinker, Steven, *The Language Instinct: The New Science of Language and Mind*, Penguin, London, 1994.

Pinker, Steven, *How the Mind Works*, Penguin Books, London, 1998.

Primack, Joel R. and Abrams, Nancy Ellen, *The View from the Center of the Universe: Discovering our Extraordinary Place in the Cosmos*, Riverhead Books, New York, 2006.

Pross, Addy, *What is Life? How Chemistry Becomes Biology*, Oxford University Press, 2012.

Raup, David M., *Extinction: Bad Genes or Bad Luck?* W.W. Norton & Co. Inc, New York, 1991.

Rees, Martin, *Just Six Numbers: The Deep Forces that Shape the Universe*, Phoenix, London, 2000.

Repcheck, Jack, *The Man Who Found Time: James Hutton and the Discovery of the Earth's Antiquity*, Pocket Books, London, 2004.

Rose, Hilary and Rose, Steven, *Genes, Cells and Brains: The Promethean Promises of the New Biology*, Verso, London, 2012.

Rose, Steven, *The Making of Memory: From Molecules to Mind*, Bantam Books, London, 1994.

Rose, Steven, *The 21st-Century Brain: Explaining, Mending and Manipulating the Mind*, Jonathan Cape, London, 2005.

Russell, Michael and Schild, Rudolph (eds.), *Origins of Life: How Life Began—Abiogenesis, Astrobiology*, papers selected from Volumes 10, 11, and 16 of the *Journal of Cosmology*, 2009–11.

Rutherford, Adam, *Creation: The Origin of Life*, Penguin Books, London, 2014.

Ryan, Sean G. and Norton, Andrew, J., *Stellar Evolution and Nucleosynthesis*, Cambridge University Press, 2010.

Searle, John R., *Intentionality*, Cambridge University Press, 1983.

Searle, John R., *The Construction of Social Reality*, Penguin, London, 1995.

Singh, Simon, *Big Bang: The Most Important Scientific Discovery of All Time and Why You Need to Know About It*, Harper Perennial, London, 2005.

Stachel, John (ed.), *Einstein's Miraculous Year: Five Papers that Changed the Face of Physics*, Princeton University Press, 2005.

Strange, Philip G., *Brain Biochemistry and Brain Disorders*, Oxford University Press, 1992.

Stringer, Chris, *The Origin of Our Species*, Penguin Books, London, 2012.

Stringer, Chris and Andrews, Peter, *The Complete World of Human Evolution*, 2nd edition, Thames & Hudson Ltd., London, 2012.

Swain, Harriett (ed.), *Big Questions in Science,* Vintage, London, 2003.

Thomas, P. J., Hicks, R. D., Chyba, C. F., and McKay, C. P. (eds.), *Comets and the Origin and Evolution of Life,* 2nd edition, Springer Verlag, Berlin, 2006.

't Hooft, Gerard, *In Search of the Ultimate Building Blocks,* Cambridge University Press, 1997.

Thompson, Richard F., *The Brain: A Neuroscience Primer,* 2nd edition, W. H. Freeman & Co., New York, 1993.

Veltman, Martinus, *Facts and Mysteries in Elementary Particle Physics,* World Scientific, London, 2003.

Waller, William H., *The Milky Way: An Insider's Guide,* Princeton University Press, 2013.

Weinberg, Steven, *The First Three Minutes: A Modern View of the Origin of the Universe,* Basic Books, New York, 1977.

Weinberg, Steven, *Dreams of a Final Theory: The Search for the Fundamental Laws of Nature,* Vintage, London, 1993.

Weinberg, Steven, *Cosmology,* Oxford University Press, 2008.

Wheeler, John Archibald, with Ford, Kenneth, *Geons, Black Holes and Quantum Foam: A Life in Physics,* W.W. Norton & Company, New York, 1998.

Zalasiewicz, Jan and Williams, Mark, *The Goldilocks Planet: The Four Billion Year Story of Earth's Climate,* Oxford University Press, 2012.

INDEX

Note: References to figures are indicated by '*f*', footnotes are indicated by '*n*', and tables by '*t*'.

HIGGS

The invention and discovery of the 'God Particle'

Jim Baggott

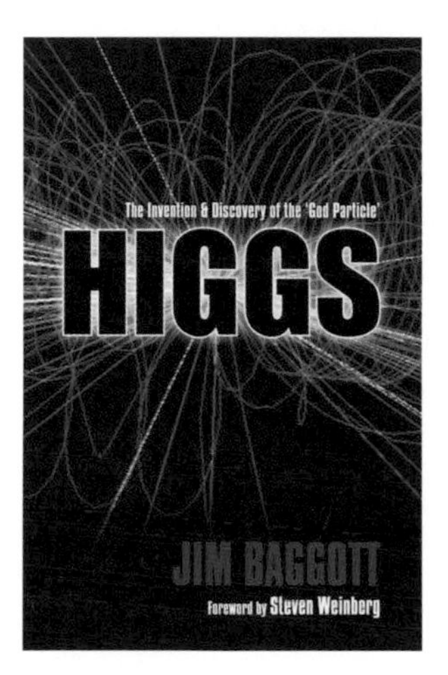

978-0-19-967957-7 | Paperback | £9.99

'A thorough and readable explanation of the lengthy hunt for the Higgs boson and why its discovery is so important.' ***New Scientist***

The hunt for the Higgs particle has involved the biggest, most expensive experiment ever. So exactly what is this particle? What does it tell us about the Universe? Did the discovery announced on 4 July 2012 finish the search? And was finding it really worth all the effort?

The short answer is yes. It's the strongest indicator yet that the Standard Model of physics really does reflect the basic building blocks of our Universe. Here, Jim Baggott explains the science behind the discovery, looking at how the concept of a Higgs field was invented, how the vast experiment was carried out, and its implications on our understanding of all mass in the Universe.

THE QUANTUM STORY

A history in 40 moments

Jim Baggott

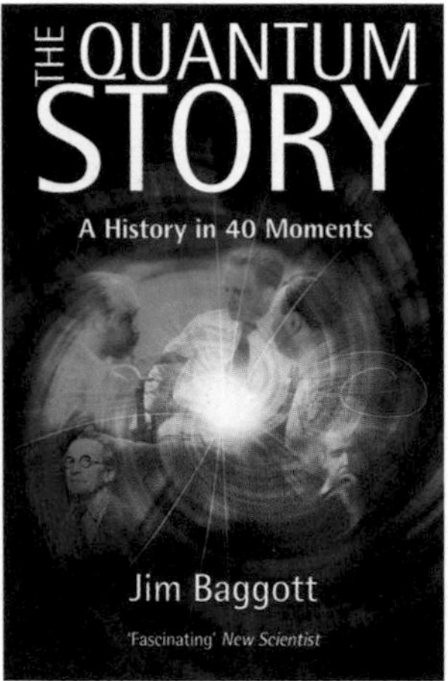

978-0-19-965597-7 | Paperback | £14.99

'Jim Baggott's survey of the history of the emergence of the twentieth century's most enigmatic but successful theory is a delight to read. It is clear, accessible, engaging, informative, and thorough. It illuminates an important, revolutionary era of modern science and the varied personalities behind it.'

Peter Atkins

Almost everything we think we know about the nature of our world comes from one theory of physics. Jim Baggott presents a celebration of this wonderful yet wholly disconcerting theory, with a history told in forty episodes—significant moments of truth or turning points in the theory's development. From its birth in the porcelain furnaces used to study black body radiation in 1900, to the promise of stimulating new quantum phenomena to be revealed by CERN's Large Hadron Collider over a hundred years later, this is the extraordinary story of the quantum world.

THE EMERALD PLANET

How plants changed Earth's history

David Beerling

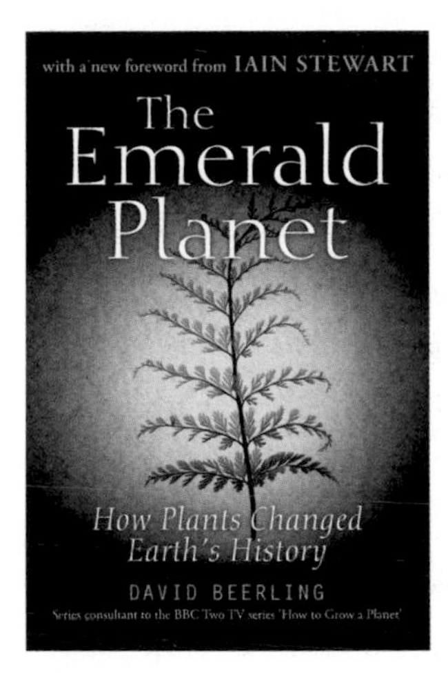

978-0-19-954814-9 | Paperback | £11.99

'My favorite nonfiction book this year. A minutely-argued but highly readable history of the last half-billion years on earth. The story Beerling tells could not have been put together even ten years ago, for it depends upon the latest insights from palaeontology, climate science, genetics, molecular biology, and chemistry, all brilliantly and beautifully integrated together. I got a special deep, quiet pleasure from reading *The Emerald Planet*—the sort of pleasure one gets from reading Darwin.'

Oliver Sacks, Book of the Year, *Observer*

Plants have profoundly moulded the Earth's climate and the evolutionary trajectory of life. David Beerling puts plants centre stage, revealing the crucial role they have played in driving global changes in the environment, in recording hidden facets of Earth's history, and in helping us to predict its future.

ONE PLUS ONE EQUALS ONE

Symbiosis and the evolution of complex life

John Archibald

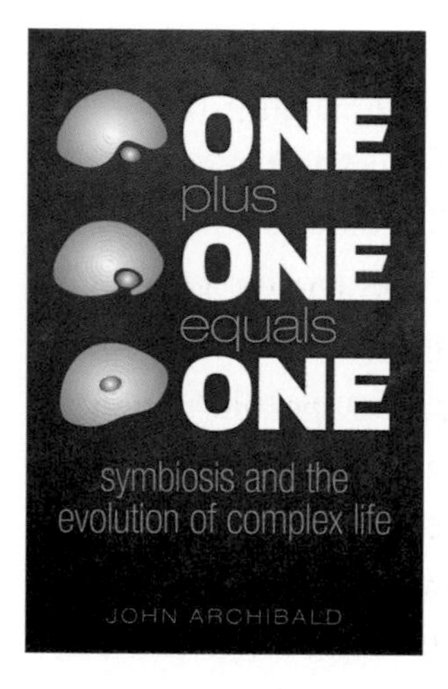

978-0-19-966059-9 | Hardback | £16.99

'*One Plus One Equals One* is an eloquent account, at times verging on the poetic. With serious scholarship, it illuminates a rare scientific endeavour.' **Nancy A. Moran, *Nature***

It is natural to look at biotechnology in the 21st century with a mix of wonder and fear. But biotechnology is not as 'unnatural' as one might think. All living organisms use the same molecular processes to replicate their genetic material and the same basic code to 'read' their genes. Here, John Archibald shows how evolution has been 'plugging-and-playing' with the subcellular components of life from the very beginning, and continues to do so today. For evidence, we need look no further than the inner workings of our own cells. Molecular biology has allowed us to gaze back more than three billion years, revealing the microbial mergers and acquisitions that underpin the development of complex life.

THE PLANET IN A PEBBLE

A journey into Earth's deep history

Jan Zalasiewicz

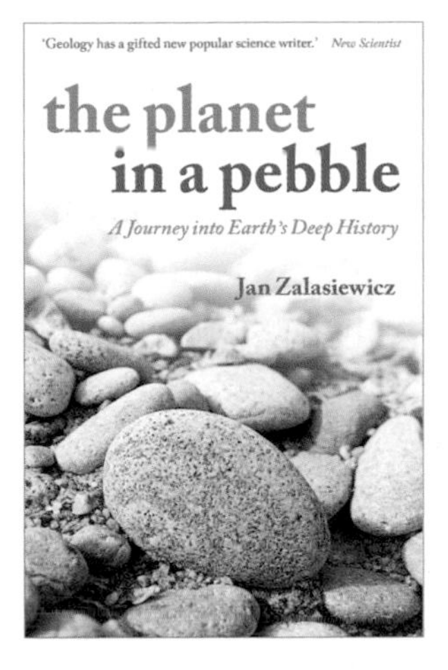

978-0-19-964569-5 | Paperback | £9.99

'A mind-expanding, awe inducing but friendly scientific exploration of the history.'

Holly Kyte, *The Sunday Telegraph*

This is a narrative of the Earth's long and dramatic history, as gleaned from a single pebble. It begins as the pebble-particles form amid unimaginable violence in distal realms of the Universe, in the Big Bang, and in supernova explosions, and continues amid the construction of the Solar System. Jan Zalasiewicz shows the incredible complexity present in such a small and apparently mundane object. It may be small, and ordinary, this pebble—but it is also an eloquent part of our Earth's extraordinary, never-ending story.

WHAT IS LIFE?

How chemistry becomes biology

Addy Pross

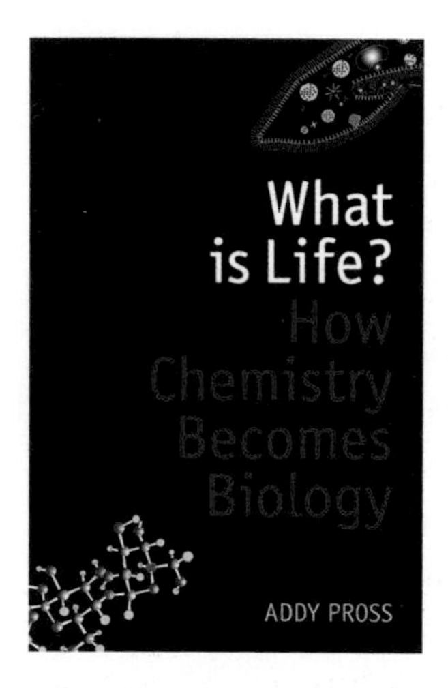

978-0-19-968777-0 | Paperback | £9.99

'Pross does an excellent job of succinctly conveying the difficulty in crafting an unambiguous general definition of life and provides a road map to much of the work on the origin of life done by chemists in the past 50 years. The book is worth the read for these discussions alone.' ***Chemical Heritage***

Living things are hugely complex and have unique properties, such as self-maintenance and apparently purposeful behaviour which we do not see in inert matter. So how does chemistry give rise to biology? What could have led the first replicating molecules up such a path? Now, developments in the emerging field of 'systems chemistry' are unlocking the problem. The gulf between biology and the physical sciences is finally becoming bridged.